VOLUME SEVENTY ONE

ADVANCES IN QUANTUM CHEMISTRY

Concepts of Mathematical Physics in Chemistry: A Tribute to Frank E. Harris - Part A

VOLUME SEVENTY ONE

Advances in QUANTUM CHEMISTRY

Concepts of Mathematical Physics in Chemistry: A Tribute to Frank E. Harris - Part A

Edited by

JOHN R. SABIN

Department of Chemistry, Physics, and Pharmacy, University of Southern Denmark, Odense, Denmark, and Quantum Theory Project, Department of Physics, University of Florida, Gainesville, Florida, USA

REMIGIO CABRERA-TRUJILLO

Instituto de Ciencias Físicas, Universidad Nacional Autónoma de México, Cuernavaca, Morelos, Mexico

ELSEVIER

AMSTERDAM • BOSTON • HEIDELBERG • LONDON
NEW YORK • OXFORD • PARIS • SAN DIEGO
SAN FRANCISCO • SINGAPORE • SYDNEY • TOKYO
Academic Press is an imprint of Elsevier

Academic Press is an imprint of Elsevier
125 London Wall, London, EC2Y 5AS, UK
525 B Street, Suite 1800, San Diego, CA 92101-4495, USA
225 Wyman Street, Waltham, MA 02451, USA
The Boulevard, Langford Lane, Kidlington, Oxford OX5 1GB, UK

First edition 2015

ISBN: 978-0-12-802824-7
ISSN: 0065-3276

For information on all Academic Press publications
visit our website at http://store.elsevier.com/

CONTENTS

PREFACE

FRANK E. HARRIS: A NAME TO BE RECKONED WITH!

It is with great pleasure that we present this volume of *Advances in Quantum Chemistry* in honor of Frank E. Harris. A workshop, "Concepts of Mathematical Physics in Chemistry," was held in Playa del Carmen, Mexico, on December 10–12, 2014, in honor of Frank and his contributions. The contents of this volume are mostly based on talks given at that workshop.

Frank is an internationally known quantum chemist who earned an A.B. in Chemistry at Harvard followed by a Ph.D. in Physical Chemistry at Berkeley, where he spent 3 years as an Assistant Professor. He joined the faculty at Stanford University in 1959, and since 1969 has been Professor of Physics and Chemistry at the University of Utah. In 1998, Frank was appointed Adjunct Professor in the UF Quantum Theory Project and Department of Chemistry, and has been a Visiting Professor of Physics during that time. Recently, he has been made Research Professor. Frank is also well known and has association with several universities such as those in Uppsala, Sweden (Uppsala University) and Odense, Denmark (The University of Southern Denmark). He has close connections with the University in Namur, Belgium, where he recently organized a symposium.

The scientific record that Frank Harris has made, and continues to expand, is impressive with more than 250 papers. His work in formal and molecular quantum mechanics is attested by his long list of research publications and invitations to talk at meetings and universities around the world. In addition, Frank has published several books, both for teaching and research. Most recently, he has completely written the standard mathematical reference for physicists—*Mathematical Methods for Physicist* by Arfkin, Weber, and Harris.

Frank has worked extensively in the quantum chemical determination of the properties of molecules as well the formal aspects of quantum chemistry. Frank was one of the earliest investigators to implement the use of Gaussian basis sets in quantum chemical calculations[1] which has become a standard method in quantum chemistry.

[1] "Gaussian Wave Functions for Polyatomic Molecules," F.E. Harris, Rev. Mod. Phys. **35**, 558–569 (1963).

The majority of the contributions in this volume are from Frank's students and collaborators at some time. The great breadth of subject matter presented in this volume is indicative of the broad range of Frank's interest and influence.

Remigio Cabrera-Trujillo and John R. Sabin
Editors

CONTRIBUTORS

Gerd Baumann
Mathematics Department, German University in Cairo, New Cairo City, Egypt, and University of Ulm, Ulm, Germany

Benedito J.C. Cabral
Instituto de Física, Universidade de São Paulo, São Paulo, Brazil; Grupo de Física Matemática da Universidade de Lisboa, and Departamento de Química e Bioquímica, Faculdade de Ciências, Universidade de Lisboa, Lisboa, Portugal

Remigio Cabrera-Trujillo
Instituto de Ciencias Físicas, Universidad Nacional Autónoma de México, Cuernavaca, Morelos, México

Sylvio Canuto
Instituto de Física, Universidade de São Paulo, São Paulo, Brazil

Marcelo Hidalgo Cardenuto
Instituto de Física, Universidade de São Paulo, São Paulo, Brazil, and Laboratoire de Chimie Théorique, Unité de Chimie Physique Théorique et Structurale, University of Namur, Namur, Belgium

Debajit Chakraborty
Quantum Theory Project, Departments of Physics and of Chemistry, University of Florida, Gainesville, Florida, USA

Benoît Champagne
Laboratoire de Chimie Théorique (UCPTS), Département de Chimie, University of Namur, Namur, Belgium

Kaline Coutinho
Instituto de Física, Universidade de São Paulo, São Paulo, Brazil

Salvador A. Cruz
Departamento de Física, Universidad Autónoma Metropolitana-Iztapalapa, México D.F., México

Jorge M. del Campo
Departamento de Física y Química Teórica, Facultad de Química, Universidad Nacional Autónoma de México, Mexico, Mexico

Fco. Javier Domínguez-Gutiérrez
Institute for Advanced Computational Science, Stony Brook University, Stony Brook, New York, USA, and Instituto de Ciencias Físicas, Universidad Nacional Autónoma de México, Cuernavaca, Morelos, México

James Dufty
Department of Physics, University of Florida, Gainesville, Fl, USA

Joseph G. Fripiat
Laboratoire de Chimie Théorique (UCPTS), Département de Chimie, University of Namur, Namur, Belgium

Frank E. Harris
Department of Physics, University of Utah, Salt Lake City, Utah, and Quantum Theory Project, University of Florida, Gainesville, Florida, USA

Xiaomin Huang
Departamento de Física y Química Teórica, Facultad de Química, Universidad Nacional Autónoma de México, Mexico, Mexico

Per Kaijser
KRI Ismaning, Germany

Valentin V. Karasiev
Quantum Theory Project, Departments of Physics and of Chemistry, University of Florida, Gainesville, Florida, USA

Andreas M. Köster
Departamento de Química, CINVESTAV, Avenida Instituto Politécnico Nacional 2508, Mexico, Mexico

Vasilios G. Koures
IISAM L3C, Cheyenne, WY, USA

Predrag S. Krstić
Institute for Advanced Computational Science, Stony Brook University, Stony Brook, New York, USA

Eugenio Ley-Koo
Instituto de Física, Universidad Nacional Autónoma de México, México D.F., México

Daniel Mejía-Rodríguez
Departamento de Química, CINVESTAV, Avenida Instituto Politécnico Nacional 2508, Mexico, Mexico

Ricardo Méndez-Fragoso
Facultad de Ciencias, Universidad Nacional Autónoma de México, Circuito Exterior S/N Delegación Coyoacán, Ciudad Universitaria, México

David A. Micha
Quantum Theory Project, Department of Chemistry and Department of Physics, University of Florida, Gainesville, Florida, USA

Jens Oddershede
Department of Chemistry, Physics, and Pharmacy, University of Southern Denmark, Odense, Denmark, and Quantum Theory Project, Department of Physics, University of Florida, Gainesville, Florida, USA

José Récamier
Instituto de Ciencias Físicas, Universidad Nacional Autónoma de México C.P. 62210 Cuernavaca, Morelos, México

Ricardo Román-Ancheyta
Instituto de Ciencias Físicas, Universidad Nacional Autónoma de México C.P. 62210 Cuernavaca, Morelos, México

John R. Sabin
Department of Chemistry, Physics, and Pharmacy, University of Southern Denmark, Odense, Denmark, and Quantum Theory Project, Department of Physics, University of Florida, Gainesville, Florida, USA

Stephan P.A. Sauer
Department of Chemistry, University of Copenhagen, Copenhagen, Denmark

Frank Stenger
SINC, LLC, School of Computing, University of Utah, Salt Lake City, Utah, USA

Samuel B. Trickey
Quantum Theory Project, Departments of Physics and of Chemistry, University of Florida, Gainesville, Florida, USA

CHAPTER ONE

Frank Harris, a Master of Mountains

Per Kaijser[1]
KRI Ismaning, Germany
[1]Corresponding author: e-mail address: per@kaijser.de

Abstract

Professor Frank Harris has excellent skills in many fields. This chapter is devoted to an area, where Frank has a special talent and where the author greatly benefitted from his knowledge during the three years 1969–1971. This guidance took place during the summer schools in Beitostølen, Norway, and has later not only helped the author conquer some rough terrains but also given him much pleasure.

This chapter is devoted to Professor Frank Harris. First time we met was in connection with the Löwdin Summer School in 1969, which took place in BeitostøIen, Norway. During 3 weeks in August that year and the years 1970 and 1971, I got to know Frank not only as an excellent teacher and scientist but also as a dear friend with many capabilities. One of these, his knowledge, skill, and love for mountains, is the topic of this chapter.

The summer schools 1969–1971 were situated in Beitostølen, Norway, a small village in Jotunheimen, the highest mountain region in Scandinavia. Professor Per-Olov Löwdin, who was in charge of these, was a very experienced climber and loved the mountains and made sure that as much of the free time as possible was spent hiking in the mountains. A hike is the perfect environment for mingling and for discussing scientific issues and ideas. Of the other staff members, the two best hikers were Professor Yngve Öhrn and, especially, Frank, who was outstanding. I was told numerous stories about his achievements in the north of Sweden and in the Dolomites, mainly from my supervisor Jean-Louis Calais from whom I learnt not only science but also much about the scientists. Besides being a technically excellent climber, Frank was most known for his speed and distances he was able to cover.

Mountains have always appealed to me, but I had no experience with alpine type of mountains with peaks several hundreds of meters above its base. I was thus very keen to get to know them better. The first mountain

Advances in Quantum Chemistry, Volume 71
ISSN 0065-3276
http://dx.doi.org/10.1016/bs.aiq.2015.02.002

lecture was how to handle large groups of inexperienced hikers. This took place at the Summer School home mountain, Bitihorn. This is a beautiful rather steep mountain but still easily accessible without advanced climbing. The students were divided into smaller groups led by one of the athletic members of the staff. These included also Osvaldo Goscinski and Erkki Brändas. The tail of each line, however, was given to less athletic staff members such as Jean-Louis Calais and Ruben Pauncz or to a junior like me. Bitihorn (see Figs. 1–3) was visited several times and during some of the later ones I had the pleasure of hiking along with Frank.

Figure 1 Erkki Brändas leading a group of participants up on Bitihorn with Jean-Louis Calais next to last in the row (August 1969).

Figure 2 Yngve Öhrn leading another group up on Bitihorn (August 1969).

Figure 3 At the top of Bitihorn with Per-Olov Löwdin at the center (August 1969).

Figure 4 Knutshøe seen from Bessvatnet on Besseggen with Gjende below. The scientist in the foreground is Gunnar Sperber (September 7, 1973).

Besides Bitihorn, we also made hikes through Svartdalen from Torfinsbu at the lake Bygdin to Gjendebu at the lake Gjende, a very nice valley surrounded by high peaks on all sides. Another mountain Knutshøe (see Fig. 4), close to Gjende, was a small and very beautiful mountain that could be climbed during a half-day tour.

Once we made a longer tour over Besseggen at the east end of lake Gjende, a famous mountain ridge 800 m above the lake and famous from the Norwegian literature. During this hike, I saw with envy how Frank was permitted to make a side trip with some of the more athletic students to Besshøe, a further away lying taller mountain, whereas I had to walk at the end making sure no one was left behind and encouraging the slower ones enough so that we would not miss the last ferry back from Memurubu.

The most advanced tour made during that first summer was the climb to the top of Galdhøpiggen, the tallest mountain in Scandinavia. Here, we were all tied up on ropes crossing the glacier up toward the peak. From the top, one could see the cracks in the ice and how the path winded between them.

The second year, 1970, I could spend more time in the mountains, since I no longer needed to attend all the lectures. This year Per-Olov also wanted to climb the east peak of Turfinstindane (see Fig. 5) along a route that Frank had found at an earlier occasion. We were a limited group, only four of us. It was a rather long and at some places exposed climb (see Fig. 6), so we sometimes used ropes. This I consider my first real full-day climb. Frank (see Fig. 7) was the excellent guide for me. The way he was carefully looking for the easiest and safest path and also how he always checked the hold or the step before making the next move, each with only one hand or one foot at a time. I observed every move with the keenest eye and had no difficulty seeing a master and follow his movements. It was a most wonderful climb giving us an impressive view from the top (see Figs. 8 and 9). It took place on Frank's 41st birthday, and was well documented by Per-Olov with his mini-camera.

Frank was an excellent hiking partner. We had the same approach to hiking. None of us was keen in bringing too much food or drinks along, and we could both walk fast and with very little need to rest. On my third season, 1971, Frank and myself, with Per-Olov's permission, set out for a long hike.

Figure 5 Turfinstindane by the lake Bygdin (August 1970).

Figure 6 The author on the climb of the eastern peak of Turfinstindane with Bygdin below. *Photo: Per-Olov Löwdin, August 26, 1970.*

Figure 7 Frank leading the climb toward the eastern peak of Turfinstindane. *Photo: Per-Olov Löwdin, August 26, 1970.*

Figure 8 Frank Harris, Per-Olov Löwdin, and Peter Lindner at the top of the eastern peak of Turfinstindane. *Photo: the author with the camera of Per-Olov Löwdin, August 26, 1970.*

Figure 9 Turfinseggi with Svartdalspiggan and Mesmogtind in the background as seen from the eastern peak of Turfinstindane. *Photo: Per-Olov Löwdin, August 26, 1970.*

This was to make the western ridge along Svartdalen (see Fig. 9) from north to south. There was no path on any map, so we had to find the way on our own. The ridge contained some peaks, all of which we of course wanted to pass. Weather was good and stable, but the trip was long, and as the start of the actual hike would be late we had to be fast in order to reach to the end before the darkness would force us to abandon it beforehand.

We left the hotel in Beitostølen with two cars early in the morning of August 25, parking one in Bygdisheim before continuing with the other to Gjendesheim. There we caught the first ferry across the lake Gjende to

Figure 10 Svartdalspiggan seen from the ridge (August 25, 1971).

Gjendebu. After having a short breakfast, we started hiking at 10:25 am. First, we climbed Svartdalspiggan (see Fig. 10) more than thousand meter above Gjende before descending down to the ridge. We conquered all the peaks along the ridge including Mesmogtind (the highest one) and Langedalstind (note that the two top pictures on page 2802 of the *Int J Quantum Chem* Vol 109 (2009) are also from this hike and shows Frank at two points along the ridge) and then followed a rather horizontal Turfinseggi to the western peak of Turfinstindane (see Fig. 11). As we reached the latter only shortly after 8 pm in the evening, we had to abandon our original goal to pass all three peaks of Turfinstindane. Notwithstanding a rather fast direct descent, it was indeed very dark when we finally reached the tourist hut in Torfinsbu close to 1000 m below. After a good night's sleep, we had an easy hike out along the lake Bygdin (see Fig. 12) to our car in Bygdisheim. This fantastic hike the author considered as taking the final mountaineering exam for Frank.

The author has since then made many hikes on all continents but Antarctica, some together with Frank (see Figs. 13 and 14). Some of these have been rather long and tough and required good speed. In Munich,

Figure 11 Turfinseggi and the three peaks of Turfinstindane (August 25, 1971).

Figure 12 Frank posing in front of an old building at Hestvollen by the shore of lake Bygdin (August 26, 1971).

Figure 13 Frank and the author's wife Kerstin Kaijser at Mirror Lake in Utah, USA (July 5, 2005).

Figure 14 Frank and Kerstin on Swedish red granite rocks (August 1, 2006).

where the author is now living, he has the reputation among some of his friends of being too fast and too seldom stopping for food and drinks, just as his teacher. But due to the excellent guidance received, the author has never got into any serious trouble while hiking anywhere in the world. And the pleasure of hiking in mountains is still immense. Thank you Frank.

CHAPTER TWO

System-Size Dependence in Grand Canonical and Canonical Ensembles

Debajit Chakraborty*, James Dufty†,[1], Valentin V. Karasiev*
*Quantum Theory Project, Departments of Physics and of Chemistry, University of Florida, Gainesville, Florida, USA
†Department of Physics, University of Florida, Gainesville, Fl, USA
[1]Corresponding author: e-mail address: dufty@phys.ufl.edu

Contents

Abstract

The thermodynamics for a system with given temperature, density, and volume is described by the Canonical ensemble. The thermodynamics for a corresponding system with the same temperature, volume, and average density is described by the Grand Canonical ensemble. In general, a chosen thermodynamic potential (e.g., free energy) is different in the two cases. Their relationship is considered here as a function of the system size. Exact expressions relating the fundamental potential for each (free energy and pressure, respectively) are identified for arbitrary system size. A formal asymptotic analysis for large system size gives the expected equivalence, but without any characterization of the intermediate size dependence. More detailed evaluation is provided for the simple case of a homogeneous, non-interacting Fermi gas. In this case, the origin of size dependence arises from only two length scales, the average interparticle distance and quantum length scale (thermal deBroglie or Fermi length). The free energies per particle calculated from each ensemble are compared for particle numbers $2 \leq N \leq 64$ for a range of temperatures above and below the Fermi temperature. The relevance of these results for applications of density functional theory is discussed briefly.

Advances in Quantum Chemistry, Volume 71
ISSN 0065-3276
http://dx.doi.org/10.1016/bs.aiq.2015.03.002

1. INTRODUCTION AND MOTIVATION

Equilibrium statistical mechanics provides the fundamental basis for the thermodynamics of a given system in terms of its Hamiltonian and the characteristics of its environment (e.g., open or closed).[1] The Canonical ensemble applies when the system is in contact with a thermal reservoir, exchanging energy at constant volume and particle number. It is parameterized by the temperature ($T \equiv 1/k_B\beta$), number density ($n \equiv N/V$), and volume (V). The fundamental thermodynamic potential associated with this ensemble is the Helmholtz free energy per particle $f_C(\beta, n, V)$. The Grand Canonical ensemble applies under the same thermodynamic conditions but with the additional exchange of particle number with its environment. It is parameterized by β, μ, and V, where μ is the chemical potential. Its thermodynamic potential is the pressure $p_G(\beta, \mu, V)$. However, the free energy per particle in the Grand Canonical ensemble $f_G(\beta, n_G, V)$ can be determined from $p_G(\beta, \mu, V)$ by a change of variables $\mu \rightarrow n_G \equiv \partial p_G/\partial \mu$ via a Legendre transform (see below). Here, n_G is the average density in the Grand Canonical ensemble. Similarly, the pressure can be defined for the Canonical ensemble by the change of variables $n \rightarrow \mu_C \equiv -\partial f_C/\partial n$ and a corresponding Legendre transform.

For large systems, it is expected on physical grounds that the system becomes extensive, in which case the free energy per particle and pressure become independent of the volume

$$f_C(\beta, n, V) \rightarrow f_C(\beta, n), \quad p_G(\beta, \mu, V) \rightarrow p_G(\beta, \mu). \tag{1}$$

Furthermore, if the two ensembles have the same β, V, and μ is chosen such that $n = n_G(\beta, \mu)$ then the thermodynamics from the two ensembles should be equivalent in this limit, e.g.,

$$f_C(\beta, n) = f_G(\beta, n_G). \tag{2}$$

It is this equivalence that allows one to choose an equilibrium ensemble for convenience of computation or simulation, rather than to fit the actual experimental conditions of interest. For example, most formulations of density functional theory are based in the Grand Canonical ensemble while actual implementations in simulation are for conditions of the Canonical ensemble, specifically for fixed density and volume. This raises the challenge of quantifying the conditions for the validity of (1) and (2), and finding relationships between properties in different ensembles. The objective here is to formulate this problem more precisely and to provide some answers for the simplest case of a non-interacting Fermi gas.

The large system limit is defined by $V \rightarrow \infty$ at constant n or n_G for the Canonical and Grand Canonical ensembles, respectively. Equivalently, this can be stated as $N \rightarrow \infty$ at constant n, or $N_G \rightarrow \infty$ at constant n_G. In detail, the shape of the system must be constrained as well, e.g., all dimensions should be of comparable size L such that L/r_0 is large, where r_0 is the average interparticle spacing defined by $4\pi n r_0^3/3 = 1$. The desired limit requires that L be large compared to all other characteristic length scales as well. One of these is the force range of interaction, a. For Coulomb systems, this is replaced by the screening length. Another length scale is the thermal de Broglie wavelength λ which becomes large at low temperatures, or the corresponding Fermi length $\tilde{\lambda}$ at temperatures near zero. There can also be a scale set by the spatial variations of an external potential. Finally, the correlation length is typically of the order of the force range, but becomes large near a critical point so that system-size dependence can be important even for macroscopic systems. In cases for which L is not the dominant length scale, the system is "small" and, while the thermodynamic formalism is universal, the details must account for the specific environment of the system being described.[2] Here, only the Canonical and Grand Canonical conditions are considered, although many other ensembles for other environments are of experimental interest.[1,2] There is a large literature on the asymptotic evaluation of the difference between properties calculated in different ensembles, e.g., fluctuations in extensive variables.[3] Much less is known away from such asymptotic conditions. However, low temperature thermodynamic properties of interacting fermions in 1D system have also been discussed in the literature (see Ref. 4).

The next section defines the ensembles and their associated thermodynamic potentials. In particular, for comparisons exact relationships between them are identified for arbitrary system size. Generally, the thermodynamic properties for the two cases are not equal. However, in Section 3, an asymptotic analysis for one of these relationships shows their equivalence for large V (or large N) at constant number density. The analysis is formal and does not expose the full dependence on V nor the cross over to the extensive limit in (1). A more detailed quantitative evaluation is provided in Section 4 for the special case of a homogeneous non-interacting gas. In that case, the only relevant length scales are r_0 and λ (or $\tilde{\lambda}$). Finally, inhomogeneous non-interacting systems with an external potential are discussed in Section 5 and related to the results of Section 4 using a local density approximation (see below). The relevance for ensemble dependence and system-size corrections to the familiar Thomas–Fermi approximation in density functional theory[5] is discussed.

2. CANONICAL AND GRAND CANONICAL ENSEMBLES AND THEIR THERMODYNAMICS

The equilibrium Canonical ensemble for a system of N particles in a volume V, coordinates $\mathbf{q}_i$, with pairwise interactions and an external single particle potential is defined by the probability density operator

$$\rho_C = e^{-\beta(H_N - Nf_C)}, \quad \beta f_C = -\frac{1}{N}\ln Tr_N\, e^{-\beta H_N}. \tag{3}$$

Here, H_N is the Hamiltonian operator for N particles

$$H_N = K_N + \Phi_N + \sum_{i=1}^{N} v(\mathbf{q}_i), \tag{4}$$

where K and Φ are the total kinetic and potential energies, respectively. The specific forms of the pair potential $\phi\left(\mathbf{q}_i, \mathbf{q}_j\right)$ and external potential $v(\mathbf{q}_i)$ are not required at this point. The equilibrium thermodynamics for this system is defined from the free energy per particle $f_C(\beta, n, V)$ which is a function of the temperature $T = 1/k_B\beta$, the density $n = N/V$, and the volume V. The trace in the definition of f_C is taken over the N particle Hilbert space with the appropriate symmetrization (Bosons or Fermions). For large systems (i.e., $V \to \infty$ at fixed finite n), it is expected that $f_C(\beta, n, V)$ becomes independent of V.

The corresponding Grand Canonical ensemble is defined by the operator

$$\rho_G = e^{-\beta(H_N - \mu N + p_G V)}, \quad \beta p_G V = \ln \sum_{N=0}^{\infty} Tr_N\, e^{-\beta(H_N - \mu N)}. \tag{5}$$

The thermodynamics now is defined from the pressure $p_G(\beta, \mu, V)$, where the density dependence of the Canonical ensemble is replaced by a dependence on the chemical potential μ. For large systems (i.e., $V \to \infty$ at fixed finite μ), it is expected that the pressure becomes independent of V.

Although the pressure is the fundamentally defined thermodynamic potential in the Grand Canonical ensemble, the corresponding free energy,

$f_G(\beta, n_G, V)$, is defined in terms of that pressure by a change of variables from μ to n_G using the Legendre transformation

$$f_G n_G = -p_G + \mu n_G. \tag{6}$$

Here, the average number density $n_G(\beta, \mu, V)$, is

$$n_G(\beta, \mu, V) \equiv \frac{\partial p_G(\beta, \mu, V)}{\partial \mu}. \tag{7}$$

Similarly, although the free energy is the fundamental potential in the Canonical ensemble, the pressure $p_C(\beta, \mu_C, V)$ is defined in terms of that free energy by a change of variables from n to μ_C using the Legendre transformation

$$p_C = -f_C n + \mu_C n, \tag{8}$$

where the chemical potential in the Canonical ensemble is

$$\mu_C(\beta, n, V) \equiv -\frac{\partial f_C(\beta, n, V)}{\partial n}. \tag{9}$$

From the forgoing definitions it is seen that the thermodynamics defined by the two ensembles are related exactly by the relation

$$e^{\beta p_G(\beta, \mu, V)V} = \sum_{N=0}^{\infty} e^{\beta \mu N} e^{-\beta f_C(\beta, n, V)N}. \tag{10}$$

The volume is the same for each term in this summation, so the density n changes accordingly. The inversion of this relationship is obtained in Appendix A:

$$e^{-\beta f_C(\beta, n, V)N} = \frac{1}{2\pi} \int_0^{2\pi} d\theta e^{i\theta N} e^{\beta p_G(\beta, \mu = -i\theta/\beta, V)V} \tag{11}$$

Note that the Grand Canonical pressure must be analytically extended to complex values of the chemical potential.

As noted above, the determination of Canonical ensemble properties from given Grand Canonical ensemble results is relevant for practical applications of density functional theory. The inversion of (10) has been discussed recently[6] where it is proposed to construct $f_C(\beta, n, V)$ from a set of linear equations obtained from evaluation of $p_G(\beta, \mu, V)$ at M discrete values of μ. In principle, this requires $M \to \infty$ but approximate values for $f_C(\beta, n, V)$ are obtained for finite M. More systematic expansions are described in Refs. 7,8. This latter work has been generalized by Lutsko[9]. Equation (11) appears to be new. A similar complex expression is given in Ref. 3, section 3.2, but under the assumption that the discrete summation over N in (10) can be replaced by an integration. In that case, it becomes a Laplace transform for which the complex Bromwich integral provides its inversion. The explicit construction of (11) for the actual discrete case is given in Appendix A.

3. THERMODYNAMIC EQUIVALENCE FOR LARGE SYSTEMS

In this section, the limit of large systems is considered. For it, the Canonical and Grand Canonical thermodynamics are expected to be equivalent. For the Canonical ensemble, large systems means the limit $N \to \infty$ at constant finite density $n = N/V$ and temperature. For the Grand Canonical ensemble, this limit is $V \to \infty$ at constant chemical potential μ and temperature. To show this equivalence, consider again (11) written as

$$\beta f_C(\beta, n, V)N = -\ln \frac{1}{2\pi}\int_0^{2\pi} d\theta e^{VA(z=\theta)}, \tag{12}$$

where now $A(z)$ is a real function of the complex variable z,

$$A(z) = izn + \beta p_G(\beta, -iz/\beta, V). \tag{13}$$

It has a stationary saddle point at the value $z \equiv z_s$ defined by $dA/dz = 0$. Using (7) this is

$$\text{Re}\ n_G(\beta, -iz_s/\beta, V) = n, \quad \text{Im}\ n_G(\beta, -iz_s/\beta, V) = 0. \tag{14}$$

Since $n_G(\beta,z,V)$ is a real function of z the solution is $z_s = i\beta\mu_s$ with real μ_s determined from

$$n_G(\beta, \mu_s, V) = n. \tag{15}$$

Now let C denote a closed contour in the z plane including the interval $[0, 2\pi]$ along the positive real axis and passing through $i\beta\mu_s$ on the complex axis. Assume that $A(z)$ is analytic on and within C, so that the integral of $\exp(VA(z))$ over the entire contour must vanish. Consequently, the integral of (12) can be replaced by an integration over that part of C complementary to the interval $[0, 2\pi]$. Denoting that part by C'

$$\beta f_C(\beta, n, V)N = -\ln \frac{\beta}{2\pi}\int_{C'} dz e^{VA(z)}, \tag{16}$$

where by definition C' passes through the stationary point μ_s tangent to the complex axis. Since $A(z)$ is multiplied by V, the contribution near $i\beta\mu_s$ gives the dominant contribution for large system size. The usual saddle-point analysis then leads to the asymptotic result

$$\begin{aligned}\beta f_C(\beta,n,V)N &\to -A(\mu_s)V - \ln\frac{\beta}{2\pi}\int_{-\infty}^{\infty} dx e^{-\frac{1}{2}V|A''|(x-\mu_s)^2} \\ &= -\mu_s\beta N + \beta p_G(\beta,\mu_s,V)V + O(\ln N)\end{aligned} \tag{17}$$

The first two terms are proportional to the free energy of the Grand ensemble evaluated at the value of the chemical potential that ensures $N_G(\beta,\ \mu_s, V\ ) = N$. The free energies are therefore the same up to small corrections of the order $(\ln N)/N$

$$f_C(\beta,n,V) = f_G(\beta,n_G,V) + O(\frac{1}{N}\ln N) \tag{18}$$

This is the expected equivalence for large systems. Note, however, that the analysis does not show that the free energy per particle is independent of V. That question is explored in more detail in the next two sections.

4. NON-INTERACTING, HOMOGENEOUS SYSTEMS AT FINITE SYSTEM SIZE

In this section, the thermodynamics for the Grand Canonical and Canonical ensembles are calculated exactly at arbitrary system size for the simplest case of non-interacting particles without external potential. The Hamiltonian for N particles is

$$H_N^0 = \sum_{i=1}^{N}\frac{\hat{p}_i^2}{2m}. \tag{19}$$

For the Canonical ensemble, the particle number and volume are fixed so the boundary conditions chosen here are a cubic box of sides L with hard walls. Then the momentum components have eigenvalues

$$p_\alpha = \frac{\pi\hbar}{L}k_\alpha, \quad \alpha = x,y,z \tag{20}$$

where k_α is a positive integer. The Grand Canonical ensemble represents an open system without fixed particle number. However, its derivation represents this as the sum of probabilities for closed systems at the same volume but different particle number. Hence the same boundary conditions can be used for calculation of components of each N within the ensemble.

4.1 Grand Canonical Ensemble

The pressure in the Grand Canonical Ensemble is given by (5). Since H_N is the sum of single-particle operators the summation and trace can be performed directly in occupation number representation with the result for spin 1/2 Fermions[10]

$$\beta p_G = \frac{2}{V}\sum_{\mathbf{k}} \ln\left(1 + e^{\beta\mu} e^{-\left(\frac{k}{\ell}\right)^2}\right). \tag{21}$$

The three-fold summation is over $\mathbf{k} = k_x, k_y, k_z$. Use has been made of

$$\beta\frac{p^2}{2m} = \frac{\beta}{2m}\left(\frac{\pi\hbar}{L}\right)^2 k^2 = \left(\frac{k}{\ell}\right)^2, \quad \ell^2 = \frac{4}{\pi}\left(\frac{L}{\lambda}\right)^2, \tag{22}$$

where $\lambda = \left(2\pi\beta\hbar^2/m\right)^{1/2}$ is the thermal de Broglie wavelength. Similarly, the average number density n_G is

$$n_G = \frac{2}{V}\sum_{\mathbf{k}} \left(e^{-\beta\mu} e^{(k/\ell)^2} + 1\right)^{-1}. \tag{23}$$

It is tempting at this point to represent the summations over $\mathbf{k}$ as integrals, i.e.,

$$\sum_{k_x} F(k/\ell) = \ell \sum_{x} \Delta x F(x) \stackrel{?}{\rightarrow} \ell \int dx F(x) \tag{24}$$

Indeed this replacement leads to the familiar textbook results in terms of Fermi integrals. However, $\Delta x = \Delta k_x/\ell = 1/\ell$ is small only for $L/\lambda \gg 1$. This is not the case for low temperatures or small system sizes. Hence for the purposes here the discrete summation must be evaluated directly.

At this point, all properties will be given a corresponding dimensionless form. The dimensionless temperature t is

$$t = \frac{1}{\beta\epsilon_F}, \quad \epsilon_F = \frac{1}{2m}\hbar^2\left(3\pi^2 n_G\right)^{2/3}. \tag{25}$$

where ϵ_F is the Fermi energy. It follows that

$$n_G\lambda^3 = \frac{8}{3\sqrt{\pi}}t^{-3/2}, \tag{26}$$

so that (23) becomes

$$t^{-3/2} = \frac{6}{\pi\ell^3}\sum_{\mathbf{k}}\left(e^{-\beta\mu}e^{(k/\ell)^2}+1\right)^{-1}, \tag{27}$$

and

$$\ell = \left(\frac{3}{\pi}N_G t^{3/2}\right)^{1/3}. \tag{28}$$

An appropriate dimensionless pressure is

$$p_G^*(t,N_G) = \frac{\beta p_G}{n_G} = \frac{2}{N_G}\sum_{\mathbf{k}}\ln\left(1+e^{\beta\mu}e^{-\left(\frac{k}{\ell}\right)^2}\right). \tag{29}$$

Here, it is understood that $\beta\mu = \beta\mu(t,N_G)$ as determined from (27). Finally, the dimensionless free energy per particle is obtained from the Legendre transformation as described in (6).

$$f_G^*(t,N_G) \equiv -p_G^*(t,N_G) + \beta\mu(t,N_G). \tag{30}$$

The dimensionless system-size parameter is now N_G. The analysis proceeds as follows: (1) choose a value for N_G and calculate $\beta\mu(t,N_G)$ as a function of t from (27). Repeat for different values of N_G. The results are shown in Fig. 1A. Also shown is the limiting value for $N_G \to \infty$ obtained from the continuum limit (i.e., 24); (2) calculate $p_G^*(t,N_G)$ from (29) as a function of t for the same set of values for N_G. The results are shown in Fig. 1B; and (3) calculate the dimensionless free energy $f_G^*(t,N_G)$ as a function of t for the same set of N_G from (30). The results are shown in Fig. 2.

These figures show that the system-size dependence is small for $N_G \geq 16$ at $t = 10$, but is more significant as the temperature is lowered. This is expected since that dependence is controlled by $\ell = \left(\frac{3}{\pi}N_G t^{3/2}\right)^{1/3}$ and vanishes only for large ℓ. Below $t = 1$, larger values of N_G are required to approach system-size independence.

4.2 Canonical Ensemble

Equation (11) shows that the Canonical ensemble free energy per particle can be obtained from the Grand Canonical pressure, extended to complex values for the chemical potential. It is written as

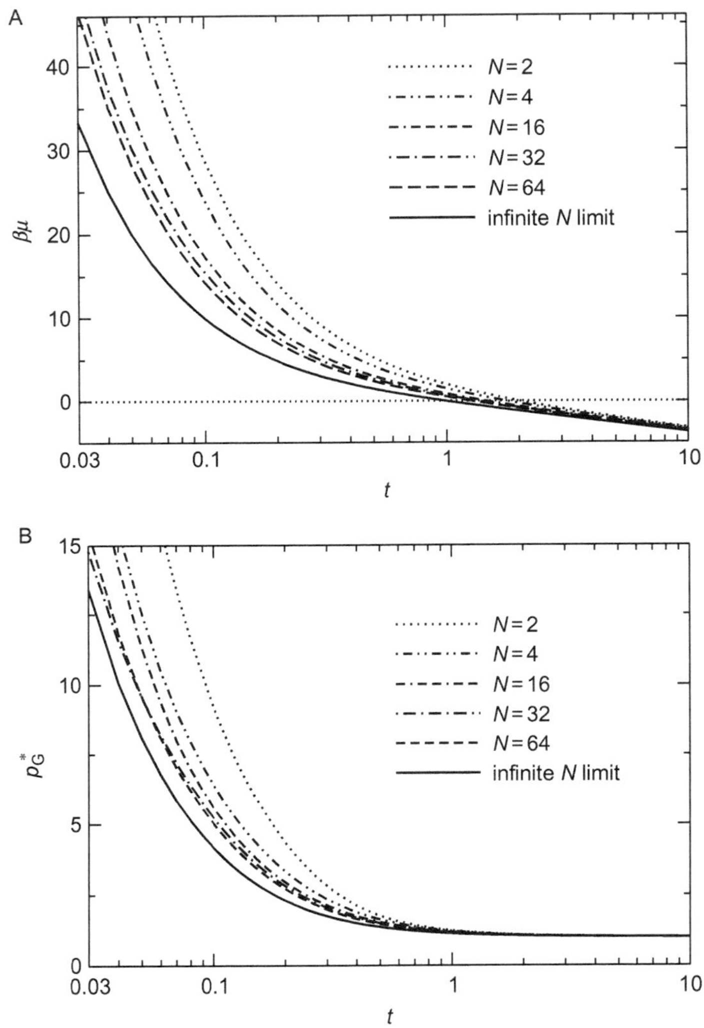

Figure 1 (A) Plot of $\beta\mu(t,N_G)$ as a function of the dimensionless temperature t for several values of $N_G = N$. Also shown is the large system-size limit. (B) Plot of the dimensionless pressure $p^*_G(t,N_G)$ as a function of the dimensionless temperature t for several values of $N_G = N$.

$$\beta f_C(\beta, n, V) = -\frac{1}{N} \ln \frac{1}{2\pi} \int_0^{2\pi} d\theta e^{(i\theta + g(\theta, t, N))N}, \tag{31}$$

with

$$g(\theta, t, N) \equiv \frac{\beta p_G(\beta, \mu = -i\theta/\beta, V)}{n} = \frac{2}{N} \sum_{\mathbf{k}} \ln\left(1 + e^{-i\theta} e^{-\left(\frac{k}{\ell}\right)^2}\right). \tag{32}$$

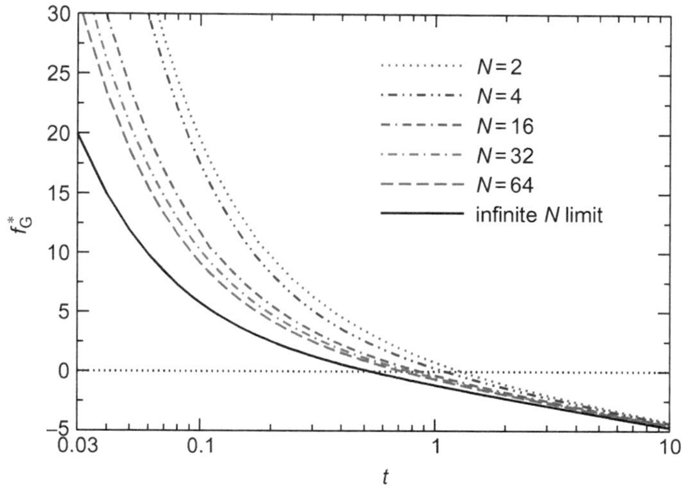

Figure 2 Plot of the dimensionless free energy per particle, $f_G^*(t, N_G) = \beta f_G(\beta, n_G, V)$, as a function of the dimensionless temperature t for several values of $N_G = N$.

The definitions of t and ℓ are the same as in (25) and (28) except with n_G and N_G replaced by n and N. The calculation of $g(\theta, t, N)$ is similar to that of $p_G^*(t, N_G)$ in (29), except that it has both real and imaginary parts. Their numerical calculation is straightforward but the final θ integration of (31) is now problematical. Due to the complex integral, it has an oscillatory integrand whose variation increases as N, and whose modulation varies between values of the order $\exp(\pm N)$. Figure 3 illustrates the problem for $N = 64$, $t = 0.63$. This is a precursor for the crossover to the asymptotic analysis of Section 3. The first difficulty of a rapidly oscillating integrand can be overcome by increasing the density of mesh points. That works in principle for both very large N and t. The second problem of huge cancellations when the integral is evaluated numerically as a sum over mesh points does not have a simple solution because of the finite precision of floating point numbers. As can be seen from the Fig. 3, the highest magnitude of the integrand is $\sim 10^9$ while the final value of the real part of the integral is $\sim 10^{-10}$. Quadruple precision is required for adequate accuracy within the restricted domain $0.1 < t < 10$ and $N \leq 64$.

Figure 4A–D shows the dimensionless Canonical ensemble free energy per particle in comparison with the corresponding Grand Canonical ensemble results of the last section. Generally, for $N > 16$ there is good agreement between the results of the two ensembles, although significant system-size

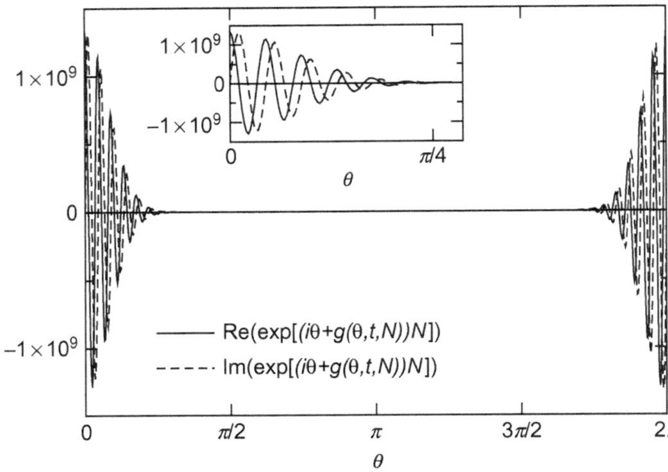

Figure 3 Illustration of the strong variation with θ for the integrand of (31), for $N = 64$, $t = 0.63$.

dependence relative to the large system-size limit remains. At smaller values of N, the discrepancies between the two ensembles decreases at lower t. On the contrary, the difference for both ensemble from the large system limit increases with smaller N and t.

For $t = 0$, the free energy for both Canonical and Grand Canonical ensembles are the same for given N as it is the sum of discrete energies up to the Fermi energy for N particles. This, however, is still different from the infinite system limit which is an integral over the density of states.

5. RELATIONSHIP TO DENSITY FUNCTIONAL THEORY

Density functional theory (DFT) describes the thermodynamics of an equilibrium, inhomogeneous system whose Hamiltonian has the form (4).[5] The external potential implies that the local density is non-uniform. DFT has a variational principle that states that the thermodynamic properties are obtained from a functional of this density at its extremum. The definition of the functional can be given as follows. First, the Grand Canonical ensemble pressure and density are computed as functionals of the external potential as in (5)

$$\beta p_G V = \ln \sum_{N=0}^{\infty} Tr_N \, e^{-\beta(H_N - \mu N)}, \quad n_G(\mathbf{r}) \equiv -\frac{\partial p_G(\beta, \mu, V)}{\partial v(\mathbf{r})}. \tag{33}$$

Next, the external potential is eliminated by inverting the second equation to give $\beta p_G V$ as a functional of the density and finally, the density functional of DFT is then given by[5]

$$F_{\mathrm{DFT}} \equiv -p_G V + \int \mathrm{d}\mathbf{r}(\mu - v(\mathbf{r}))n_G(\mathbf{r}), \tag{34}$$

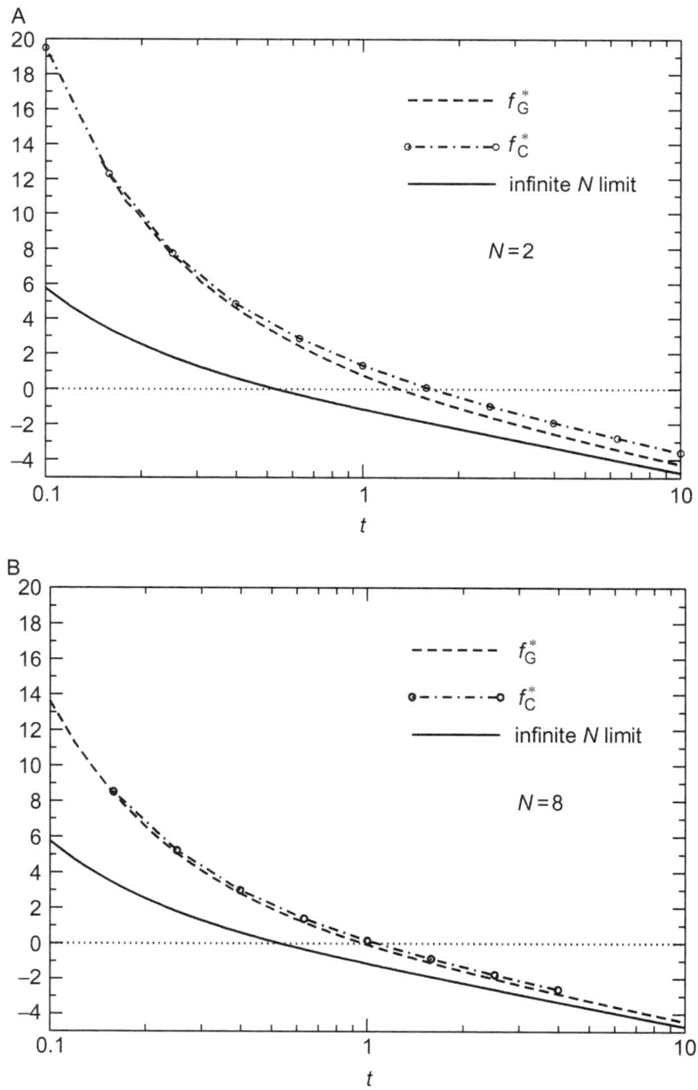

Figure 4—Cont'd

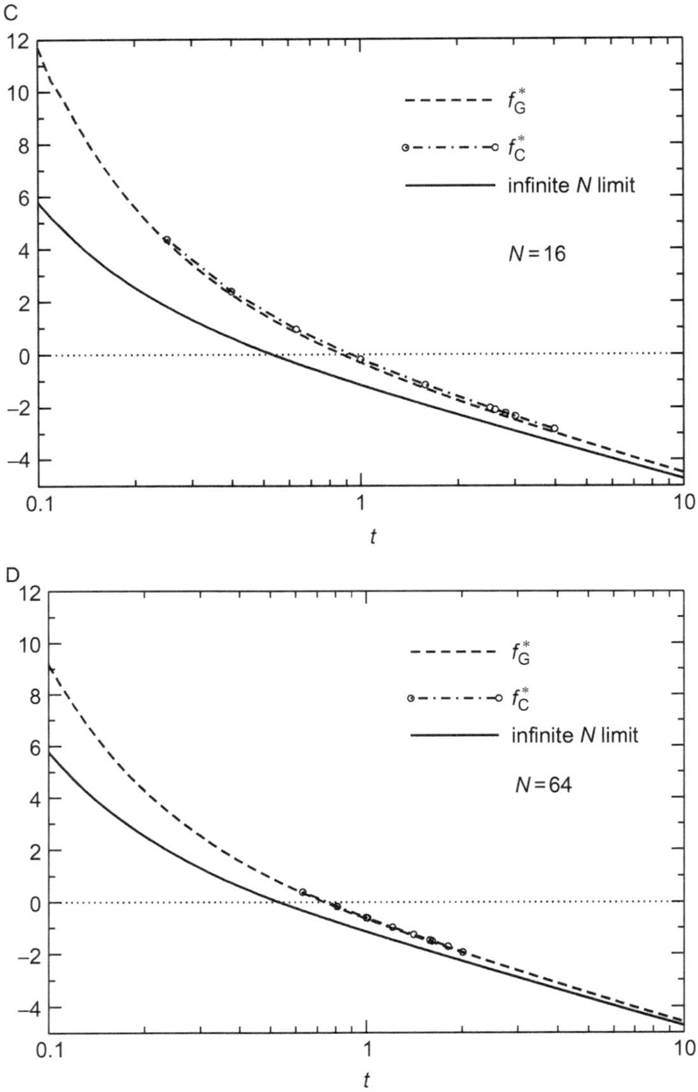

Figure 4—Cont'd Panel (A) compares the dimensionless free energy per particle for Canonical and Grand Canonical ensemble ($f_G^*(t, N_G)$, $fC^*(t, N)$ as a function of the dimensionless temperature t for $N = 2$. Also shown is the large system-size limit. Panels (B)–(D) show the same comparison for $N = 8$, $N = 16$, and $N = 64$, respectively.

It is understood that the density and external potential in the second term are now independent functions. They become related by the extremum condition that provides the equilibrium density in terms of the external potential. Finally, with that relationship established, evaluation of F_{DFT} at its extremum gives the Legendre transform (6) (extended to the inhomogeneous case) and hence the equilibrium Grand Canonical free energy.

It is clear from this brief description of DFT that its theoretical formulation is tied to the Grand Canonical ensemble. However, in practice construction of approximate functionals often presumes the large system-size limit (e.g., Thomas–Fermi and local density approximations). Calculations almost always fix the total number of particles, N, as in the Canonical ensemble. Consequently, system-size corrections and ensemble dependencies are overlooked or ignored. The analysis of the previous sections is therefore quite relevant for current problems of DFT.

To illustrate this, consider the non-interacting part of the DFT functional constructed as above

$$\beta p_{\mathrm{G}}^{(0)} V = \int d\mathbf{r} \left\langle \mathbf{r} \left| \ln\left(1 + e^{\beta\mu} e^{-\beta\left(\frac{\hat{p}^2}{2m} + v(\hat{\mathbf{q}})\right)}\right) \right| \mathbf{r} \right\rangle \tag{35}$$

$$n_{\mathrm{GC}}^{(0)}(\mathbf{r}) = \left\langle \mathbf{r} \left| \left(e^{-\beta\mu} e^{\beta\left(\frac{\hat{p}^2}{2m} + v(\hat{\mathbf{q}})\right)} + 1\right)^{-1} \right| \mathbf{r} \right\rangle \tag{36}$$

$$\begin{aligned} \beta F_{\mathrm{DFT}}^{(0)} = & -\int d\mathbf{r} \left\langle \mathbf{r} \left| \ln\left(1 + e^{\beta\mu} e^{-\beta\left(\frac{\hat{p}^2}{2m} + v^{(0)}(\hat{\mathbf{q}} n_{\mathrm{GC}})\right)}\right) \right| \mathbf{r} \right\rangle \\ & + \int d\mathbf{r} \beta(\mu - v(\mathbf{r})) n_{\mathrm{GC}}^{(0)}(\mathbf{r}). \end{aligned} \tag{37}$$

On the right side of (36) $v^{(0)}(\hat{\mathbf{q}} \mid n_{\mathrm{GC}}^{(0)})$ denotes the inversion of (35) to obtain $v(\mathbf{r})$ as a functional of $n_{\mathrm{GC}}^{(0)}(\mathbf{r})$. Further construction of $\beta F_{\mathrm{DFT}}^{(0)}$ is nontrivial for general external potential and entails diagonalization of the single particle Hamiltonian $\hat{p}^2/2m + v(\hat{\mathbf{q}})$ and self-consistent inversion of the expression for $n_{\mathrm{GC}}^{(0)}(\mathbf{r})$ (the Kohn–Sham approach).[11,12] A simpler method is the local density approximation that replaces the operator dependence of the external potential by its value at the point of interest, $(v(\hat{\mathbf{q}})) \rightarrow v(\mathbf{r}))$. Then for instance the density equation can be evaluated in momentum representation using the same boundary conditions as above

$$n_{\mathrm{GC}}^{(0)}(\mathbf{r}) \to \frac{2}{V}\sum_{\mathbf{k}}\left(\mathrm{e}^{-\beta(\mu-v(r))}\mathrm{e}^{(k/\ell)^2}+1\right)^{-1}|\psi_{\mathbf{k}}(\mathbf{r})|^2 \tag{38}$$

In the large system-size limit, the summation can be represented as an integration and becomes the familiar finite temperature Thomas–Fermi approximation

$$n_{\mathrm{TF}}^{(0)}(\mathbf{r}) \to h^{-3}\int d\mathbf{p}\left(\mathrm{e}^{-\beta(\mu-v(r))}e^{\beta\frac{p^2}{2m}}+1\right)^{-1}. \tag{39}$$

Equations (38) and (39) are the same results as for the homogenous system analysis of the last section, with only the replacement $\mu \to \mu - v(r)$. Hence the system-size corrections found there for small N,t apply here as well, and those corrections for the free energy per particle identified in Figs. 2 and 4A–D are required for the DFT functional as well. Notwithstanding those corrections, it is expected that differences between the results for the two ensembles are small for $N = N_{\mathrm{G}} > 16$. Further discussion of system-size and ensemble dependence of the DFT functional will be given elsewhere.

ACKNOWLEDGMENTS

The authors thank J. Lutsko for sharing unpublished research and for helpful comments. Work supported by U.S. Dept. of Energy grant DE-SC0002139.

A. APPENDIX

A.1 Determination of $f_{\mathrm{C}}(\beta, n, V)$ from $p_{\mathrm{G}}(\beta, \mu, V)$

The definition of $f_{\mathrm{C}}(\beta,n,V)$ in (3) can be written in the equivalent form

$$\beta f_{\mathrm{C}} = -\frac{1}{N}\ln Tr_N\,\mathrm{e}^{-\beta H_N} = -\frac{1}{N}\ln\sum_{M=0}^{\infty}Tr_M\delta_{N,M}\mathrm{e}^{-\beta H_M}, \tag{A.1}$$

with a representation for the Kronecker delta $\delta_{N,M}$ to get

$$\begin{aligned}\beta f_{\mathrm{C}}(\beta,n,V) &= -\frac{1}{N}\ln\frac{1}{2\pi}\int_0^{2\pi}d\theta\mathrm{e}^{i\theta N}\sum_{M=0}^{\infty}Tr_M\mathrm{e}^{-i\theta M}\mathrm{e}^{-\beta H_M}\\ &= -\frac{1}{N}\ln\frac{1}{2\pi}\int_0^{2\pi}d\theta\mathrm{e}^{i\theta N}e^{\beta p_{\mathrm{G}}(\beta,\mu=-i\theta/\beta,V)V}.\end{aligned} \tag{A.2}$$

The last line follows from the definition of p_G in (5). This gives the relationship (11) quoted in the text

$$e^{-\beta f_C(\beta, n, V)N} = \frac{1}{2\pi}\int_0^{2\pi} d\theta e^{i\theta N} e^{\beta p_G(\beta, \mu=-i\theta/\beta, V)V}. \quad (A.3)$$

The consistency of this result with its inverse (10) can be demonstrated by substituting the latter into the right side of (11)

$$\begin{aligned} e^{-\beta f_C(\beta, n, V)N} &= \frac{1}{2\pi}\int_0^{2\pi} d\theta e^{i\theta N} \sum_{M=0}^{\infty} e^{-i\theta M} e^{-\beta f_C(\beta, n=M/V, V)M} \\ &= \sum_{M=0}^{\infty} \frac{1}{2\pi}\int_0^{2\pi} d\theta e^{i\theta(N-M)} e^{-\beta f_C(\beta, n=M/V, V)M} \\ &= e^{-\beta f_C(\beta, n, V)N}. \end{aligned} \quad (A.4)$$

REFERENCES

1. Hill, T. *Statistical Mechanics*; Dover: New York, 1987.
2. Hill, T. *Thermodynamics of Small Systems*; Dover: New York, 1994.
3. Munster, A. *Statistical Thermodynamics*; Vol. 3(1); Springer-Verlag: Berlin, 1969.
4. Schönhammer, K.; Meden, V. Fermion-Boson Transmutation and Comparison of Statistical Ensembles in One Dimension. *Am. J. Phys.* **1996**, *64*, 1168.
5. Mermin, N. Thermal Properties of Inhomogeneous Electron Gas. *Phys. Rev.* **1965**, *137*, A1441.
6. de las Heras, D.; Schmidt, M. Full Canonical Information from Grand-Potential Density-Functional Theory. *Phys. Rev. Lett.* **2014**, *113*, 238304.
7. Kosov, D.; Gelen, M.; Vdovin, A. Calculations of Canonical Averages from the Grand Canonical Ensemble. *Phys. Rev. E* **2008**, 77, 021120.
8. Gonzalez, A.; White, J.; Roman, A.; Evans, R. How the Structure of a Confined Fluid Depends on the Ensemble: Hard Spheres in a Spherical Cavity. *J. Chem. Phys.* **1998**, *109*, 3637.
9. Lutsko, J. Relations Between Averages Computed in Different Ensembles. Unpublished.
10. Mazenko, G. *Equilibrium Statistical Mechanics*; Wiley: New York, 2000.
11. Kohn, W.; Sham, L. Self-Consistent Equations Including Exchange and Correlation Effects. *Phys. Rev.* **1965**, *140* (4A), A1133.
12. Parr, R.; Yang, W. *Density-Functional Theory of Atoms and Molecules*; Clarendon Press: Oxford [England], 1994.

CHAPTER THREE

The Mean Excitation Energy of Atomic Ions

Stephan P.A. Sauer*, Jens Oddershede†,‡, John R. Sabin†,‡,1
*Department of Chemistry, University of Copenhagen, Copenhagen, Denmark
†Department of Chemistry, Physics, and Pharmacy, University of Southern Denmark, Odense, Denmark
‡Quantum Theory Project, Department of Physics, University of Florida, Gainesville, Florida, USA
[1]Corresponding author: e-mail address: sabin@qtp.ufl.edu

Contents

Abstract

A method for calculation of the mean excitation energies of atomic ions is presented, making the calculation of the energy deposition of fast ions to plasmas, warm, dense matter, and complex biological systems possible. Results are reported to all ions of helium, lithium, carbon, neon, aluminum, and argon.

1. INTRODUCTION

The sorts of materials treated by theoretical methods in physics have changed radically in recent years, from perfect solids and gasses at zero temperature to much more complicated systems such as plasmas,[1] warm dense matter,[2] and large and complex biochemical structures.[3] For example, the components in plasmas are almost always highly ionized, depending on the plasma energy being higher than component ionization potentials, with electrons, atoms, and ions composing a quasineutral soup or mixture. It is frequently desired to pump energy into these systems, and this is often done by focusing a fast ion beam on the target mixture, where collisions, perhaps the result of complicated motions depending on what E and B fields are

ISSN 0065-3276
http://dx.doi.org/10.1016/bs.aiq.2015.02.001

present, result in conversion of some of the kinetic energy of the fast ion to kinetic and electronic energy of the components of the target mixture.

Understanding of the details of the deposition of energy by a swift ion in materials is important to the understanding of many processes, from interaction of molecules in space with the solar wind to radiation treatment of tumors. In all such cases, projectile kinetic energy is converted to target electronic energy on collision. The quantity describing such energy transfer is the energy deposited by the projectile per unit length of the projectile trajectory, known as the stopping power, $-\mathrm{d}E(v)/\mathrm{d}x$, of the target, which depends on the projectile velocity v, and the scatterer density, n. Frequently the stopping power is normalized with respect to scatterer density in order to make comparisons among systems of different densities

$$S(v) = -\frac{1}{n}\frac{\mathrm{d}E(v)}{\mathrm{d}x} \tag{1}$$

Here, $S(v)$ is referred to as the stopping cross-section of the target. The stopping cross-section, with appropriate constants removed, is referred to in Bethe-like theories[4] as the stopping number $L(v)$:

$$S(v) = \frac{4\pi e^4 Z_1^2 Z_2}{mv^2} L(v) \tag{2}$$

where Z_1 and Z_2 are the projectile charge and target electron number. The stopping number is further expanded in powers of the projectile charge:

$$L(v) = \sum_{i=0} Z_1^i L_i \approx L_0 + Z_1 L_1 + Z_1^2 L_2 \tag{3}$$

of which, the first, or Bethe,[5] term is by far the most important and is the subject of this chapter. The Bethe term consists of two terms, the Bethe logarithm and the so-called shell corrections:

$$L(v) = \ln\frac{2mv^2}{I_0} - \frac{C}{Z_1} \tag{4}$$

Further small corrections to the stopping number L_1 and L_2 are reported by Lindhard[6,7] and Bloch.[8]

Of the quantities in Eq. (4), the critical quantity is the mean excitation energy of the target, I_0, which is defined[9] as the first energy weighted moment of the target dipole oscillator strength distribution:

$$\ln I_0 = \frac{\int \frac{\mathrm{d}f}{\mathrm{d}E} \ln E \mathrm{d}E}{\int \frac{\mathrm{d}f}{\mathrm{d}E} \mathrm{d}E} \tag{5}$$

The mean excitation energy describes how easily a target, typically a molecule or atom can absorb kinetic energy from the projectile, primarily as electronic (including ionization) and vibrational (including fragmentation) excitation.

The mean excitation energy is the characteristic of the target only and has no dependence on the properties of the projectile ion.

The next section will describe how we are calculating the electronic excitation energies and corresponding transition moments need to obtain this material constant.

2. THE POLARIZATION PROPAGATOR AND MEAN EXCITATION ENERGY

The vertical electronic excitation energies and associated electronic transition dipole moments of a molecule are conveniently extracted from the linear response function or polarization propagator,[10] as evident from its spectral representation in the basis of eigenstates $\{|\Psi_0\rangle, |\Psi_k\rangle\}$ of the electronic Hamiltonian $\hat{H}$:

$$\langle\langle \hat{P}; \hat{Q} \rangle\rangle_E = \sum_{k \neq 0} \left[\frac{\langle \Psi_0 | \hat{P} | \Psi_k \rangle \langle \Psi_k | \hat{Q} | \Psi_0 \rangle}{E - E_k + E_0} + \frac{\langle \Psi_0 | \hat{Q} | \Psi_k \rangle \langle \Psi_k | \hat{P} | \Psi_0 \rangle}{E + E_k - E_0} \right] \tag{6}$$

in which the sum extends over all excited states $\{|\Psi_k\rangle\}$ of the system, both bound and continuum states. The poles and residues of the propagator give the excitation energies, $E_{0k} = E_k - E_0$, and transition matrix elements, $\langle \Psi_0 | \hat{P} | \Psi_k \rangle$, of the operators $\hat{P}$ and $\hat{Q}$. If we choose $\hat{P} = \hat{Q} = \sum_i \vec{r}_i$ (using atomic units), the residues and poles are usable to compute the oscillator strengths of the system in the dipole-length formulation. Similarly, the choice of $\hat{P} = \hat{Q} = \sum_i \vec{p}_i$ yields the dipole velocity and the choice of $\hat{Q} = \sum_i \vec{r}_i$ and $\hat{P} = \sum_i \vec{p}_i$ the mixed formulations.

This approach yields a finite number of excitations. As in previous work, we approximate the continuum with discrete excitations (pseudostates) of finite number placed such that they represent the continuum. We have found that this discretization of the continuum works satisfactorily when sums over the entire excitation spectrum are taken,[11] but no significance attaches to an individual pseudostate. The energy-weighted DOSD sum rules and mean excitation energies in Eq. (5) are obtained on explicit summation of the oscillator strengths to all bound states and to the continuum pseudostates.

Experience shows[12–15] that inclusion of electron correlation to some extent is required to calculate reliable spectral moments of the DOSD. One must calculate the propagator at the level of, at least, the linearized time-dependent Hartree–Fock level, also called the random phase approximation (RPA); this RPA adds correlation to both ground and excited states in a balanced way. In a complete computational basis set, the RPA, used in most of our propagator calculations, the dipole oscillator strengths, calculated in velocity, length, or mixed representation, and all sum rules, would be identical. The Thomas–Reiche–Kuhn (TRK) sum rule

$$S_0 = \sum_{n>0} f_{0n} = N \tag{7}$$

would be fulfilled exactly, i.e., the sum of the dipole oscillator strengths would be equal to the number of electrons, if the computational basis was complete. Comparison of the oscillator strengths calculated in various formulations thus provides a measure of the completeness of the computational basis in addition to the fulfillment of the TRK sum rule. Both criteria are extensively employed in our work to ensure the completeness of the employed one-electron basis sets. Further, inclusion of correlation, e.g., at the coupled cluster, level turned out to lead only to minor changes of the RPA results.[16]

3. COMPUTATIONAL DETAILS

Calculations were carried out for the neutral atom and all ions for He, Li, C, Ne, Al, and Ar. All calculations have been carried out with the 2013 version of the Dalton program package[17] apart from the calculations on the one-electron systems, for which the Turbomole program was employed.[18,19] In practice, the RPA calculations of the open-shell ions were in the Dalton program carried out as MCSCF calculations with one configuration corresponding to one of the states of the atomic term. Based on our previous basis set studies,[16,20] we have adopted the largest correlation-consistent basis sets of Dunning and coworkers[21] as one-electron basis sets.

In particular, we have employed the aug-cc-pCV5Z basis set for Ne, C, Al, and Ar ions, the aug-cc-pV5Z basis set for the He ions, and the aug-cc-pCVQZ basis set for Li ions. For all of these atoms and their ions, the TRK sum rule was found to be fulfilled to at least to decimal places in the length, mixed, and velocity representations. Also, the mean excitation energies are the same to three digits in the three representations. Although this equivalence is not a sufficient condition, it is nevertheless a necessary condition for basis set completeness.

4. RESULTS

The calculated mean excitation energies and values of the TRK sums (S_0) for several lighter atoms and their ions are given in the Tables 1–6. In all

Table 1 I_0 (eV) and S_0 of He and Its Ion Calculated with the aug-cc-pV5Z Basis Set

	S_0	I_0 (in eV)
He	2.00	42.68
He^{1+}	1.00	59.88

Table 2 I_0 (eV) and S_0 of Li and Its Ions Calculated with the aug-cc-pCVQZ Basis Set

	S_0	I_0 (in eV)
Li	3.00	33.1
Li^{1+}	2.00	108.1
Li^{2+}	1.00	134.1

Table 3 I_0 (eV) and S_0 of C and Its Ions Calculated with the aug-cc-pCV5Z Basis Set

	S_0	I_0 (in eV)
C	6.00	65.9
C^{1+}	5.00	92.6
C^{2+}	4.00	134.8
C^{3+}	3.00	214.2
C^{4+}	2.00	486.2
C^{5+}	1.00	539.5

Table 4 I_0 (eV) and S_0 of Ne and Its Ions Calculated with the aug-cc-pCV5Z Basis Set

	S_0	I_0 (in eV)	OS[22]	MST[23]
Ne	10.00	137.2	137.3	92.5
Ne^{1+}	9.00	165.2	145.5	95.3
Ne^{2+}	8.00	196.9	156.5	98.9
Ne^{3+}	7.00	235.2	171.8	103.7
Ne^{4+}	6.00	282.8	194.6	110.6
Ne^{5+}	5.00	352.6	231.7	120.9
Ne^{6+}	4.00	475.0	301.0	138.1
Ne^{7+}	3.00	696.8	446.5	211.8
Ne^{8+}	2.00	1409.2	982.7	497.8
Ne^{9+}	1.00	1498.4	982.7	497.8

Table 5 I_0 (eV) and S_0 of Al and Its Ions Calculated with the aug-cc-pCV5Z Basis Set

	S_0	I_0 (in eV)	MPP[24]
Al	13.00	132.0	122.8
Al^{1+}	12.00	172.4	160.0
Al^{2+}	11.00	225.4	223.7
Al^{3+}	10.00	310.4	310.9
Al^{4+}	9.00	350.5	325.5
Al^{5+}	8.00	398.1	390.4
Al^{6+}	7.00	458.4	457.5
Al^{7+}	6.00	536.3	530.8
Al^{8+}	5.00	654.8	643.0
Al^{9+}	4.00	867.1	817.9
Al^{10+}	3.00	1238.3	1236.0
Al^{11+}	2.00	2406.9	2373.0
Al^{12+}	1.00	2523.2	-

Table 6 I_0 (eV) and S_0 of Ar and Its Ions Calculated with the aug-cc-pCV5Z Basis Set

	S_0	I_0 (in eV)
Ar	18.00	188.5
Ar^{1+}	17.00	219.4
Ar^{2+}	16.00	253.8
Ar^{3+}	15.00	293.4
Ar^{4+}	14.00	339.1
Ar^{5+}	13.00	394.5
Ar^{6+}	12.00	463.4
Ar^{7+}	11.00	568.0
Ar^{8+}	10.00	728.0
Ar^{9+}	9.00	795.9
Ar^{10+}	8.00	879.8
Ar^{11+}	7.00	989.9
Ar^{12+}	6.00	1138.1
Ar^{13+}	5.00	1369.5
Ar^{14+}	4.00	1791.2
Ar^{15+}	3.00	2497.0
Ar^{16+}	2.00	4677.2
Ar^{17+}	1.00	4838.2

cases, the TRK sum rule is fulfilled to two decimal places, so that the one-electron basis sets used are nearly complete in all cases. For the Al and Ne series, we compare with literature values for the mean excitation energies.

5. DISCUSSION

There are very few measurements or calculations of the mean excitation energy of ions to compare against. However, the accuracy of the calculated mean excitation energies may be estimated from the favorable fulfillment of the TRK sum rule results (Eq. 7) for each of the ions is reported.

Values calculated for the mean excitation energies of aluminum ions agree well with those reported by McGuire, Peek, and Pitchford,[24] whereas those calculated for neon from orbital mean excitation energies[22] are not as good as the ones reported here. Calculations based on DFT[23] for Ne give much less reliable results.

Mean excitation energies of the hydrogenic ions with charge Z follow the Bohr scheme:

$$I_{1\text{e}^-\text{ion}} = Z^2 I_\text{H} \tag{8}$$

where $I_\text{H} = 14.9916\,\text{eV}$.[25] It is readily verified that our calculated mean excitation energies for the one-electron ions in Tables 1–6 agree with those calculated from Eq. (8).

The behavior of the mean excitation energy as a function of ionicity is similar for each of the atoms calculated, namely I_0 increases with ionicity, leading to a strong loss of stopping power for increasingly ionic targets. It is clear that the lower charged ions, having more electrons to absorb projectile energy than the more highly charged species, will have lower mean excitation energies. Another viewpoint would be that lower charged ions will have occupied atomic orbitals which are not available to electronic transitions in higher charged ions of the same atom, thus allowing more transitions to enter into the mean excitation energy for the more highly charged ions.

The increase in I_0 with increasing ionicity in dominantly quadratic in agreement with Eq. (8), however, with clear deviations from the quadratic behavior that can be attributed to the shell structure of the ion. The largest change in mean excitation energy among adjacent charged ions comes from ions with the $1s^2$ and the $1s^2 2s$ configurations and the difference increases with atomic number. This is due to the $1s \rightarrow 2s$ energy difference being so large. On the other hand, the mean excitation energies of ions with a $1s^1$ configuration and ions of the same atom with a configuration of $1s^2$ are small, as the orbitals available for transition and the respective orbital energies are very similar, leading to similar sets of transition moments in the two cases.

In plasma technology, the plasma is generally composed of a variety of atoms and ions. In order to describe the characteristics of energy absorption of the plasma from energetic ions, such as protons or alpha particles, the mean excitation energy and abundance of the component ions must be known. Assuming that the electronic structure of the individual constituents of the plasma may be treated independently the stopping power for the mixture as a whole for a projectile of charge Z_1 would then be

$$\left(-\frac{dE}{dx}\right)_{mix} = n_{mix}\frac{4\pi e^4 Z_1^2 Z_{mix}}{mv^2}\ln\frac{2mv^2}{I_0^{mix}} \tag{9}$$

Here, n_{mix} is a density of scattering centers, where $n_{mix} = \sum_i n_i$. The weighted average of the number of electrons per scatterer is given by $Z_{mix} = \sum_i n_i Z_i / n_{mix}$, and I_0^{mix} is the mean excitation energy appropriate to the mixture. Such treatment would derive from a sum of stopping powers of the components, weighted by their relative density of scattering centers

$$\begin{aligned}\left(-\frac{dE}{dx}\right)_{mix} &= \sum_i n_i \frac{4\pi e^4 Z_1^2 Z_i}{mv^2}\ln\frac{2mv^2}{I_0^i}\\ &= \frac{4\pi e^4 Z_1^2}{mv^2}\sum_i n_i Z_i \ln\frac{2mv^2}{I_0^i}\end{aligned} \tag{10}$$

Equating Eqs. (9) and (10), one obtains

$$\ln I_0^{mix} = \frac{\sum_i n_i Z_i \ln I_0^i}{\sum_i n_i Z_i} \tag{11}$$

Thus, the mean excitation energy of the mixture of nonbonding components is simply the appropriate weighted average of the mean excitation energies of those components, analogous to the Bragg Rule.[26]

In a plasma, there are many charged ions of the various components. Thus, the target atom or ion will not be in zero filed, as are the ions in Tables 1–6 above. In order to estimate the effect an electric field would have on the mean excitation energy of these target species, the mean excitation energies of several Ne ions in electric fields were calculated and are reported in Table 7.

Table 7 I_0 (eV) of Ne and Two of Its Ions Versus External Electric Field

E (au)	Ne	Ne^{6+}	Ne^{8+}
0	137.2	475.0	1402.2
0.1	136.7	474.9	1402.2
0.5	-	471.7	1409.2
1.0	-	462.6	1409.2

It is clear that it is not necessary to treat external fields from nearby ions explicitly, as they influence the mean excitation energy, and thus the stopping power of plasma component ions only marginally.

The conclusion is that the method described here can calculate the mean excitation energies of atomic ions reliably, and thus the energy absorption properties of complex systems such as plasmas, warm dense matter, and large and complex biochemical structures may be reliable treated.

ACKNOWLEDGMENT

We thank DCSC for financial support of this project.

REFERENCES

1. Chen, F. F. Introduction to Plasma Physics and Controlled Fusion, 2nd ed.; *Plasma Physics*, Vol. 1; Plenum: New York, NY, 1984.
2. Karasiev, V. V.; Chakraborty, D.; Trickey, S. B. Progress on New Approaches to Old Ideas: Orbital-Free Density Functionals. In: *Many-Electron Approaches in Physics, Chemistry, and Mathematics*; Delle Site, L., Bach, V., Eds.; Springer: Heidelberg, 2014; pp 113–134; Karasiev, V. V.; Sjostrom, T.; Chakraborty, D.; Dufty, J. W.; Harris, F. E.; Runge, K.; Trickey, S. B. Innovations in Finite-Temperature Density Functionals. In: *Frontiers and Challenges in Warm Dense Matter*; Graziani, F., et al., Eds.; Springer: Heidelberg, 2014; pp 61–85.
3. Sauer, S. P. A.; Oddershede, J.; Sabin, J. R. Mean Excitation Energies for Biomolecules: Glycine to DNA. *Adv. Quantum Chem.* **2011**, *62*, 215–242.
4. Inokuti, M. Inelastic Collisions of Fast Charged Particles with Atoms and Molecules—The Bethe Theory Revisited. *Rev. Mod. Phys.* **1971**, *43*, 297–347.
5. Bethe, H. Zur Theorie des Durchgangs schneller Korpuskularstrahlendurch Materie. *Ann. Phys. (Leipzig)* **1930**, *5*, 325–400.
6. Barkas, W. H.; Dyer, J. N.; Heckman, H. H. Resolution of the Σ^--Mass Anomaly. *Phys. Rev. Lett.* **1963**, *11*, 26–28.
7. Lindhard, J. Barkas Effect—or Z_{31}, Z_{41}—Corrections to Stopping of Swift Charged Particles. *Nucl. Inst. Meth.* **1976**, *132*, 1–5.
8. Bloch, F. The Slow Down of Rapidly Moving Particles in the Their Passing through Solid Matter. *Ann. Phys. (Leipzig)* **1933**, *16*, 285–320.
9. Defined in Ref. [4].
10. For a review of the theory and implementation of the polarization propagator method, see Oddershede, J.; Jørgensen, P.; Yaeger, D. L. Polarization Propagator Methods in Atomic and Molecular Calculations. *Comput. Phys. Rep.* **1984**, *2*, 33–92; Oddershede, J. Propagator Methods. *Adv. Chem. Phys.* **1987**, *69*, 201–239; Sauer, S. P. A. *Molecular Electromagnetism: A Computational Chemistry Approach*; Oxford University Press: Oxford, 2011.
11. See Bichsel, H. Shell Corrections in Stopping Powers. *Phys. Rev. A* **2002**, *65*, 052709, for a review of the calculation of shell corrections.
12. Sauer, S. P. A.; Sabin, J. R.; Oddershede, J. Calculated Mean Excitation Energies for Some Small Molecules. *Nucl. Inst. Meth. B* **1995**, *100*, 458–463.
13. Fano, U. Penetration of Protons, Alpha Particles, and Mesons. *Ann. Rev. Nuc. Sci.* **1963**, *13*, 2–66.

14. Oddershede, J.; Sabin, J. R. Effect of Correlation on the Mean Excitation Energy of Beryllium. *Phys. Rev. A* **1989**, *39*, 5565–5571.
15. Bruun-Ghalbia, S.; Sauer, S. P. A.; Oddershede, J.; Sabin, J. R. Comparison of the Directional Characteristics of Swift Ion Excitation for Two Small Biomolecules: Glycine and Alanine. *Eur. Phys. J. D* **2010**, *60*, 71–76.
16. Sauer, S. P. A.; Ul Haq, I.; Sabin, J. R.; Oddershede, J.; Christiansen, O.; Coriani, S. Coupled Cluster Calculations of Mean Excitation Energies of the Noble Gas Atoms He, Ne and Ar and of the H_2 Molecule. *Mol. Phys.* **2014**, *3*, 751–761.
17. Aidas, K.; Angeli, C.; Bak, K. L.; Bakken, V.; Boman, L.; Christiansen, O.; Cimiraglia, R.; Coriani, S.; Dahle, P.; Dalskov, E. K.; Ekstrøm, U.; Enevoldsen, T.; Eriksen, J. J.; Ettenhuber, P.; Fernandez, B.; Ferrighi, L.; Fliegl, H.; Frediani, L.; Hald, K.; Halkier, A.; Hättig, C.; Heiberg, H.; Helgaker, T.; Hennum, A. C.; Hettema, H.; Høst, S.; Høyvik, I.-M.; Iozzi, M. F.; Jansik, B.; A. Jensen, H. J.; Jonsson, D.; Jørgensen, P.; Kauczor, J.; Kirpekar, S.; Kjærgaard, T.; Klopper, W.; Knecht, S.; Kobayashi, R.; Koch, H.; Kongsted, J.; Krapp, A.; Kristensen, K.; Ligabue, A.; Lutnæs, O. B.; Melo, J. I.; Mikkelsen, K. V.; Myhre, R. H.; Neiss, C.; Nielsen, C. B.; Norman, P.; Olsen, J.; Olsen, J. M. H.; Osted, A.; Packer, M. J.; Pawlowski, F.; Pedersen, T. B.; Provasi, P. F.; Reine, S.; Rinkevicius, Z.; Ruden, T. A.; Ruud, K.; Rybkin, V.; Salek, P.; Samson, C. C. M.; Sánchez de Merás, A.; Saue, T.; Sauer, S. P. A.; Schimmelpfennig, B.; Sneskov, K.; Steindal, A. H.; Sylvester-Hvid, K. O.; Taylor, P. R.; Teale, A. M.; Tellgren, E. I.; Tew, D. P.; Thorvaldsen, A. J.; Thøgersen, L.; Vahtras, O.; Watson, M.; Wilson, D. J.; Ziolkowski, M.; Ågren, H. The DALTON Quantum Chemistry Program System. *WIREs Comput. Mol. Sci.* **2014**, *4*, 269–284.
18. Ahlrichs, R.; Bär, M.; Häser, M.; Horn, H.; Kölmel, C. Electronic Structure Calculations on Workstation Computers: The Program System Turbomole. *Chem. Phys. Lett.* **1989**, *162*, 165–169; Häser, M.; Ahlrichs, R. Improvements on the Direct SCF Method. *J. Comput. Chem.* **1989**, *10*, 104–111; Treutler, O.; Ahlrichs, R. Efficient Molecular Numerical Integration Schemes. *J. Chem. Phys.* **1995**, *102*, 346–354.
19. *TURBOMOLE V6.6 2014 a Development of University of Karlsruhe and Forschungszentrum Karlsruhe GmbH, 1989–2007, TURBOMOLE GmbH, Since 2007*; http://www.turbomole.com, Available from.
20. Sauer, S. P. A.; Oddershede, J.; Sabin, J. R. Directional Dependence of the Mean Excitation Energy and Spectral Moments of the Dipole Oscillator Strength Distribution of Glycine and Its Zwitterion. *J. Phys. Chem. A* **2006**, *110*, 8811–8817.
21. Dunning, T. H. Gaussian Basis Sets for the Use in Correlated Molecular Calculations. I. The Atoms Boron Through Neon and Hydrogen. *J. Chem. Phys.* **1989**, *90*, 1007–1023; Kendall, R. A.; Dunning, T. H.; Harrison, R. J. Electron Affinities of the First Row Atoms Revisited. Systematic Basis Sets and Wave Functions. *J. Chem. Phys.* **1992**, *96*, 6796–6806; Woon, D. E.; Dunning, T. H. Gaussian Basis Sets for Use in Correlated Molecular Calculations. III. The Atoms Aluminum Through Argon. *J. Chem. Phys.* **1993**, *98*, 1358–1371; Woon, D. E.; Dunning, T. H. Gaussian Basis Sets for Use in Correlated Molecular Calculations. IV. Calculation of Static. Electrical Response Properties. *J. Chem. Phys.* **1994**, *100*, 2975–2988; Woon, D. E.; Dunning, T. H. Gaussian Basis Sets for Use in Correlated Molecular Calculations. V. Core-Valence Basis Sets for Boron Through Neon. *J. Chem. Phys.* **1995**, *103*, 4572–4585.
22. Oddershede, J.; Sabin, J. R. Orbital and Whole Atom Stopping Power and Shell Corrections for Atoms with $Z < 36$. *At. Dat. Nuc. Dat. Tables* **1984**, *31*, 275–297.
23. Meltzer, D. E.; Sabin, J. R.; Trickey, S. B. Calculation of Mean Excitation Energies and Stopping Cross Sections in the Orbital Local Plasma Approximation. *Phys. Rev. A* **1990**, *41*, 220–232. (Erratum, 42, 666 (1990)).

24. McGuire, E. J.; Peek, J. M.; Pitchford, L. C. Proton Stopping of Aluminum Ions. *Phys. Rev. A* **1982**, *26*, 1318–1325.
25. Rosendorff, S.; Schlaile, H. G. Logarithmic Mean Excitation Energies—Hydrogen and Helium. *Phys. Rev. A* **1989**, *40*, 6892.
26. Bragg, W. H.; Kleeman, H. On the Alpha Particles of Radium and Their Loss of Range When Passing Through Various Atoms and Molecules. *Philos. Mag.* **1905**, *10*, 318–340.

CHAPTER FOUR

Hybrid Functionals with Variationally Fitted Exact Exchange

Daniel Mejía-Rodríguez*, Xiaomin Huang†, Jorge M. del Campo†, Andreas M. Köster*,[1]

*Departamento de Química, CINVESTAV, Avenida Instituto Politécnico Nacional 2508, Mexico, Mexico
†Departamento de Física y Química Teórica, Facultad de Química, Universidad Nacional Autónoma de México, Mexico, Mexico
[1]Corresponding author: e-mail address: akoster@cinvestav.mx

Contents

Abstract

This chapter describes the use of the variationally fitted exact exchange for the computation of hybrid density-functional approximations. Accuracy and performance of the resulting methodology, which combines auxiliary density functional theory with the local-density fitted exact exchange algorithm, are presented.

1. INTRODUCTION

Density functional theory (DFT)[1,2] permits the application of electronic structure theory to complex chemical problems at a computation cost characteristic to mean-field approximations. In theory, DFT is an exact formulation, but in practice one has to rely on density functional approximations (DFAs). The most popular formulation of DFT is the one given by

Advances in Quantum Chemistry, Volume 71
ISSN 0065-3276
http://dx.doi.org/10.1016/bs.aiq.2015.03.009

Kohn and Sham (KS),[3] in which the electronic density is obtained from a single Slater determinant of orthonormal molecular orbitals (MOs). The accuracy of KS-DFT depends crucially on the chosen approximation for the exchange-correlation functional, i.e., how close the chosen DFA models the exact functional. The different DFAs developed during the last decades are usually sorted in groups according to their increasing accuracy and complexity. A well-known example for this type of sorting is Perdew's Jacob's ladder.[4] Here, the lowest rung is given by the local density approximation (LDA) which depends only on the electronic density $\rho(\mathbf{r})$. On the next rung is the generalized gradient approximation (GGA) that, in addition to $\rho(\mathbf{r})$, also depends on the density gradient, $\nabla \rho(\mathbf{r})$. Several other rungs may follow which combine either higher derivatives of $\rho(\mathbf{r})$, explicit dependency on MOs and their eigenvalues, or a combination of them. In particular, the so-called hybrid DFAs combine some fraction of exact exchange with any given "pure" exchange-correlation functional, usually of the GGA or higher rungs. This mixing is motivated by the observation that both DFAs, LDA and GGA, tend to overbind molecules[5] whereas Hartree–Fock binding energies are usually too low due to the missing correlation contributions.[6] The formal justification for building hybrid functionals is given by the adiabatic connection.[7–9] In fact, one of the most popular DFAs in chemistry is the B3LYP[10] hybrid functional. The main disadvantage of hybrid DFAs is that the simplicity of KS-DFT is partially lost by the incorporation of the more complex and time-consuming computation of exact exchange.

In KS-DFT, the functional form of the exact exchange energy is equivalent to the exchange term in Hartree–Fock theory; therefore, Fock exchange and exact exchange are sometimes used as synonyms. The functional form of exact exchange is given by:

$$E_X = -\sum_{i}^{\text{occ}} \iint \frac{\psi_i(\mathbf{r}_1)\psi_j(\mathbf{r}_1)\psi_i(\mathbf{r}_2)\psi_j(\mathbf{r}_2)}{|\mathbf{r}_1 - \mathbf{r}_2|} \, d\mathbf{r}_1 \, d\mathbf{r}_2, \tag{1}$$

where $\psi_i(\mathbf{r})$ and $\psi_j(\mathbf{r})$ represent occupied KS MOs. In the linear combination of the Gaussian-type orbital (LCGTO) approach, the expansion of the KS MOs into the atomic orbitals (AOs) yields for Eq. (1):

$$E_X = -\frac{1}{4}\sum_{\mu,\nu}\sum_{\sigma,\tau} P_{\mu\nu}\, P_{\sigma\tau} \iint \frac{\mu(\mathbf{r}_1)\sigma(\mathbf{r}_1)\tau(\mathbf{r}_2)\nu(\mathbf{r}_2)}{|\mathbf{r}_1 - \mathbf{r}_2|} \, d\mathbf{r}_1 \, d\mathbf{r}_2. \tag{2}$$

Here, **P** is the closed-shell density matrix with elements:

$$P_{\mu\nu} = 2\sum_{i}^{\text{occ}} c_{\mu i}\, c_{\nu i}, \tag{3}$$

and **c** is the MO coefficient matrix with elements $c_{\mu i}$ and $c_{\nu i}$. Note that Eq. (2) can be subject to efficient electron repulsion integral (ERI) screening based on density matrix elements and AO overlap integrals. This leads to the well-known linear-scaling algorithms like ONX[11] and LinK.[12] These algorithms are based on the short-range nature of exact exchange as seen from Eq. (2). The two-electron four-center ERIs appearing in Eq. (2) have nonnegligible values only if $\mu(\mathbf{r})$ is close to $\sigma(\mathbf{r})$ and $\nu(\mathbf{r})$ is close to $\tau(\mathbf{r})$. Furthermore, for insulating systems the density matrix elements $P_{\mu\nu}$ and $P_{\sigma\tau}$ vary significantly during the self-consistent field (SCF) iterations only if $\mu(\mathbf{r})$ is close to $\nu(\mathbf{r})$ and $\sigma(\mathbf{r})$ is close to $\tau(\mathbf{r})$, respectively. As a result, all four AOs must be localized somewhat close in space and the computational effort can be reduced to $\mathcal{O}(N)$.

However, the linear-scaling exact exchange is achieved by algorithms like ONX and LINK only with respect to system size. It is based on the fact that the number of quartets $\mu\nu\sigma\tau$ grows as $\mathcal{O}(N^4)$, whereas the number of nonvanishing contributions in Eq. (2) grows only as $\mathcal{O}(N)$. On the other hand, computing exact exchange in the same system with enlarging basis set size will still result in the unfavorable $\mathcal{O}(N^4)$ scaling. Therefore, the development of algorithms with low scaling for both system size and basis set enlargement has been a very active research field over the last decades. Two prominent examples for this development are the pseudo-spectral method from Friesner[13,14] that avoids the explicit computation of four-center ERIs, and the chain-of-spheres exchange (COSX)[15-17] algorithm that exploits the short-range nature of exact exchange by performing a semi-numerical integration on a grid.

Another group of algorithms aiming in the same direction are based on the approximation of the orbital distributions $\rho_{ij}(\mathbf{r}) = \psi_i(\mathbf{r})\psi_j(\mathbf{r})$. Similar to the variational fitting of the Coulomb potential,[18-22] a reduction of the formal scaling from $\mathcal{O}(N^4)$ to $\mathcal{O}(N^3)$ can be achieved in this way. Examples are the so-called resolution of the identity (RI),[23-25] Cholesky decomposition (CD),[26,27] atomic RI (ARI),[28] local density-fitting (LDF),[29,30] and the concentric atomic density-fitting (CADF) [31,32] algorithms that work with almost the same basic equations. Computational speed-ups achieved by these methods can become significant, especially when large basis sets are used.

In this chapter, we present the development and implementation of LDF exact exchange energy and potential within the auxiliary density functional theory (ADFT) framework[33] in order to obtain a fast and low-scaling algorithm useful for the application of hybrid DFAs in large systems and in Born–Oppenheimer molecular dynamics (BOMD) simulations. To this end, we first introduce the ADFT framework as one of the most efficient KS-DFT methods available. We then continue to discuss the LDF exact exchange algorithm recently developed in our group.[30] Finally, we show how the combination of these two methods performs for three popular hybrid DFAs: B3LYP,[10] PBE0,[34,35] and M06-2X.[36]

2. AUXILIARY DENSITY FUNCTIONAL THEORY

ADFT[33,37] is based on density-fitting density functional theory (DF-DFT). Therefore, we review both methodologies in this section. In the LCGTO ansatz, the Kohn–Sham energy expression can be written as

$$E = \sum_{\mu,\nu} P_{\mu\nu} H_{\mu\nu} + \frac{1}{2} \sum_{\mu,\nu} \sum_{\sigma,\tau} P_{\mu\nu} P_{\sigma\tau} \langle \mu\nu \parallel \sigma\tau \rangle + E_{\mathrm{xc}}[\rho] \,, \tag{4}$$

with

$$H_{\mu\nu} = \left\langle \mu \middle| -\frac{1}{2}\nabla^2 \middle| \nu \right\rangle - \sum_{C}^{\text{Atoms}} \left\langle \mu \middle| \frac{Z_C}{|\mathbf{r}-\mathbf{C}|} \middle| \nu \right\rangle \tag{5}$$

and

$$\langle \mu\nu \parallel \sigma\tau \rangle = \iint \frac{\mu(\mathbf{r}_1)\nu(\mathbf{r}_1)\sigma(\mathbf{r}_2)\tau(\mathbf{r}_2)}{|\mathbf{r}_1 - \mathbf{r}_2|} \, d\mathbf{r}_1 \, d\mathbf{r}_2 \tag{6}$$

where the symbol $\parallel$ represents the $1/|\mathbf{r}_1 - \mathbf{r}_2|$ operator. Computational complexity of Eq. (4) can be analyzed term by term. Construction of the core Hamiltonian matrix **H**, appearing in Eqs. (4) and (5), has a formal scaling of $\mathcal{O}(N^2)$ with respect to the number of basis functions N. The second term in Eq. (4), which represents the electron–electron coulombic repulsion, introduces a formal $\mathcal{O}(N^4)$ scaling. The last term is given by the exchange-correlation energy E_{xc}, generally obtained by a numerical integration. Since the Kohn–Sham orbital density $\rho(\mathbf{r})$ has a quadratic dependency on the number of basis functions, the numerical integration has a formal scaling of $\mathcal{O}(N^2) \times \mathcal{O}(G)$, where G denotes the number of grid points used in the quadrature.

It can be seen that, in standard LCGTO-KS-DFT, the calculation of the Coulomb repulsion term represents the most demanding computational task to be performed. The introduction of the variational approximation of the Coulomb energy and potential,[18-20] along with the corresponding integral recurrence relations,[21,22,38] reduces the formal scaling of this term to $\mathcal{O}(N^2) \times \mathcal{O}(M)$ and the asymptotic ones to $\mathcal{O}(N)$, where M is the number of auxiliary functions used to approximate the density. This technique was pioneered by the early LCGTO-KS-DFT programs DGAUSS[39] and deMon[40] and is used in most LCGTO-DFT software packages today. It is formally identical to the so-called RI approach[41] that is also employed in wave function methods.[24,42,43]

The variational approximation of the Coulomb potential is based on the minimization of the self-interaction term:

$$\varepsilon_2 = \frac{1}{2} \iint \frac{[\rho(\mathbf{r}_1) - \tilde{\rho}(\mathbf{r}_1)][\rho(\mathbf{r}_2) - \tilde{\rho}(\mathbf{r}_2)]}{|\mathbf{r}_1 - \mathbf{r}_2|} \, d\mathbf{r}_1 \, d\mathbf{r}_2 \tag{7}$$

In deMon2k,[44,45] the auxiliary density $\tilde{\rho}(\mathbf{r})$ is expanded in primitive Hermite Gaussians[22] which are centered on atoms

$$\tilde{\rho}(\mathbf{r}) = \sum_{\bar{c}} x_{\bar{c}} \bar{c}(\mathbf{r}) \tag{8}$$

Throughout the text, primitive Hermite Gaussian auxiliary functions are indicated by a bar. Thus, an (unnormalized) auxiliary function $\bar{c}(\mathbf{r})$ centered on atom C with exponent $\zeta_{\bar{c}}$ has the general form

$$\bar{c}(\mathbf{r}) = \left(\frac{\partial}{\partial C_x}\right)^{\bar{c}_x} \left(\frac{\partial}{\partial C_y}\right)^{\bar{c}_y} \left(\frac{\partial}{\partial C_z}\right)^{\bar{c}_z} e^{-\zeta_{\bar{c}}(\mathbf{r}-\mathbf{C})^2} \tag{9}$$

where the sum of the partial derivative orders equals the angular momentum index of the auxiliary function $\bar{c}(\mathbf{r})$.

Expanding both $\rho(\mathbf{r})$ and $\tilde{\rho}(\mathbf{r})$, we obtain the following representation for ε_2:

$$\varepsilon_2 = \frac{1}{2} \sum_{\mu,\nu} \sum_{\sigma,\tau} P_{\mu\nu} P_{\sigma\tau} \langle \mu\nu \| \sigma\tau \rangle - \sum_{\mu,\nu} \sum_{\bar{c}} P_{\mu\nu} \langle \mu\nu \| \bar{c} \rangle x_{\bar{c}} + \frac{1}{2} \sum_{\bar{c},\bar{d}} x_{\bar{c}} x_{\bar{d}} \langle \bar{c} \| \bar{d} \rangle \tag{10}$$

The fitting coefficients $x_{\bar{c}}$ of the auxiliary density are determined by minimization of ε_2:

$$\frac{\partial \varepsilon_2}{\partial x_{\bar{c}}} = -\sum_{\mu,\nu} P_{\mu\nu} \langle \mu\nu \| \bar{c} \rangle + \sum_{\bar{d}} x_{\bar{d}} \langle \bar{d} \| \bar{c} \rangle = 0 \; \forall \, \bar{c} \tag{11}$$

The set of equations, one for each auxiliary function, with the same form as Eq. (11) defines the inhomogeneous equation system

$$\mathbf{Gx} = \mathbf{J} \tag{12}$$

The fitting coefficients are collected in the vector $\mathbf{x}$. The Coulomb matrix $\mathbf{G}$ is defined as

$$\mathbf{G} = \begin{pmatrix} \langle \bar{1} \| \bar{1} \rangle & \langle \bar{1} \| \bar{2} \rangle & \cdots & \langle \bar{1} \| \bar{c} \rangle \\ \langle \bar{2} \| \bar{1} \rangle & \langle \bar{2} \| \bar{2} \rangle & \cdots & \langle \bar{2} \| \bar{c} \rangle \\ \vdots & \vdots & \ddots & \vdots \\ \langle \bar{c} \| \bar{1} \rangle & \langle \bar{c} \| \bar{2} \rangle & \cdots & \langle \bar{c} \| \bar{c} \rangle \end{pmatrix} \tag{13}$$

and the Coulomb vector $\mathbf{J}$ as

$$\mathbf{J} = \begin{pmatrix} \sum_{\mu,\nu} P_{\mu\nu} \langle \mu\nu \| \bar{1} \rangle \\ \sum_{\mu,\nu} P_{\mu\nu} \langle \mu\nu \| \bar{2} \rangle \\ \vdots \\ \sum_{\mu,\nu} P_{\mu\nu} \langle \mu\nu \| \bar{c} \rangle \end{pmatrix} \tag{14}$$

The Coulomb matrix $\mathbf{G}$ is symmetric and positive definite which guarantees the minimization in Eq. (11). In principle, the very efficient Cholesky decomposition algorithm can be used to solve the fitting equation system. However, if very large auxiliary function sets are used, $\mathbf{G}$ tends to become ill-conditioned. In order to avoid numerical instabilities by solving the fitting equations with the Cholesky decomposition approach, deMon2k performs a singular value decomposition (SVD) at the beginning of the SCF and uses a robust numerical solver[46] that acts only on the nonredundant space of the SVD for subsequent solutions. Performing a full SVD is time consuming; however, the stability gained during the initial SCF and the application of the numerical solver even for different geometries (for example, in a geometry optimization or along a BOMD trajectory) overcompensate the computational load of the SVD.

Because ε_2 is positive definite, an approximate SCF energy which is based on Eq. (4) can be derived and takes the form:

$$E = \sum_{\mu,\nu} P_{\mu\nu} H_{\mu\nu} + \sum_{\bar{c}} \sum_{\mu,\nu} P_{\mu\nu} \langle \mu\nu \parallel \bar{c} \rangle x_{\bar{c}} - \frac{1}{2} \sum_{\bar{c},\bar{d}} x_{\bar{c}} x_{\bar{d}} \langle \bar{c} \parallel \bar{d} \rangle + E_{\mathrm{xc}}[\rho] \quad (15)$$

The variation of this energy expression with respect to the MO coefficients, constraining the Kohn–Sham MOs to be orthonormal, leads to the standard Roothaan–Hall equation

$$\mathbf{Kc} = \mathbf{Sc}\boldsymbol{\epsilon} \quad (16)$$

The elements of the Kohn–Sham matrix **K** are given by:

$$K_{\mu\nu} = H_{\mu\nu} + \sum_{\bar{c}} \langle \mu\nu \parallel \bar{c} \rangle x_{\bar{c}} + \langle \mu | v_{\mathrm{xc}}[\rho] | \nu \rangle \quad (17)$$

As can be seen from Eq. (17), the Kohn–Sham matrix depends on both the fitted density and the Kohn–Sham density. We refer to this method as density-fitted DFT (DF-DFT). As already discussed, the formal scaling of the Coulomb repulsion term in DF-DFT is lowered to $\mathcal{O}(N^3)$ and approaches asymptotically the linear regime. As a consequence, the numerical integration of the exchange-correlation energy and potential becomes the computationally most demanding task in this approach.

To circumvent this computational bottleneck, the direct use of the auxiliary density for the calculation of the exchange-correlation energy and potential has been investigated over the last years.[33,47-51] The resulting energy expression is variational and has the form[33]:

$$E = \sum_{\mu,\nu} P_{\mu\nu} H_{\mu\nu} + \sum_{\bar{c}} \sum_{\mu,\nu} P_{\mu\nu} \langle \mu\nu \parallel \bar{c} \rangle x_{\bar{c}} - \frac{1}{2} \sum_{\bar{c},\bar{d}} x_{\bar{c}} x_{\bar{d}} \langle \bar{c} \parallel \bar{d} \rangle + E_{\mathrm{xc}}[\tilde{\rho}] \quad (18)$$

Because the auxiliary density is a linear combination of functions, the density calculation at each grid point scales linearly. Obviously, this represents a simplification of the grid work from $\mathcal{O}(N^2) \times \mathcal{O}(G)$ to $\mathcal{O}(N) \times \mathcal{O}(G)$. At this point, it is also important to mention that the special structure of the primitive Hermite Gaussian auxiliary function sets used in deMon2k facilitates significantly the calculation of analytical and numerical molecular integrals. Because the primitive Hermite Gaussian auxiliary functions are grouped together into *s*, *spd*, and *spdfg* sets with common exponents, very efficient recurrence relations for the ERI calculation can be derived.[21,22] The shared exponent structure also reduces considerably the number of expensive exponential function evaluations at each grid point in the numerical integration of the auxiliary density.[33] Last, but not least, the primitive Hermite Gaussian form of the auxiliary function facilitates asymptotic integral expansions[38] and permits the use of Hermite polynomial recurrence relations[52] on the grid. Nowadays, this methodology is known as ADFT and can be used with LDA and GGA functionals.

Also, the variation of the ADFT energy with respect to the MO coefficients yields a Roothaan–Hall equation. The difference to Eq. (16) is in the calculation of the exchange-correlation contribution to the Kohn–Sham matrix. In ADFT, the variation of the exchange-correlation energy is given by:

$$\frac{\partial E_{\mathrm{xc}}[\tilde{\rho}]}{\partial c_{\mu i}} = \int \frac{\delta E_{\mathrm{xc}}[\tilde{\rho}]}{\delta \tilde{\rho}(\mathbf{r})} \frac{\partial \tilde{\rho}(\mathbf{r})}{\partial c_{\mu i}} \, d\mathbf{r} \tag{19}$$

where the resulting functional derivative defines the exchange-correlation potential calculated with the auxiliary density

$$v_{\mathrm{xc}}[\tilde{\rho};\mathbf{r}] \equiv \frac{\delta E_{\mathrm{xc}}[\tilde{\rho}]}{\delta \tilde{\rho}(\mathbf{r})} \tag{20}$$

The derivative of the auxiliary density with respect to the MO coefficients is given by:

$$\frac{\partial \tilde{\rho}(\mathbf{r})}{\partial c_{\mu i}} = \sum_{\bar{d}} \frac{\partial x_{\bar{d}}}{\partial c_{\mu i}} \bar{d}(\mathbf{r}) \tag{21}$$

which leads to

$$\frac{\partial E_{\mathrm{xc}}[\tilde{\rho}]}{\partial c_{\mu i}} = 4 \sum_{\bar{c}} \sum_{\nu} c_{\nu i} \langle \mu\nu \,\|\, \bar{c} \rangle z_{\bar{c}} \tag{22}$$

with

$$z_{\bar{c}} = \sum_{\bar{d}} G^{-1}_{\bar{c}\bar{d}} \langle \bar{d} | v_{\mathrm{xc}}[\tilde{\rho}] \rangle \tag{23}$$

The corresponding Kohn–Sham matrix elements are defined as:

$$K_{\mu\nu} = H_{\mu\nu} + \sum_{\bar{c}} \langle \mu\nu \,\|\, \bar{c} \rangle (x_{\bar{c}} + z_{\bar{c}}) \tag{24}$$

As Eq. (24) shows, the ADFT Kohn–Sham matrix depends only on the auxiliary density. To obtain the spin-dependent exchange-correlation fitting coefficients **z**, Eq. (23) is transformed into the inhomogeneous equation system

$$\mathbf{Gz} = \mathbf{L} \tag{25}$$

where the vector **L** is given by

$$\mathbf{L} = \begin{pmatrix} \langle v_{xc}[\tilde{\rho}] \mid \bar{1} \rangle \\ \langle v_{xc}[\tilde{\rho}] \mid \bar{2} \rangle \\ \vdots \\ \langle v_{xc}[\tilde{\rho}] \mid \bar{c} \rangle \end{pmatrix} \tag{26}$$

This equation system can be solved efficiently with a preconditioned conjugate gradient algorithm previously described.[46]

3. VARIATIONALLY FITTED EXACT EXCHANGE

The significant reduction in the computational load achieved by using ADFT may be hidden in hybrid ADFT calculations due to the time-consuming computation of exact exchange contributions via four-center ERIs. Also, standard RI or DF exact exchange algorithms are not improving this situation due to the cumbersome handling of three-index quantities.[24] For this reason, a new local density-fitting (LDF) approach for the exact exchange (EXX) energy was recently developed in our group.[30]

Following the variational fitting of the Coulomb potential, the LDF-EXX approach finds the variationally best fit for the orbital products

$$\rho_{ij}(\mathbf{r}) = \psi_i(\mathbf{r})\psi_j(\mathbf{r}) \tag{27}$$

appearing in the exact exchange functional in Eq. (1). The fitting is performed with the same Hermite Gaussian auxiliary function sets used for the variationally Coulomb fitting. Thus, we find for the auxiliary orbital products:

$$\tilde{\rho}_{ij}(\mathbf{r}) = \sum_{\bar{c}} x_{\bar{c}ij}\, \bar{c}(\mathbf{r}) \tag{28}$$

In this case, the functional to be maximized is

$$\varepsilon_2^{\mathrm{EXX}} = -\sum_{i,j}^{\mathrm{occ}} \iint \frac{[\rho_{ij}(\mathbf{r}_1) - \tilde{\rho}_{ij}(\mathbf{r}_1)][\rho_{ij}(\mathbf{r}_2) - \tilde{\rho}_{ij}(\mathbf{r}_2)]}{|\mathbf{r}_1 - \mathbf{r}_2|}\, d\mathbf{r}_1\, d\mathbf{r}_2 \tag{29}$$

It can be shown that the functional $\varepsilon_2^{\mathrm{EXX}}$ is negative definite. Note the difference to ε_2 which is positive definite. Therefore, a maximization of $\varepsilon_2^{\mathrm{EXX}}$

will correspond to a minimization of the approximate exact exchange energy. On the other hand, it was shown that the minimization of ε_2 for the variational Coulomb fitting, Eq. (11), leads to an energy maximization.[53] Therefore, significant error compensation is expected if Coulomb and exact exchange variational fittings are performed with the same auxiliary function set. Moreover, in the case of Hartree–Fock, the approximated energy expression is self-interaction-free.[30] After expanding the orbital products in Eq. (29), we find:

$$\varepsilon_2^{\mathrm{EXX}} = -\sum_{i,j}^{\mathrm{occ}} \langle ij \parallel ij \rangle + 2\sum_{i,j}^{\mathrm{occ}} \sum_{\bar{c}} \langle ij \parallel \bar{c} \rangle x_{\bar{c}ij} - \sum_{i,j}^{\mathrm{occ}} \sum_{\bar{c},\bar{d}} x_{\bar{c}ij} x_{\bar{d}ij} \langle \bar{c} \parallel \bar{d} \rangle \tag{30}$$

Again, the expansion coefficients $x_{\bar{c}ij}$ are determined by maximization of $\varepsilon_2^{\mathrm{EXX}}$

$$\frac{\partial \varepsilon_2^{\mathrm{EXX}}}{\partial x_{\bar{c}ij}} = \langle ij \parallel \bar{c} \rangle - \sum_{\bar{d}} x_{\bar{d}ij} \langle \bar{c} \parallel \bar{d} \rangle = 0 \ \forall \ \bar{c}, i, j \tag{31}$$

which defines a set of linear equation systems, one for each ρ_{ij} product:

$$\mathbf{G}\mathbf{x}_{ij} = \mathbf{J}_{ij} \tag{32}$$

with

$$\mathbf{J}_{ij} = \begin{pmatrix} \langle ij \parallel \bar{1} \rangle \\ \langle ij \parallel \bar{2} \rangle \\ \vdots \\ \langle ij \parallel \bar{m} \rangle \end{pmatrix} \tag{33}$$

After the fitting equations have been solved, the fitted exact exchange energy is given by:

$$E_{\mathrm{EXX}} = -\sum_{i,j}^{\mathrm{occ}} \sum_{\bar{c},\bar{d}} x_{\bar{c}ij} G_{\bar{c}\bar{d}} x_{\bar{d}ij} \tag{34}$$

which can be further expanded to yield the following expression:

$$E_{\mathrm{EXX}} = -\sum_{i,j}^{\mathrm{occ}} \sum_{\bar{c},\bar{d}} \langle ij \parallel \bar{c} \rangle G_{\bar{c}\bar{d}}^{-1} \langle \bar{d} \parallel ij \rangle \tag{35}$$

Inserting Eq. (35) directly into the corresponding energy expression avoids the explicit calculation of the three-index fitting coefficients appearing in the expansion of the auxiliary orbital products (Eq. 28). Note, however, that this is the result of a variational fitting as described above. If the approximated exact exchange energy expression of Eq. (35) is added to the self-interaction including Hartree energy with Coulomb fitting, the variationally fitted self-interaction-free Hartree–Fock energy expression is obtained.[30] On the other hand, if the fitted exact exchange energy is combined with exchange-correlation functionals, hybrid ADFT energy expressions without four-center ERIs are obtained. In particular, we find the following ADFT or DF-DFT energy expressions for the here discussed B3LYP (Eq. 36a), PBE0 (Eq. 36b), and M06-2X (Eq. 36c) functionals:

$$\begin{aligned} E^{\mathrm{B3LYP}} = & \sum_{\mu,\nu} P_{\mu\nu} H_{\mu\nu} + \sum_{\bar{c}} \sum_{\mu,\nu} P_{\mu\nu} \langle \mu\nu \| \bar{c} \rangle x_{\bar{c}} - \frac{1}{2} \sum_{\bar{c},\bar{d}} x_{\bar{c}} x_{\bar{d}} \langle \bar{c} \| \bar{d} \rangle \\ & -0.20 \sum_{i,j}^{\mathrm{occ}} \sum_{\bar{c},\bar{d}} \langle ij \| \bar{c} \rangle G^{-1}_{\bar{c}\bar{d}} \langle \bar{d} \| ij \rangle + 0.80 E_{\mathrm{X}}^{\mathrm{Slater}}[\tilde{\rho}] + 0.72 E_{\mathrm{X}}^{\mathrm{B88}}[\tilde{\rho}] \\ & + 0.19 E_{\mathrm{c}}^{\mathrm{VWN}}[\tilde{\rho}] + 0.81 E_{\mathrm{c}}^{\mathrm{LYP}}[\tilde{\rho}] \end{aligned} \tag{36a}$$

$$\begin{aligned} E^{\mathrm{PBE0}} = & \sum_{\mu,\nu} P_{\mu\nu} H_{\mu\nu} + \sum_{\bar{c}} \sum_{\mu,\nu} P_{\mu\nu} \langle \mu\nu \| \bar{c} \rangle x_{\bar{c}} - \frac{1}{2} \sum_{\bar{c},\bar{d}} x_{\bar{c}} x_{\bar{d}} \langle \bar{c} \| \bar{d} \rangle \\ & -0.25 \sum_{i,j}^{\mathrm{occ}} \sum_{\bar{c},\bar{d}} \langle ij \| \bar{c} \rangle G^{-1}_{\bar{c}\bar{d}} \langle \bar{d} \| ij \rangle + 0.75 E_{\mathrm{X}}^{\mathrm{PBE96}}[\tilde{\rho}] + E_{\mathrm{c}}^{\mathrm{PBE96}}[\tilde{\rho}] \end{aligned} \tag{36b}$$

$$\begin{aligned} E^{\mathrm{M06\text{-}2X}} = & \sum_{\mu,\nu} P_{\mu\nu} H_{\mu\nu} + \sum_{\bar{c}} \sum_{\mu,\nu} P_{\mu\nu} \langle \mu\nu \| \bar{c} \rangle x_{\bar{c}} - \frac{1}{2} \sum_{\bar{c},\bar{d}} x_{\bar{c}} x_{\bar{d}} \langle \bar{c} \| \bar{d} \rangle \\ & -0.54 \sum_{i,j}^{\mathrm{occ}} \sum_{\bar{c},\bar{d}} \langle ij \| \bar{c} \rangle G^{-1}_{\bar{c}\bar{d}} \langle \bar{d} \| ij \rangle + E_{\mathrm{X}}^{\mathrm{M06\text{-}2X}}[\rho] + E_{\mathrm{c}}^{\mathrm{M06\text{-}2X}}[\rho] \end{aligned} \tag{36c}$$

Bear in mind that the contribution from the Becke 88 exchange functional[54] appearing in the B3LYP energy expression is only from the nonlocal part of the functional. The variation of any of these energy expressions with respect to the MO coefficients, under the MO orthonormality constraints, leads to Roothaan–Hall equation systems. The hybrid ADFT Kohn–Sham matrix has contributions from the variation of E_{EXX},

$$\frac{\partial E_{\mathrm{EXX}}}{\partial c_{\mu i}} = -4\sum_{j}^{\mathrm{occ}}\sum_{\nu}\sum_{\bar{c},\bar{d}} \langle \mu j \,\|\, \bar{c}\rangle G_{\bar{c}\bar{d}}^{-1} \langle \bar{d} \,\|\, j\nu\rangle c_{\nu i}, \tag{37}$$

which leads to the definition of the following general form for hybrid ADFT Kohn–Sham matrix elements:

$$K_{\mu\nu} = H_{\mu\nu} + \sum_{\bar{c}} \langle \mu\nu \,\|\, \bar{c}\rangle (x_{\bar{c}} + z_{\bar{c}}) - \alpha \sum_{i}^{\mathrm{occ}}\sum_{\bar{c},\bar{d}} \langle \mu i \,\|\, \bar{c}\rangle G_{\bar{c}\bar{d}}^{-1} \langle \bar{d} \,\|\, i\nu\rangle \tag{38}$$

Here α denotes the fraction of exact exchange to be used in the hybrid DFAs. Note that exchange-correlation fitting coefficients appearing in Eq. (38) absorb the contribution from all exchange-correlation functionals wired into the hybrid functional. Also note that Eq. (38) does not hold for the meta-GGA M06-2X functional which is only available with the DF-DFT methodology due to its explicit dependency on KS MOs via the kinetic energy density.

Straightforward implementation of Eqs. (36) and (38) yields computationally unfavorable algorithms that can be outperformed by exact exchange algorithms with four-center ERIs. This is particularly true for smaller basis sets of double-zeta or lesser quality.[23,25,28-30] In order to overcome this situation, the usual strategy has been to define some rather local fitting domains that reduce the computational demand of calculating the last term in Eq. (38) and the exact exchange energy in Eq. (35). The rationale behind this argument is the already discussed local nature of exact exchange. In the term

$$-\sum_{i}^{\mathrm{occ}}\sum_{\bar{c},\bar{d}} \langle \mu i \,\|\, \bar{c}\rangle G_{\bar{c}\bar{d}}^{-1} \langle \bar{d} \,\|\, i\nu\rangle \tag{39}$$

the MO $\psi_i(\mathbf{r})$ appears in both three-center ERIs. Therefore, only ERIs with nonvanishing overlaps between the AOs $\mu(\mathbf{r})$ or $\nu(\mathbf{r})$ with the MO $\psi_i(\mathbf{r})$ will contribute to the expression in Eq. (39). Thus, the use of tightly localized MOs is very important in order to diminish the number of atomic orbitals overlapping to any given MO. Furthermore, due to the variational nature of the procedure, one can restrict the auxiliary function set and still obtain very good approximations, provided that the auxiliary functions chosen describe appropriately the Hilbert space spanned by the products $\psi_i(\mathbf{r})\mu(\mathbf{r})$ and $\psi_i(\mathbf{r})\nu(\mathbf{r})$.

With this in mind, we developed the LDF-EXX approach. The selection of the local fitting domains in this approach is described in Algorithm 1. The first step is to localize the MOs. To this end, we use the very fast pivoted

ALGORITHM 1 Selection of Local Domains

Localize molecular orbitals.
Löwdin orthogonalize the local MOs: $\breve{c} = \mathbf{S}^{1/2}\mathbf{c}$.
for $i = 1$ to $i = \text{occ}$ **do**
Find Löwdin atomic population $Q_A = \sum_{\mu \in A} \breve{c}_{\mu i}\ \forall\ A$.
Sort Q_A's from largest to smallest.
Sum Q_A's until a predefined threshold, $\tau = 0.99$.
These atoms define the local atomic domain for MO i.
All $\bar{c}$, centered on these atoms, define the local auxiliary function set $\mathcal{A}_i$.
All μ, centered on any atom which overlaps significantly to the local atomic domain, define the local basis set $\mathcal{B}_i$.
end for

Cholesky decomposition of the density matrix[55] followed by a tighter Foster-Boys localization[56,57] with some of the linear scaling modifications proposed by Guo et al.[58] This procedure has demonstrated to be very efficient and produces tightly localized MOs.[30] Then a local set of atoms around each localized MO $\psi_i(\mathbf{r})$ is defined according to the atomic Löwdin populations. All auxiliary functions centered on these atoms define the local auxiliary function set $\mathcal{A}_i$ which is used to approximate all $\psi_i(\mathbf{r})\mu(\mathbf{r})$ products of the given MO. Additionally, the number of $\psi_i(\mathbf{r})\mu(\mathbf{r})$ products is restricted to AOs centered on atoms with significant overlap to any atom in the local set. This defines the local basis set $\mathcal{B}_i$. These restricted auxiliary function and basis sets are called the local domain for $\psi_i(\mathbf{r})$. Once the local domains of the occupied KS MOs are selected, the actual computation of the exact-exchange contribution to the Kohn–Sham matrix is performed. Algorithm 2 shows the steps followed in order to obtain the exact exchange contribution to the Kohn–Sham matrix. Note that for each MO, one has to compute and invert its corresponding Coulomb matrix $\mathbf{G}_i$. At first glance, this seems like a serious disadvantage of the proposed algorithm. However, the local $\mathbf{G}_i$ are relatively small, symmetric, positive definite, and well conditioned. Thus, their inverse can be always calculated by fast Cholesky decomposition algorithms. This is different from other local approaches like CADF[31,32] where positive definiteness cannot be guaranteed. Another important difference of the here described LDF-EXX procedure to other local approaches[28,29] is the definition of a local basis set $\mathcal{B}_i$ for each MO and the explicit localization of MOs during each SCF iteration. The MO localization is of primary importance for obtaining the smallest local auxiliary and basis sets possible.

ALGORITHM 2 Computation of the exact exchange contribution to the Kohn–Sham matrix

Select local domains (Algorithm 1).
for $i = 1$ to $i = \mathrm{occ}$ **do**
Compute $E^{i}_{\mu\bar{c}} = \langle \mu i \| \bar{c} \rangle \; \forall \; \mu \in \mathcal{B}_i$ and $\bar{c} \in \mathcal{A}_i$
Compute $G^{i}_{\bar{c}\bar{d}} = \langle \bar{c} \| \bar{d} \rangle \; \forall \; \bar{c}, \bar{d} \in \mathcal{A}_i$
Obtain the inverse Cholesky factor $\mathbf{G}_i^{-1} = \mathbf{Z}_i\mathbf{Z}_i^T$
Compute $\mathbf{H}_i = \mathbf{E}_i\mathbf{Z}_i$
Compute $\mathbf{K}_i = -\mathbf{H}_i\mathbf{H}_i^T$
Sum the elements of $\mathbf{K}_i$ into the appropriate elements of $\mathbf{K}$.
end for

4. COMPUTATIONAL DETAILS

The hybrid DFAs B3LYP, PBE0, and M06-2X were implemented in a developer version of the deMon2k code.[44,45] For B3LYP, the VWN formula 1 with RPA parameters[59] for the local correlation contribution was used. Note that B3LYP implementations may vary among different codes, either by the VWN functional form or by the parameters used (Monte Carlo or RPA). In particular, the current implementation coincides with the one implemented in the NWChem[60] software package.

The hybrid GGAs B3LYP and PBE0 calculations were performed using either DF-DFT or ADFT methodologies as described in Section 2. The meta-GGA M06-2X calculations were performed only with the DF-DFT methodology due to the explicit dependency of the meta-GGA exchange-correlation functional on MOs through the kinetic energy density.

The exchange-correlation potential was numerically integrated on an adaptive grid.[61,62] The grid accuracy was set to 10^{-5} a.u. in all calculations. The Coulomb and exact exchange energies were calculated by the variational fitting procedures presented in Sections 2 and 3. The automatically generated GEN-A2*[63] auxiliary function set was used for both fittings. The default MinMax[53] SCF convergence stabilization and acceleration techniques were used, i.e., initially dynamical damping of the auxiliary density (ADFT) or Kohn–Sham density (DF-DFT) followed by Pulay's direct inversion of the iterative subspace[53,64,65] once the MinMax error dropped below a 10^{-2} a.u. threshold. The MinMax SCF convergence criteria were set to 10^{-5} a.u. in all calculations. NWChem computations were performed using default settings.

The Def2-TZVPP basis set[66] was used for validation calculations. This basis set was retrieved from the EMSL basis set exchange site.[67,68] For timing benchmarks, the DFT-optimized DZVP basis set[69] was employed. The spherical representation for atomic orbitals was used for all calculations.

All reported timings refer to serial direct SCF calculations either on an Intel®Core™i5-2400 CPU @ 3.10 GHz with 2 GB of maximum allocatable memory (linear alkane chains) or on an Intel® Xeon® X5675 @ 3.07 GHz with a maximum of 4 GB of RAM (amylose chains).

5. VALIDATION

In order to validate our hybrid DFA implementations, we compared several properties that depend on energy differences. First, we performed calculations of standard enthalpies of formation ΔH_f^{298} using the G3/99X test set.[70–72] The protocol for calculating ΔH_f^{298} was the one established by Curtiss and collaborators.[73] The values of mean absolute error (MAE) with respect to experiment[74] are reported in Tables 1, 2, and 3 for B3LYP, PBE0, and M06-2X, respectively. They are listed under the entry "Heats of formation." Standard DFT results from NWChem calculations for these three DFAs are also included in each table under the "DFT" column. As Tables 1–3 show, all LDF-DFT MAEs are in excellent agreement with the DFT reference. The largest deviation is only 0.05 kcal/mol for the M06-2X functional. Although M06-2X uses a significant fraction of exact exchange (54%), part of the observed small deviations may also arise from sources other than the LDF-EXX computation, for example, integration errors specific to the M06 suite.[75] The use of LDF-ADFT increases the MAE for heats of formation with B3LYP, but decreases the same quantity for PBE0. This difference between LDF-DFT and LDF-ADFT is mainly

Table 1 Mean Absolute Errors (MAEs) of the B3LYP Hybrid Exchange-Correlation Functional for Heats of Formation and Reaction Barrier Heights from Different DFT Approaches

Property	DFT	LDF-DFT	LDF-ADFT
Heats of formation	5.55	5.53	6.40
Hydrogen transfer forward	4.29	4.28	4.15
Hydrogen transfer backward	4.65	4.65	4.53
Nonhydrogen transfer forward	6.09	6.17	6.10
Nonhydrogen transfer backward	5.24	5.39	5.34

All energies are in kcal/mol.

Table 2 Mean Absolute Errors (MAEs) of the PBE0 Hybrid Exchange-Correlation Functional for Heats of Formation and Reaction Barrier Heights from Different DFT Approaches

Property	DFT	LDF-DFT	LDF-ADFT
Heats of formation	5.80	5.82	5.43
Hydrogen transfer forward	4.05	4.03	3.69
Hydrogen transfer backward	4.90	4.89	4.68
Nonhydrogen transfer forward	4.60	4.69	4.63
Nonhydrogen transfer backward	4.45	4.55	4.53

All energies are in kcal/mol.

Table 3 Mean Absolute Errors (MAEs) of the M06-2X Hybrid Exchange-Correlation Functional for Heats of Formation and Reaction Barrier Heights from Different DFT Approaches

Property	DFT	LDF-DFT
Heats of formation	2.20	2.25
Hydrogen transfer forward	1.04	1.08
Hydrogen transfer backward	1.38	1.41
Nonhydrogen transfer forward	1.76	1.79
Nonhydrogen transfer backward	1.92	1.89

All energies are in kcal/mol.

due to the use of the auxiliary density for the calculation of the exchange-correlation energy and potential in LDF-ADFT.[33] The overall small MAE deviations ($<$ 1 kcal/mol) of LDF-DFT and LDF-ADFT with respect to standard DFT confirm that the LDF-EXX algorithm is well suited for thermochemical applications. In particular, it preserves the intrinsic accuracy of the underlying hybrid DFAs.

For the validation of reaction barrier heights, the HTBH38/08 and NHTBH38/08 databases[76–79] were used. The HTBH database contains 19 hydrogen transfer reactions with values for the forward and reverse classical reaction barrier heights. The NHTBH database contains 19 reactions, including heavy-atom transfer, bimolecular nucleophilic substitution, association, and unimolecular reactions. For the purpose of this chapter, the "best estimates" provided in the Minnesota databases[80] are used as experimental references. MAEs with respect to these reference data are reported in rows two to four of Tables 1–3. Again, standard DFT results from NWChem calculations are reported in the first column of these tables under "DFT."

For all three studied hybrid functionals, the LDF-DFT reaction barrier height MAEs are within 0.15 kcal/mol to the DFT ones. The LDF-ADFT reaction barrier height MAEs are always smaller than the LDF-DFT ones. This underlies the intrinsic reliability of the ADFT approach. In this context, it is also important to note that other statistical parameters, like maximum and minimum errors, show deviations from standard DFT implementations in a similar energy range of 0.15 kcal/mol. Therefore, LDF-EXX-based hybrid functionals are also well suited for the calculation of activation barriers in chemical reactions. In particular, the MO localization and restricted auxiliary and basis function expansion in the LDF-EXX methodology do not alter the improved energetical description of hybrid functionals.

6. TIMINGS

The computational performance of the new hybrid DFA implementations was benchmarked on the linear alkane chains C_nH_{2n+2} from $n=10$ to $n=80$. Since timings are almost identical for the two hybrid GGAs, B3LYP and PBE0, only B3LYP results will be discussed. Figures 1 and 2 show the CPU timing breakdown of single-point deMon2k calculations of alkane chains with the LDF-ADFT and LDF-DFT methodologies, respectively. It is evident from these figures that the initial SVD calculation becomes increasingly dominant with system size and takes up to 28% of the total time for the largest molecule. Note that the SVD timing (height of the bar) will remain the same even when multiple single-point energies have to be calculated, e.g., in geometry optimizations or molecular dynamics runs, since this step is performed only once at the beginning of the first SCF.

For the LDF-ADFT approach, shown in Fig. 1, the second most time-consuming step is given by the two-electron Coulomb calculation (28%), closely followed by the LDF-EXX (27%) contribution in the largest alkane. It has been noted before that LDF-EXX can take only a fraction of the time needed to compute the Coulomb contribution in large one-dimensional systems employing the direct SCF scheme.[30] Finally, the exchange-correlation (XC) contribution takes only 13% of the total time for the largest alkane due to the ADFT approach. In all cases, convergence was achieved

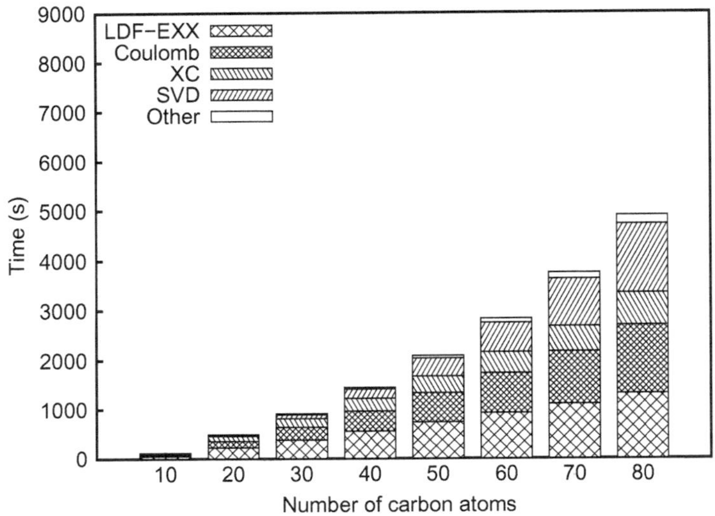

Figure 1 CPU timing breakdown for single-point calculations of linear alkane chains employing the B3LYP hybrid functional and the auxiliary density functional theory (ADFT) methodology.

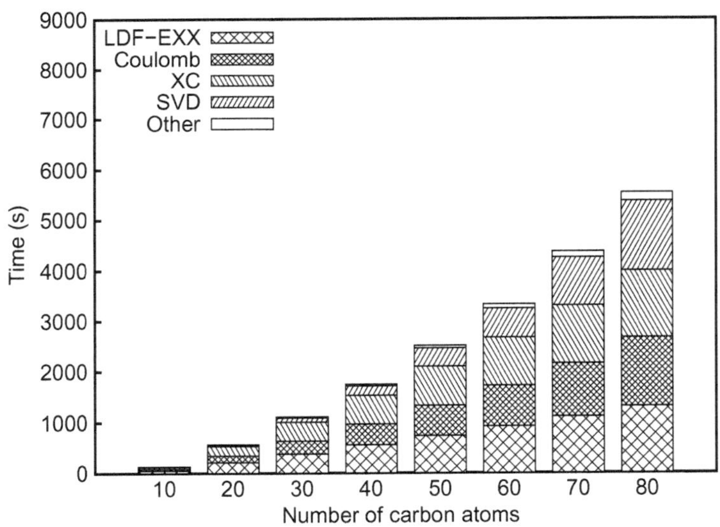

Figure 2 CPU timing breakdown for single-point calculations of linear alkane chains employing the B3LYP hybrid functional and the density-fitting density functional theory (DF-DFT) methodology.

after 18 SCF iterations. This means that the serial average direct SCF iteration takes only 3.1 min for the $C_{80}H_{162}$ chain, described with 2010 basis functions and 11,004 auxiliary functions.

In Fig. 2, CPU timings for the LDF-DFT approach are shown. By and large, this graphic is very similar to the one for LDF-ADFT except that, in this case, the time needed to compute the exchange-correlation contribution is almost the same as the one needed for the Coulomb or LDF-EXX contributions. In fact, for $C_{80}H_{162}$ the total CPU time is almost equally partitioned between SVD (25%), Coulomb (25%), exchange-correlation (24%), and LDF-EXX (24%). Convergence was also achieved after 18 SCF iterations with an average of 3.7 min per iteration for the largest alkane chain.

Timings are very different for the hybrid meta-GGA M06-2X. Figure 3 shows that the most demanding computational task here is the calculation of the exchange-correlation contribution. For the largest alkane, this task accounts for almost 50% of the total CPU time. The increased workload arises from the computation of the kinetic energy density $\tau(\mathbf{r})$ and the τ-dependent part of the Kohn–Sham matrix on the grid that prevents the use of ADFT. Nevertheless, the difference between LDF-DFT and LDF-ADFT timings is rather small for linear alkane chains (compare

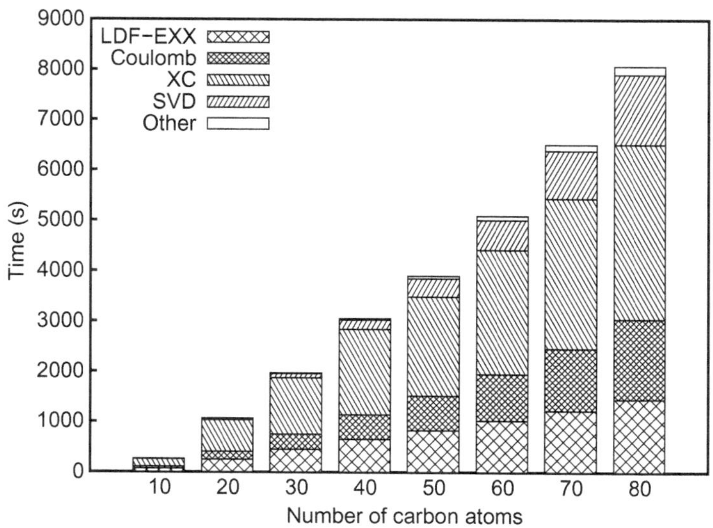

Figure 3 CPU timing breakdown for single-point calculations of linear alkane chains employing the M06-2X hybrid functional and the density-fitting density functional theory (DF-DFT) methodology.

Fig. 1 with Fig. 2). The reason is the ideal one-dimensional expansion of these systems that facilitates screening on the grid and in the local domains of the LDF-EXX approach.[30] To demonstrate this in more detail, we now discuss benchmark calculations on amylose chains with up to 16 d-glucose units. These systems are still rather sparse as their use for benchmarking other low-scaling algorithms[81] shows. The largest amylose chain is composed of 96 carbon, 81 oxygen, and 162 hydrogen atoms. It has almost the same extension as the $C_{40}H_{82}$ alkane chain, depicted in Fig. 4, but four times more heavy atoms (C and O). These larger calculations were performed with the GEN-A2/GEN-A2* strategy described in Ref. 30. In short, the first auxiliary function set (GEN-A2) is used during the SCF, while the second one (GEN-A2*) is used to calculate a non-self-consistent energy with the converged Kohn–Sham density. This approach has the best cost–accuracy ratio and is recommended for large-scale exact exchange computations. Figures 5 and 6 show the CPU timing breakdown of the first two SCF iterations with the PBE0 hybrid functional for the LDF-ADFT and LDF-DFT approaches, respectively. Most notable in the comparison of Figs. 5 and 6 is the large difference in the exchange-correlation timings due to the ADFT approach. In fact, in LDF-DFT (Fig. 6) the XC contribution dominates the timings of the two largest amylose chains with 8 and 16 d-glucose units. Note that the LDF-EXX time is smaller than the time needed for the numerical integration of the exchange-correlation energy and potential in LDF-DFT. On the other hand, LDF-ADFT timings are dominated by the Coulomb and LDF-EXX times. This qualitative difference to the alkane chain benchmark results arises from the more packed structure of amylose. The difference further increases for more dense packing, for example, in diamond cell calculations with the PBE functional reported in Ref. 30, ADFT was almost one order of magnitude faster than DF-DFT. These findings are the motivation for our search of even more efficient algorithms for the calculation of Coulomb[38] and exact exchange contributions.[82]

7. CONCLUSION

In this chapter, we present the implementation of hybrid DFAs in the software package deMon2k. The new implemented functionals can be used with both density-fitted- and auxiliary density functional theory. The exact-exchange contribution is calculated with the LDF approach proposed in our group[30] which has been demonstrated to be robust and efficient.

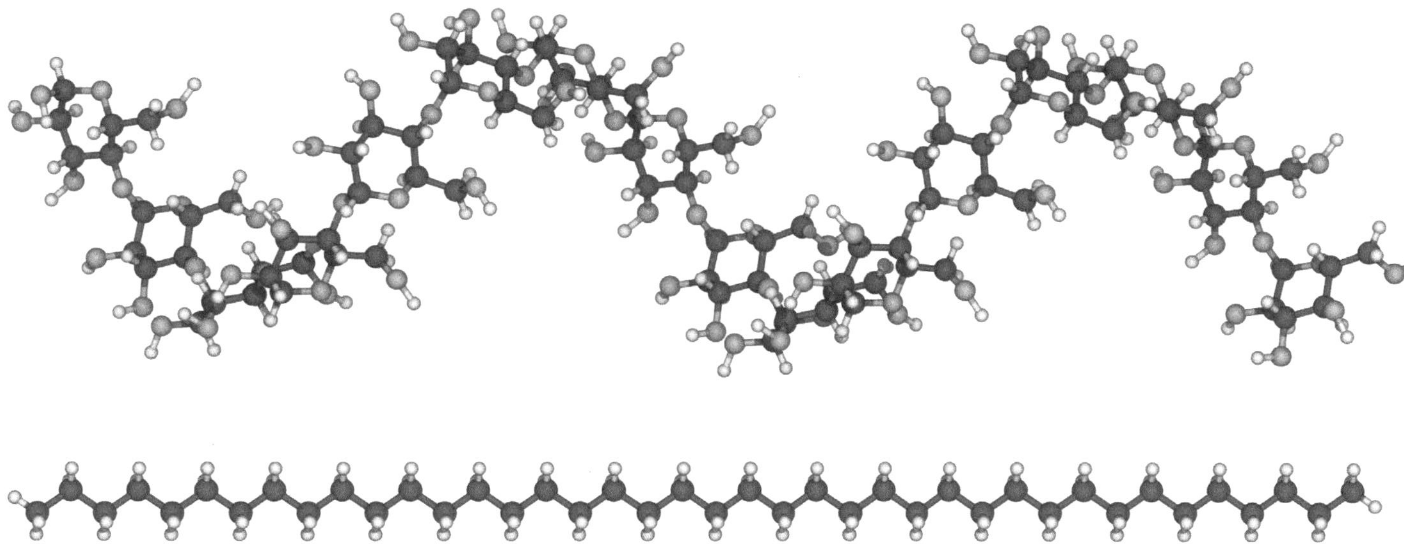

Figure 4 Comparison of amylose$_{16}$ helix (top) and linear alkane chain $C_{40}H_{82}$ (bottom) extension. End-to-end distance is about 50 Å for the amylose helix and 49 Å for the alkane chain.

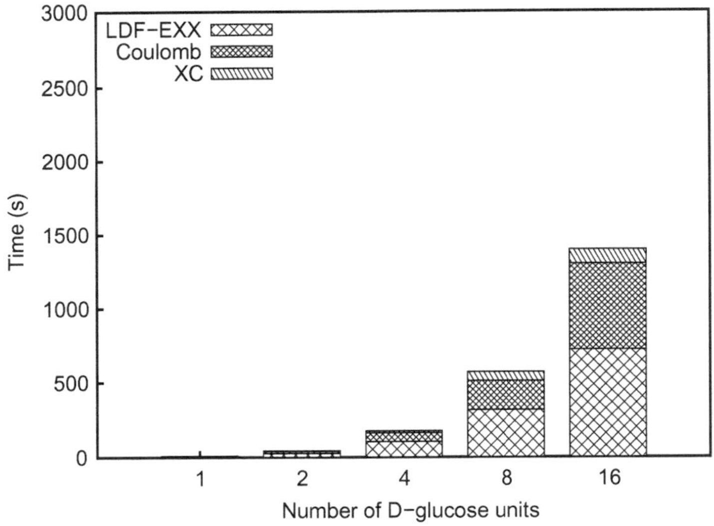

Figure 5 CPU timing breakdown for the first two SCF iterations for amylose helices described with the PBE0 hybrid functional and the auxiliary density functional theory (ADFT) methodology.

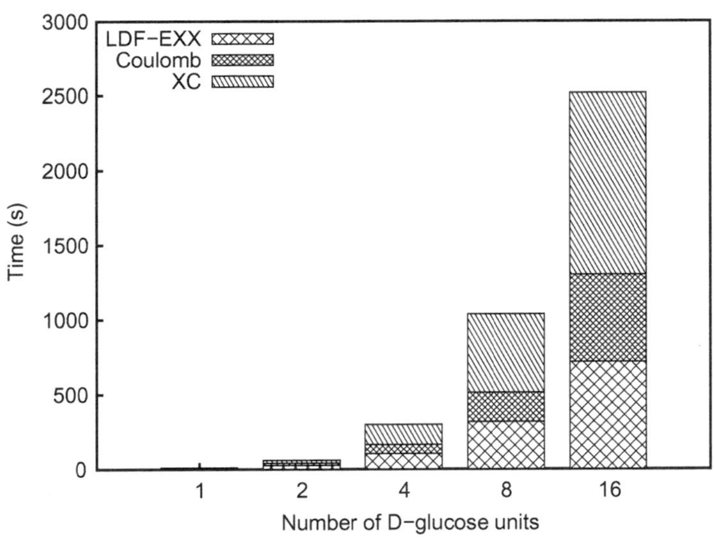

Figure 6 CPU timing breakdown for the first two SCF iterations for amylose helices described with the PBE0 hybrid functional and the density-fitting density functional theory (DF-DFT) methodology.

The newly implemented DFAs B3LYP, PBE0, and M06-2X were validated using 223 standard enthalpies of formation,[70–72] 19 hydrogen-transfer reaction barrier heights, and 19 nonhydrogen transfer reaction barrier heights.[76–79] The mean absolute deviations with respect to experimental data for the standard enthalpies of formation of LDF-DFT and LDF-ADFT hybrid functional calculations are within 1 kcal/mol identical to conventional Kohn–Sham calculations. This shows that the here presented density-fitting methodology for two-electron Coulomb and exact exchange contributions is well suited for thermochemical applications. Moreover, chemical reaction barriers obtained with the new LDF-DFT and LDF-ADFT are virtually indistinguishable to conventional DFT results. The high accuracy and robustness of the variational fitting approach comes along with a substantial improvement in computational performance. In particular, the LDF-ADFT methodology employing the GEN-A2/GEN-A2* strategy permits hybrid DFT energy calculations of large systems in very reasonable computation times, fairly independent of the system structure. Currently, we extend the here described methodology to energy gradients, aiming for large-scale molecular dynamics simulations with hybrid functionals. In addition, further efforts to test hybrid LDF-ADFT on response properties already implemented in deMon2k are being carried out in our groups.

ACKNOWLEDGMENTS

This work was performed in the framework of the CONACyT group project CB-173409. D.M.-R. gratefully acknowledges a Ph.D. CONACyT fellowship (CVU 47922). J.M.C. and X.H. thank DGTIC-UNAM for the use of their facilities and support provided by DGAPA-UNAM Grant No. IA102114.

REFERENCES

1. Hohenberg, P.; Kohn, W. Inhomogeneous Electron Gas. *Phys. Rev.* **1964**, *136*, B864–B871.
2. Kohn, W. Nobel Lecture Electronic Structure of Matter-Wave Functions and Density Functionals. *Rev. Mod. Phys.* **1999**, *71*, 1253–1266.
3. Kohn, W.; Sham, L. J. Self-Consistent Equations Including Exchange and Correlation Effects. *Phys. Rev.* **1965**, *140*, A1133–A1138.
4. Perdew, J. P. Jacob's Ladder of Density Functional Approximations for the Exchange-Correlation Energy. AIP Conference Proceedings, 577; AIP, 2001; pp 1–20.
5. Becke, A. D. Perspective: Fifty Years of Density-Functional Theory in Chemical Physics. *J. Chem. Phys.* **2014**, *140*, 18A301.
6. Clementi, E.; Corongin, G. Computational Chemistry: The Correlation Energy and the Global Simulations. In Clementi, E., Corongin, G., Eds.; METECC-95; STEF: Cagliari, Italy, 1995; p 32.
7. Parr, R. G.; Yang, W. *Density Functional Theory of Atoms and Molecules*. Oxford University Press: New York; 1989; 186 pp.

8. Becke, A. D. A New Mixing of Hartree-Fock and Local Density-Functional Theories. *J. Chem. Phys.* **1993**, *98*, 1372–1377.
9. Becke, A. D. Density-Functional Thermochemistry. III. The Role of Exact Exchange. *J. Chem. Phys.* **1993**, *98*, 5648–5652.
10. Stephens, P. J.; Devlin, F. J.; Chabalowski, C. F.; Frisch, M. Ab Initio Calculations of Vibrational Absorption and Circular Dichroism Spectra Using Density Functional Force Fields. *J. Phys. Chem.* **1994**, *98*, 11623–11627.
11. Schwegler, E.; Challacombe, M.; Head-Gordon, M. Linear Scaling Computation of the Fock Matrix. II. Rigorous Bounds on Exchange Integrals and Incremental Fock Build. *J. Chem. Phys.* **1997**, *106*, 9708–9717.
12. Ochsenfeld, C.; White, C. A.; Head-Gordon, M. Linear and Sublinear Scaling Formation of Hartree-Fock Type Exchange Matrices. *J. Chem. Phys.* **1998**, *109*, 1663–1669.
13. Friesner, R. A. Solution of Self-Consistent Field Electronic Structure Equations by a Pseudospectral Method. *Chem. Phys. Lett.* **1985**, *116*, 39–45.
14. Friesner, R. A. Solution of the Hartree-Fock Equations by a Pseudospectral Method: Application to Diatomic Molecules. *J. Chem. Phys.* **1986**, *85*, 1462–1468.
15. Kossmann, S.; Neese, F. Comparison of Two Efficient Approximate Hartree-Fock Approaches. *Chem. Phys. Lett.* **2009**, *481*, 240–243.
16. Neese, F.; Wennmohs, F.; Hansen, A.; Becker, U. Efficient, Approximate and Parallel Hartree-Fock and Hybrid DFT Calculations. A "Chain-of-Spheres" Algorithm for the Hartree-Fock Exchange. *Chem. Phys.* **2009**, *356*, 98–109.
17. Izsák, R.; Neese, F. An Overlap Fitted Chain of Spheres Exchange Method. *J. Chem. Phys.* **2011**, *135*, 144105.
18. Dunlap, B. I.; Conolly, J. W. D.; Sabin, J. R. On First-Row Diatomic Molecules and Local Density Models. *J. Chem. Phys.* **1979**, *71*, 4993–4999.
19. Mintmire, J. W.; Dunlap, B. I. Fitting the Coulomb Potential Variationally in Linear-Combination-of-Atomic-Orbitals Density Functional Calculations. *Phys. Rev. A* **1982**, *25*, 88–95.
20. Mintmire, J. W.; Sabin, J. R.; Trickey, S. B. Local-Density Functional Methods in Two-Dimensional Periodic Systems. Hydrogen and Beryllium Monolayers. *Phys. Rev. B* **1982**, *26*, 1743–1753.
21. Köster, A. M. Efficient Recursive Computation of Molecular Integrals for Density Functional Methods. *J. Chem. Phys.* **1996**, *104*, 4114–4124.
22. Köster, A. M. Hermite Gaussian Auxiliary Functions for the Variational Fitting of the Coulomb Potential in Density Functional Methods. *J. Chem. Phys.* **2003**, *118*, 9943–9951.
23. Kendall, R. A.; Früchtl, H. A. The Impact of the Resolution of the Identity Approximate Integral Method on Modern Ab Initio Algorithm Development. *Theor. Chem. Acc.* **1997**, *97*, 158–163.
24. Hamel, S.; Casida, M. E.; Salahub, D. R. Assessment of the Quality of Orbital Energies in Resolution-of-the-Identity Hartree-Fock Calculations Using deMon Auxiliary Basis Sets. *J. Chem. Phys.* **2001**, *114*, 7342–7350.
25. Weigend, F. A Fully Direct RI-HF Algorithm: Implementation, Optimised Auxiliary Basis Sets, Demonstration of Accuracy and Efficiency. *Phys. Chem. Chem. Phys.* **2002**, *4*, 4285–4291.
26. Aquilante, F.; Pedersen, T.; Lindh, R. Low-Cost Evaluation of the Exchange Fock Matrix from Cholesky and Density Fitting Representations of the Electron Repulsion Integrals. *J. Chem. Phys.* **2007**, *126*, 194106.
27. Aquilante, F.; Lindh, R.; Pedersen, T. Analytic Derivatives for the Cholesky Decomposition of the Two-Electron Integrals. *J. Chem. Phys.* **2008**, *129*, 034106.
28. Sodt, A.; Head-Gordon, M. Hartree-Fock Exchange Computed Using the Atomic Resolution of the Identity Approximation. *J. Chem. Phys.* **2008**, *128*, 104106.

29. Polly, R.; Werner, H. J.; Manby, F. R.; Knowles, P. J. Fast Hartree-Fock Theory Using Local Density Fitting Approximations. *Mol. Phys.* **2004**, *102*, 2311–2321.
30. Mejía-Rodríguez, D.; Köster, A. M. R. Robust and Efficient Variational Fitting of Fock Exchange. *J. Chem. Phys.* **2014**, *141*, 124114.
31. Merlot, P.; Kjærgaard, T.; Helgaker, T.; Lindh, R.; Aquilante, F.; Reine, S.; Pedersen, T. B. Attractive Electron-Electron Interactions within Robust Local Fitting Approximations. *J. Comput. Chem.* **2013**, *34*, 1486–1496.
32. Hollman, D.; Schaefer, H.; Valeev, E. Semi-Exact Concentric Atomic Density Fitting: Reduced Cost and Increased Accuracy Compared to Standard Density Fitting. *J. Chem. Phys.* **2014**, *140*, 064109.
33. Köster, A. M.; Reveles, J. U.; del Campo, J. M. Calculation of Exchange-Correlation Potentials with Auxiliary Function Densities. *J. Chem. Phys.* **2004**, *121*, 3417–3424.
34. Perdew, J. P.; Ernzerhof, M.; Burke, K. Rationale for Mixing Exact Exchange with Density Functional Approximations. *J. Chem. Phys.* **1996**, *105*, 9982–9985.
35. Adamo, C.; Barone, V. Toward Reliable Density Functional Methods Without Adjustable Parameters: The PBE0 Model. *J. Chem. Phys.* **1999**, *110*, 6158–6170.
36. Zhao, Y.; Truhlar, D. G. The M06 Suite of Density Functionals for the Main Group Thermochemistry, Thermochemical Kinetics, Noncovalent Interactions, Excited States, and Transition Elements: Two New Functionals and Systematic Testing of Four M06-Class Functionals and 12 Other Functionals. *Theor. Chem. Acc.* **2007**, *120*, 215–241.
37. Calaminici, P.; Domínguez-Soria, V. D.; Flores-Moreno, R.; Gamboa-Martinez, G. U.; Geudtner, G.; Goursot, A.; Salahub, D. R.; Köster, A. M. Auxiliary Density Functional Theory: From Molecules to Nanostructures. In: *Handbook of Computational Chemistry*; Leszczynski, J. Ed, 1st; Springer: Netherlands, 2012, pp 573–610.
38. Alvarez-Ibarra, A.; Köster, A. M. Double Asymptotic Expansion of Three-Center Electronic Repulsion Integrals. *J. Chem. Phys.* **2013**, *139*, 024102.
39. Andzelm, J.; Wimmer, E. Density Functional Gaussian-Type Orbital Approach to Molecular Geometries, Vibrations, and Reaction Energies. *J. Chem. Phys.* **1992**, *96*, 1280–1303.
40. St-Amant, A.; Salahub, D. New Algorithm for the Optimization of Geometries in Local Density Functional Theory. *Chem. Phys. Lett.* **1990**, *169*, 387–392.
41. Dunlap, B. I.; Rösch, N.; Trickey, S. B. Variational Fitting Methods for Electronic Structure Calculations. *Mol. Phys.* **2010**, *108*, 3167–3180.
42. Vahtras, O.; Almlöf, J.; Feyereisen, M. Integral Approximations for LCAO-SCF Calculations. *Chem. Phys. Lett.* **1993**, *213*, 514–518.
43. Flores-Moreno, R.; Ortiz, J. V. Integral Approximations in Ab Initio Electron Propagator Calculations. *J. Chem. Phys.* **2009**, *131*, 124110.
44. Köster, A. M.; Geudtner, G.; Calaminici, P.; Casida, M. E.; Domínguez, V. D.; Flores-Moreno, R.; Gamboa, G. U.; Goursot, A.; Heine, T.; Ipatov, A.; Janetzko, F.; del Campo, J. M.; Reveles, J. U.; Vela, A.; Zuniga-Gutierrez, B.; Salahub, D. R. *deMon2k Version, 3;* The deMon Developers, Cinvestav: Mexico City, 2011.
45. Geudtner, G.; Calaminici, P.; Carmona-Espíndola, J.; del Campo, J. M.; Domínguez-Soria, V. D.; Flores-Moreno, R.; Gamboa, G. U.; Goursot, A.; Köster, A. M.; Reveles, J. U.; Mineva, T.; Vázquez-Pérez, J. M.; Vela, A.; Zuniga-Gutierrez, B.; Salahub, D. R. deMon2k. *WIREs Comput. Mol. Sci.* **2011**, *2*, 548–555.
46. Dominguez-Soria, V.; Geudtner, G.; Morales, J.; Calaminici, P.; Köster, A. M. Robust and Efficient Density Fitting. *J. Chem. Phys.* **2009**, *131*, 124102.
47. Laikov, D. Fast Evaluation of Density Functional Exchange-Correlation Terms Using the Expansion of the Electron Density in Auxiliary Basis Sets. *Chem. Phys. Lett.* **1997**, *281*, 151–156.
48. Köster, A. M. Entwicklung einer LCGTO-Dichtefunktionalmethode mit Hilfsfunktionen. Habilitation Thesis; Universität Hannover, 1998.

49. Birkenheuer, U.; Gordienko, A.; Nasluzov, V.; Fuchs-Rohr, M.; Rösch, N. Model Density Approach to the Kohn-Sham Problem: Efficient Extension of the Density Fitting Technique. *Int. J. Quantum Chem.* **2005**, *102*, 743–761.
50. Belpassi, L.; Tarantelli, F.; Sgamellotti, A.; Quiney, H. M. Electron Density Fitting for the Coulomb Problem in Relativistic Density Functional Theory. *J. Chem. Phys.* **2006**, *124*, 124104.
51. Janetzko, F.; Köster, A. M.; Salahub, D. R. Development of the Cyclic Cluster Model Formalism for Kohn-Sham Auxiliary Density Functional Theory Methods. *J. Chem. Phys.* **2008**, *128*, 024102.
52. Saunders, V. R. Molecular Integrals for Gaussian Type Functions. In: *Methods in Computational Molecular Physics*; Diercksen, G. H. F., Wilson, S., Eds., Springer: Netherlands, 1983; pp 1–36.
53. Köster, A. M.; del Campo, J. M.; Janetzko, F.; Zuniga-Gutierrex, B. MinMax Self-Consistent Field Approach for Auxiliary Density Functional Theory. *J. Chem. Phys.* **2009**, *130*, 114106.
54. Becke, A. D. Density-Functional Exchange-Energy Approximation with Correct Asymptotic Behaviour. *Phys. Rev. A* **1988**, *38*, 3098–3100.
55. Aquilante, F.; Pedersen, T. B.; de Merás, A. S.; Koch, H. Fast Noniterative Orbital Localization for Large Molecules. *J. Chem. Phys.* **2006**, *125*, 174101.
56. Boys, S. Construction of Some Molecular Orbitals to be Approximately Invariant for Changes from One Molecule to Another. *Rev. Mod. Phys.* **1960**, *32*, 296–299.
57. Foster, J. M.; Boys, S. F. Canonical Configuration Interaction Procedure. *Rev. Mod. Phys.* **1960**, *32*, 300–302.
58. Guo, Y.; Li, W.; Li, S. An Efficient Linear Scaling Procedure for Constructing Localized Orbitals of Large Molecules Based on the One-Particle Density Matrix. *J. Chem. Phys.* **2011**, *135*, 134107.
59. Vosko, S. H.; Wilk, L.; Nusair, M. Accurate Spin-Dependent Electron Liquid Correlation Energies for Local Spin Density Calculations: A Critical Analysis. *Can. J. Phys.* **1980**, *58*, 1200–1211.
60. Valiev, M.; Bylaska, E. J.; Govind, N.; Kowalski, K.; Straatsma, T. P.; Van Dam, H. J. J.; Wang, D.; Nieplocha, J.; Apra, E.; Windus, T. L.; deJong, W. A. N. A Comprehensive and Scalable Open-Source Solution for Large Scale Molecular Simulations. *Comput. Phys. Commun.* **2010**, *181*, 1477–1489.
61. Krack, M.; Köster, A. M. An Adaptive Numerical Integrator for Molecular Integrals. *J. Chem. Phys.* **1998**, *108*, 3226–3234.
62. Köster, A. M.; Flores-Moreno, R.; Reveles, J. U. Efficient and Reliable Numerical Integration of Exchange-Correlation Energies and Potentials. *J. Chem. Phys.* **2004**, *121*, 681–690.
63. Calaminici, P.; Janetzko, F.; Köster, A. M.; Mejía-Olvera, R.; Zuniga-Gutierrez, B. Density Functional Theory Optimized Basis Sets for Gradient Corrected Functionals: 3d Transition Metal Systems. *J. Chem. Phys.* **2007**, *126*, 044108.
64. Pulay, P. Convergence Acceleration of Iterative Sequences, the Case of SCF Iteration. *Chem. Phys. Lett.* **1980**, *73*, 393–398.
65. Pulay, P. Improved SCF Convergence Acceleration. *J. Comput. Chem.* **1982**, *3*, 556–560.
66. Weigend, F.; Ahlrichs, R. Balanced Basis Sets of Split Valence, Triple Zeta Valence and Quadruple Zeta Valence Quality for H to Rn: Design and Assessment of Accuracy. *Phys. Chem. Chem. Phys.* **2005**, 7, 3297–3305.
67. Schuchart, K. L.; Didier, B. T.; Elsethagen, T.; Sun, L.; Gurumoorthi, V.; Chase, J.; Li, J.; Windus, T. L. Basis Set Exchange: A Community Database for Computational Sciences. *J. Chem. Inf. Model.* **2007**, *47*, 1045–1052.

68. Feller, D. The Role of Databases in Support of Computational Chemistry Calculations. *J. Comput. Chem.* **1996**, *17*, 1571–1586.
69. Godbout, N.; Salahub, D. R.; Andzelm, J.; Wimmer, E. Optimization of Gaussian-Type Basis Sets for Local Spin Density Functional Calculations. Part I. Boron Through Neon, Optimization Technique and Validation. *Can. J. Chem.* **1992**, *70*, 560–571.
70. Curtiss, L. A.; Redfern, P. C.; Raghavachari, K.; Pople, J. A. Assessment of Gaussian-2 and Density-Functional Theories for the Computation of Ionization Potentials and Electron Affinities. *J. Chem. Phys.* **1998**, *109*, 42–55.
71. Curtiss, L. A.; Raghavachari, K.; Redfern, P. C.; Rassolov, V.; Pople, J. A. Gaussian-3 (G3) Theory for Molecules Containing First and Second-Row Atoms. *J. Chem. Phys.* **1998**, *109*, 7764–7776.
72. Curtiss, L. A.; Redfern, P. C.; Raghavachari, K.; Pople, J. A. Gaussian-3X (G3X) Theory: Use of Improved Geometries, Zero-Point Energies and Hartree-Fock Basis Sets. *J. Chem. Phys.* **2001**, *114*, 108–117.
73. Curtiss, L. A.; Raghavachari, K.; Redfern, P. C.; Pople, J. A. Assessment of Gaussian-2 and Density Functional Theories for the Computation of Enthalpies of Formation. *J. Chem. Phys.* **1997**, *106*, 1063–1079.
74. Chase, M. W., Jr.; Davies, C. A.; Downey, J. R., Jr.; Frurip, D. J.; McDonald, R. A.; Syverud, A. JANAF Thermochemical Tables. *J. Phys. Chem. Ref. Data* **1985**, *14,* Suppl. 1; Pts. I and II.
75. Wheeler, S. E.; Houk, K. N. Integration Grid Errors for Meta-GGA-Predicted Reaction Energies: Origin of Grid Errors for the M06 Suite of Functionals. *J. Chem. Theory Comput.* **2010**, *6*, 395–404.
76. Zhao, Y.; González-García, N.; Truhlar, D. G. Benchmark Database of Barrier Heights for Heavy Atom Transfer, Nucleophilic Substitution, Association, and Unimolecular Reactions and Its Use to Test Theoretical Methods. *J. Phys. Chem. A* **2005**, *109*, 2012–2018.
77. Zhao, Y.; Lynch, B. J.; Truhlar, D. G. Multi-Coefficient Extrapolated Density Functional Theory for Thermochemistry and Thermochemical Kinetics. *Phys. Chem. Chem. Phys.* **2005**, 7, 43–52.
78. Zhao, Y.; Schultz, N. E.; Truhlar, D. G. Design of Density Functionals by Combining the Method of Constraint Satisfaction with Parametrization for Thermochemistry, Thermochemical Kinetics, and Noncovalent Interactions. *J. Chem. Theory Comput.* **2006**, *2*, 364–382.
79. Zheng, J.; Zhao, Y.; Truhlar, D. G. The DBH24/08 Database and Its Use to Assess Electronic Structure Model Chemistries for Chemical Reaction Barrier Heights. *J. Chem. Theory Comput.* **2009**, *5*, 808–821.
80. Peverati, R.; Truhlar, D. G. Quest for a Universal Density Functional: The Accuracy of Density Functionals Across a Broad Spectrum of Databases in Chemistry and Physics. *Philos. Trans. R. Soc. A* **2014**, *372*, 20120476.
81. Kussmann, J.; Ochsenfeld, C. Linear-Scaling Methods for Calculating Nuclear Magnetic Resonance Chemical Shifts Using Gauge-Including Atomic Orbitals Within Hartree-Fock and Density-Functional Theory. *J. Chem. Phys.* **2007**, *127*, 054103.
82. Köster, A. M.; Mejía-Rodríguez, D. *Robust and Efficient Evaluation of Hartree-Fock Exchange, Concepts of Mathematical Physics in Chemistry: A Tribute to Frank E. Harris, Playa del Carmen*; Quintana Roo: Mexico, 2014.

CHAPTER FIVE

The Hydrogen H_2^+ and HeH^{2+} Molecular Ions Confined in Dihedral Angles

Salvador A. Cruz*[,1], Eugenio Ley-Koo[†]
*Departamento de Física, Universidad Autónoma Metropolitana-Iztapalapa, México D.F., México
[†]Instituto de Física, Universidad Nacional Autónoma de México, México D.F., México
[1]Corresponding author: e-mail address: cruz@xanum.uam.mx

Contents

Advances in Quantum Chemistry, Volume 71
ISSN 0065-3276
http://dx.doi.org/10.1016/bs.aiq.2015.03.001

Abstract

Exact solutions of the Schrödinger equation for the H_2^+ and HeH^{2+} molecular ions, confined in dihedral angles and in the Born–Oppenheimer approximation, are constructed. The confinement in the dihedral angle breaks the rotational symmetry around the internuclear axis, with the consequent breaking of the parity inversion symmetry relative to the plane bisecting the internuclear axis of the homonuclear molecule; the need and the way to restore the second symmetry are analyzed and implemented. The Schrödinger equation is separable in prolate spheroidal coordinates, with the nuclei at the positions of the foci; the confinement in a dihedral angle is modeled by the boundary condition on the electronic wavefunction of vanishing at the meridian planes defining the angle. The electronic wavefunctions are the products of trigonometric functions in the azimuthal angle with noninteger magnetic numbers, superpositions of Legendre polynomials with definite η-parity in the hyperboloidal coordinate, and Jaffé functions in the spheroidal coordinate. They are evaluated in the standard way using matrix diagonalizations to determine their eigenvalues for the electronic and molecular energies as functions of the internuclear distance. The potential energy curves for the ground, first, and second excited states of the homonuclear H_2^+ molecule and for the metastable state of HeH^{2+} are reported illustrating their variations with the confining angle. In the first case, the Dunham parameters needed for the description of the molecular vibrations are also evaluated; the electric dipole moment is also illustrated as a manifestation of the breaking of the rotational symmetry due to the confinement. For the heteronuclear HeH^{2+} molecule, the variations of the metastability of its first excited state are also analyzed as functions of the angle of confinement. Comparisons between the ions confined in the largest angle and the free ones, based on their wavefunctions, are found to be illustrative of their common and different properties.

1. INTRODUCTION

This is the written version of our contribution to the Frank Harris Workshop on Concepts of Mathematical Physics in Chemistry. There, the oral presentation was limited to preliminary results on the homonuclear hydrogen molecular ion confined in dihedral angles; here, we include additional results about the same ion and also for its heteronuclear counterpart HeH^{2+}.

The molecular properties of the free H_2^+ and HeH^{2+} ions have been extensively studied both theoretically and experimentally. These two systems are the simplest examples of homo- and heteronuclear one-electron diatomic molecules and serve as important reference test cases to explore the effect of external constrictions—such as quantum confinement—on their properties. The H_2^+ molecular ion under different confinement conditions has been the subject of increasing research interest since the pioneering work of Cottrell,[1] who investigated variationally the effects of pressure on its ground-state energy using the molecule-in-a-box model with a hard spheroidal wall. Since then, different approaches using hard and soft box models to study confinement effects on H_2^+ have been put forward, ranging from variational,[2–5] perturbative,[6,7] and accurate calculations.[8–10] Also, more sophisticated treatments of H_2^+ subjected to different confining barrier potentials have been developed.[11,12] To the authors' knowledge, similar studies for the confined HeH^{2+} system are scarce[4,8] and necessary. Note that all the foregoing treatments have as a common characteristic, the conservation of rotational symmetry around the internuclear axis, imposed by the confining potential. It is therefore important to explore the situation where the confinement conditions lead to a rotational symmetry-breaking effect, such as the case of a molecule confined by meridian planes forming a dihedral angle, which is the aim of this work. The physical situation for which this study could have relevance is that of a diatomic molecule sitting on the edge of two planar boundaries in the surface of a given material.

The study reported in this chapter was anticipated in Ref. 13 on "Surface effects in the hydrogen atom confined in dihedral angles," in its section identifying several physical systems in the same situation of confinement. The confinement in dihedral angles presents a situation of breaking of the rotational symmetry around the z-axis, with the consequent and unexpected breaking of the $z \rightarrow -z$ parity symmetry. Reference 14 analyzed such symmetry breakings for the hydrogen atom, or any central potential physical system, and the restoration of the z-parity symmetry.

The hydrogen molecular ion H_2^+ exhibits gerade and ungerade states in the free situation and is expected to do the same for the confinement in dihedral angles. However, the situation encountered initially in the hydrogen atom also occurs for the homonuclear molecular ion. In both cases, the magnetic quantum numbers are in general no longer integer under dihedral confinement. The Legendre functions for the spherical and spheroidal harmonics, with noninteger associativity, contain both even and odd powers of their arguments: their parities are lost. Fortunately, these functions exist in two forms as different hypergeometric functions with arguments $(1-\eta)/2$ and $(1+\eta)/2$, which can be summed or subtracted to convert into a solution with only even or odd powers of η: the parity is restored. The explicit construction of such associated Legendre functions with a well-defined parity is illustrated in the Appendix, together with the evaluation of the matrix elements of η and η^2 in the complete orthonormal basis of such functions, needed for the solution of the Schrödinger equation for the molecular ions.

The chapter is structured in the following way. In Section 2, the exact solutions of the Schrödinger equation in the Born–Oppenheimer approximation for the one-electron diatomic molecular ions, in the situation of confinement in dihedral angles, are constructed. This task includes considering constants of motion, boundary conditions, separation, and integration of the eigenvalue ordinary differential equations in the azimuthal, hyperboloidal, and spheroidal coordinates and their proper matching.

Section 3 presents the results for the homonuclear H_2^+ molecular ion: electronic energies, potential energy curves including Dunham parametrization, analysis of the hyperboloidal wavefunctions including a comparison with those of the free ions, and electric dipole moment for successive values of the angles of confinement. Section 4 includes the corresponding analysis and results for the heteronuclear HeH^{2+} molecular ion, with no binding states and an interesting metastable state. The main results and conclusions are discussed in Section 5.

2. THE SCHRÖDINGER EQUATION AND ITS EXACT SOLUTIONS WITHIN THE BORN–OPPENHEIMER APPROXIMATION

This section deals with the exact solutions of the Schrödinger equation for the H_2^+ and HeH^{2+} molecular ions, within the Born–Oppenheimer approximation, for the situation of confinement in dihedral angles. In fact, the equation is known to be separable and integrable in prolate spheroidal coordinates for the free ions, and it also holds for the situation of confinement of our interest, as presented in Section 2.1. The modeling of the latter in a dihedral angle with impenetrable meridian planes is implemented by the boundary condition that the electronic wavefunction vanishes for $\phi = 0$ and $\phi = \phi_0$ in Section 2.2. The effect of the confinement is manifested as the breaking of the rotational symmetry around the internuclear axis. Its consequence is that, for homonuclear molecules, the symmetry of reflection in the plane bisecting the internuclear axis of the electronic function is also broken. Before the calculations for the solution of the equation in the hyperboloidal coordinate can be implemented, it is necessary to construct the spheroidal harmonics in the hyperboloidal coordinate with a definite parity as explained in Section 2.3, with the details in Appendix A. Section 2.4 constructs the matrix solution of the hyperboloidal equation separated from the Schrödinger equation using the matrix elements of the hyperboloidal coordinate η and its square η^2, on the basis of spheroidal harmonics with definite parity obtained explicitly in Appendices B and C; the solutions are valid for both homo- and heteronuclear molecules. The solution of the corresponding equation in the spheroidal coordinate has the Jaffé form[15] and can be evaluated in a matrix form also, as shown in Section 2.5. The equations in the hyperboloidal and spheroidal coordinates share the parameters corresponding to the confining angle, the internuclear distance and common eigenvalues of the electronic energy, and the separation constant; the matching of these two eigenvalues from the respective solutions must be implemented before the product of the solutions, including the factor in the azimuthal angle, becomes a bona fide solution of the Schrödinger equation, as discussed in the same subsection.

2.1 Separation of the Schrödinger Equation in Prolate Spheroidal Coordinates

The spheroidal coordinates are chosen with the foci coinciding with the fixed positions of the nuclei with charge numbers Z_1 and Z_2 at the positions $(x_1 = 0, y_1 = 0, z_1 = -R/2)$ and $(x_2 = 0, y_2 = 0, z_2 = -R/2)$, respectively, so that the z-axis coincides with the internuclear axis. The midpoint between the nuclei is the origin of the coordinate system, and R is the separation between the nuclei. The coordinates $(\xi = (r_1 + r_2)/R$, $\eta = (r_1 - r_2)/R$, $\phi)$ correspond to the spheroidal coordinate $(1 \leq \xi < \infty)$, hyperboloidal coordinate $(-1 \leq \eta \leq 1)$, and the azimuthal coordinate $(0 \leq \phi \leq 2\pi)$, respectively, involving the distances r_1 and r_2 of any point from the respective nuclei. Fixed values of ξ define prolate spheroids and measure the reciprocal of their eccentricities; fixed values of η define two-sheeted hyperboloids and also measure the reciprocal of their eccentricities; fixed values of ϕ define the positions of meridian half-planes meeting at the z-axis.

The Hamiltonian for the diatomic molecules of our interest with a single electron, in the approximation of fixed nuclei, includes the electron kinetic energy operator, its Coulomb energies of interaction with each nucleus, and the Coulomb energy for both nuclei

$$H = -\frac{\hbar^2}{2m_e}\nabla^2 - \frac{Z_1 e^2}{r_1} - \frac{Z_2 e^2}{r_2} + \frac{Z_1 Z_2 e^2}{R}. \tag{1}$$

By using the expression for the Laplace operator in the spheroidal coordinates,[16] as well as $r_1 = R(\xi+\eta)/2$ and $r_2 = R(\xi-\eta)/2$, the eigenvalue problem for the molecular Hamiltonian leads to the Schrödinger equation

$$\left\{-\frac{\hbar^2}{2m_e}\left[\frac{4}{R^2(\xi^2-\eta^2)}\left(\frac{\partial}{\partial\xi}(\xi^2-1)\frac{\partial}{\partial\xi}+\frac{\partial}{\partial\eta}(1-\eta^2)\frac{\partial}{\partial\eta}\right)+\frac{4}{R^2(\xi^2-1)(1-\eta^2)}\frac{\partial^2}{\partial\phi^2}\right.\right.$$
$$\left.-\frac{2Z_1e^2}{R(\xi+\eta)}-\frac{2Z_2e^2}{R(\xi-\eta)}+\frac{Z_1Z_2e^2}{R}\right\}\Psi(\xi,\eta,\phi;R)=E(R)\Psi(\xi,\eta,\phi;R), \tag{2}$$

where the parametric dependence on the nuclear separation R is explicitly indicated.

The second derivative with respect to the azimuthal angle is identified as the square of the z-component of the angular momentum, and a constant of the motion, due to the rotational invariance around the internuclear axis of the Hamiltonian. Erikson and Hill[17] identified the other constant of motion involving the contraction of the angular momenta of the electron from the positions of the nuclei, plus a combination of the unit vectors from those positions to the position of the electron projected along the internuclear axis: $\Omega = \mathbf{L_1}\cdot\mathbf{L_2} + m_e e^2 R(Z_1\cos\vartheta_1 - Z_2\cos\vartheta_2)$.

The existence of these two constants of motion is behind the separation of the Schrödinger equation, Eq. (2), admitting the factorizable wavefunction

$$\Psi(\xi,\eta,\phi) = \Xi(\xi)H(\eta)\Phi(\phi) \tag{3}$$

and leading to the ordinary differential equations for the respective factors

$$\frac{d^2\Phi}{d\Phi^2} = -\mu^2\Phi \tag{4}$$

$$\left[-\frac{d}{d\eta}(1-\eta^2)\frac{d}{d\eta}+\frac{\mu^2}{1-\eta^2}-(Z_2-Z_1)\rho\eta-p^2\eta^2\right]H(\eta)=KH(\eta) \tag{5}$$

$$\left[-\frac{d}{d\xi}(\xi^2-1)\frac{d}{d\xi}+\frac{\mu^2}{\xi^2-1}-(Z_1+Z_2)\rho\xi+p^2\xi^2\right]\Xi(\xi)=-K\Xi(\xi). \tag{6}$$

The adimensional parameters ρ and p^2 are, respectively, the nuclear separation in units of the Bohr radius, $\rho = R/a_0$, and the square of the latter multiplied by the electronic energy W in units of $e^2/2a_0$, $p^2 = -\rho^2 W/4 > 0$, where $W = 2a_0E/e^2 - 2Z_1Z_2e^2/\rho$, the difference between the total energy of the molecule E and the Coulomb internuclear repulsive energy.

2.2 Boundary Conditions and Breaking of Rotational Symmetry Around the Internuclear Axis

The confinement in dihedral angles with impenetrable meridian planes at $\phi = 0$ and $\phi = \phi_0$ is modeled by the boundary conditions on the azimuthal eigenfunctions of Eq. (4) vanishing at both planes:

$$\Phi(\phi = 0) = 0, \quad \Phi(\phi = \phi_0) = 0. \tag{7}$$

The solutions of Eq. (4) are of the types $\sin \mu\phi$ and $\cos \mu\phi$. The sine function satisfies the first boundary condition, while the cosine function does not and must be eliminated. The second boundary condition becomes $\sin \mu\phi_0 = 0$ requiring the quantization restriction on the eigenvalues

$$\mu = \frac{n_\phi \pi}{\phi_0} \quad \text{for } n_\phi = 1, 2, 3, \ldots. \tag{8}$$

The eigenvalues of the z-component of the angular momentum are no longer integer, in general, depending on the value of the confining angle and the order of excitation $(n_\phi - 1)$, $n_\phi = 1$ being the ground state with no nodes in the interval $(0, \phi_0)$ of ϕ. This is in contrast with the integer magnetic numbers and is a manifestation of the breaking of the rotational symmetry around the internuclear axis due to the presence of the confining meridian planes.

2.3 Breaking and Restoration of the Reflection Parity Symmetry $\eta \to -\eta$ for Homonuclear Molecules

Before attempting to solve Eq. (5), let us examine its version for homonuclear molecules with $Z_1 = Z_2$, for which the linear term in η vanishes. Correspondingly, the remaining terms in the equation are quadratic, and consequently, it can be anticipated that its solutions have well-defined parities.

Now, let us begin the attempt to solve Eq. (5). It is recognized that the first two terms on the l.h.s. correspond to the Legendre operator. Dismissing the quadratic term for the time being, by making $p^2 = 0$, the equation reduces to Legendre's equation. Its solutions are the familiar associated Legendre polynomials, with a singularity factor and a polynomial which can be written in its hypergeometric function representation:

$$\left(1 - \eta^2\right)^{\mu/2} {}_2F_1\left(-n_\vartheta, n_\vartheta + \mu + 1; \mu + 1; \frac{1 - \eta}{2}\right). \tag{9}$$

The consecutive powers of the argument of the function, $(1 - \eta)/2$, contain both even and odd powers of η. In the case of $\mu = m$ an integer, the hypergeometric function itself contains only the even or odd powers of η because the terms with the other powers cancel out. For the situation of confinement in dihedral angles, μ is no longer an integer and the hypergeometric function contains both even and odd powers of η: its parity symmetry is broken.

The first paragraph in this section demands solutions with a definite parity for homonuclear molecules. Can we restore the parity symmetry? The answer is yes. In fact, Eq. (5) is a second-order differential equation and admits a second solution with argument $(1+\eta)/2$. For $\mu = m$ an integer, both solutions coincide. But for μ noninteger, they are linearly independent. Their linear combinations with the coefficient $(-1)^{n_\eta}$ include only the powers of η with this parity, while the terms of the other parity cancel out.

In the case of the hydrogen atom in the same situation of confinement,[13] the same breaking and restoration of the parity symmetry for spherical harmonics occur. Here, the details for the spheroidal harmonics are presented in Appendix A. The methodology of ladder operators developed for the hydrogen atom[14] is also applied in Appendices B and C to use recurrence relations to evaluate normalization constants and matrix elements of η and η^2 on the basis of the Legendre polynomials of noninteger associativity μ and well-defined parity.

2.4 Matrix Solution of the Hyperboloidal Coordinate Equation with the Basis of Legendre Polynomials with Associativity μ and Definite Parity

We use the ket notation to represent the orthonormal basis of μ-associated Legendre polynomials with definite parity $|n_\eta\mu\rangle$ with a fixed value of μ from Eq. (8). This is a good basis to expand the solutions of Eq. (5):

$$H(\eta) = \sum_{n_\eta} a_{n_\eta} |n_\eta \mu\rangle. \tag{10}$$

Then Eq. (5) is converted into its matrix form

$$\left\langle n'_\eta\mu \middle| \left(\hat{l}^2 - K\right) \middle| n_\eta\mu \right\rangle - \rho(Z_2 - Z_1)\left\langle n'_\eta\mu \middle| \eta \middle| n_\eta\mu \right\rangle - p^2\left\langle n'_\eta\mu \middle| \eta^2 \middle| n_\eta\mu \right\rangle = 0. \tag{11}$$

The first term involves diagonal terms with eigenvalues $\left(n_\eta + \mu\right)\left(n_\eta + \mu + 1\right)$ and K, respectively; the second one has the selection rules $n'_\eta = n_\eta \pm 1$ and the values from Appendix C; and the third one has the selection rules $n'_\eta = n_\eta$, $n'_\eta \pm 2$, and the values from the same Appendix.

The explicit form of the matrix eigenvalue equation is:

$$\left[(n_\eta+\mu)(n_\eta+\mu+1)-p^2\frac{(2n_\eta+2\mu+3)(n_\eta+2\mu)(n_\eta)+(2n_\eta+2\mu-1)(n_\eta+2\mu+1)(n_\eta+1)}{(2n_\eta+2\mu+1)(2n_\eta+2\mu-1)(2n_\eta+2\mu+3)}-K\right]\delta_{n'_\eta,n_\eta}$$
$$-\rho(Z_2-Z_1)\left[\sqrt{\frac{(n_\eta+2\mu)(n_\eta)}{(2n_\eta+2\mu+1)(2n_\eta+2\mu-1)}}\delta_{n'_\eta,n_\eta-1}+\sqrt{\frac{(n_\eta+2\mu+1)(n_\eta+1)}{(2n_\eta+2\mu+1)(2n_\eta+2\mu+3)}}\delta_{n'_\eta,n_\eta+1}\right]$$
$$-p^2\left[\frac{1}{(2n_\eta+2\mu-1)}\sqrt{\frac{(n_\eta+2\mu+1)(n_\eta+2\mu-1)n_\eta(n_\eta-1)}{(2n_\eta+2\mu+1)(2n_\eta+2\mu-3)}}\delta_{n'_\eta,n_\eta-2}\right.$$
$$\left.+\frac{1}{(2n_\eta+2\mu+3)}\sqrt{\frac{(n_\eta+2\mu+1)(n_\eta+2\mu+2)(n_\eta+1)(n_\eta+2)}{(2n_\eta+2\mu+1)(2n_\eta+2\mu+5)}}\delta_{n'_\eta,n_\eta+2}\right]. \tag{12}$$

For homonuclear molecules, $Z_1 = Z_2$, the middle term drops out. Then the summation in Eq. (10) involves only even or odd values of the hyperboloidal quantum number, corresponding to gerade or ungerade states, respectively. The matrix in Eq. (12) becomes tridiagonal. For heteronuclear molecules, $Z_1 \neq Z_2$, and the matrix elements of η must be included. Their eigenstates in Eq. (10) include even and odd quantum numbers because the system no longer has the symmetry under $\eta \to -\eta$. The corresponding matrix in Eq. (12) is pentadiagonal.

For chosen values of the parameters ρ and p^2, the diagonalization of large enough matrices yields the eigenvalues of the separation constant K_i and its associated eigenvectors $a_{n_\eta}(K_i)$ in the expansion of Eq. (10). In practice, for a given nuclear separation ρ is fixed, and p^2 is determined by the square of ρ and the electronic energy W (see Section 2.1), which must be negative and is assigned changing values. The sizes of the matrices must be sufficiently large to ensure stability in the convergence and accuracy of the eigenvalues K_i and eigenvectors $a_{n_\eta}(K_i)$ as functions of W.

2.5 Jaffé Solutions for the Spheroidal Coordinate Equation and Their Matching with the Solutions in the Hyperboloidal Coordinate

Note that Eq. (6) has the same structure as Eq. (5); except for the different sign in the coefficient of their linear terms, both equations are almost the same. The important difference resides in the different domains of their independent variables η and ξ, and the corresponding boundary conditions.

They also share the removable singularity at the value one, but from different sides in their respective domains. The spheroidal function must vanish for $\xi \to \infty$ in order to guarantee its quadratic integrability. Here, we choose Jaffé solutions[15] with the following form:

$$\Xi(\xi) = \left(\xi^2 - 1\right)^{\mu/2} e^{-p\xi} \left[(\xi + 1)^{\sigma} \sum_{t=0}^{\infty} b_t \left(\frac{\xi - 1}{\xi + 1} \right)^t \right], \tag{13}$$

where $\sigma = (Z_1 + Z_2)\rho/2p - (\mu + 1)$, determined by the differential equation satisfied by the third factor after removal of the singularities at one and infinity; in the series expansion, the variable is the fraction in the domain [0, 1]. The expansion coefficients satisfy a three-term recurrence relation:

$$\alpha_t b_{t+1} + \beta_t b_t + \gamma_t b_{t-1} = 0, \tag{14}$$

where

$$\begin{aligned} \alpha_t &= (t+1)(t+\mu+1) \\ \beta_t &= -2t^2 - (4p - 2\sigma)t - (K + p^2) \\ \gamma_t &= (t - 1 - \sigma)(t - 1 - \sigma - \mu) \end{aligned} \tag{15}$$

with the same p^2 introduced in Section 2.1.

The quadratic and linear terms in the summation index t in the α_t and γ_t coefficients can be traced to the second and first derivatives of the two factors inside the brackets in Eq. (13); the respective terms in β_t are associated with the quadratic and linear terms in ξ of Eq. (6), after the change of variable $\xi = (\xi + 1) - 1$, and its constant term combines the separation constant and the coefficient of the quadratic term in Eq. (6).

The recurrence relations can be cast into an eigenvalue problem in matrix form, following the method of Hunter and Pritchard.[18] The matrix is tridiagonal and nonsymmetric, the b_t coefficients in a column matrix form the eigenvectors, and $K + p^2$ is the eigenvalue. The diagonalization of the matrix yields, as in the previous subsection, the eigenvalues of the separation constant K_i and the eigenvectors $b_t(K_i)$, as functions of the parameters ρ and p^2. In this case, the diagonalization must take into account the nonsymmetric character of the matrix. Also the sizes of the matrices can be enlarged to test for convergence and accuracy in the eigenvalues and eigenvectors.

The matching of the hyperboloidal and spheroidal coordinate solutions of Eqs. (10) and (13) requires the same values of the parameters ρ and p^2 as well as the same values of the separating constants from the diagonalizations of Eqs. (12) and (14) $K_\eta(R, W) = K_\xi(R, W)$. The electronic energy must be changed until this equality is achieved for each nuclear separation. Then the complete eigenfunctions of Eq. (2) can be formed $\Psi_{N_\xi,N_\eta,n_\phi}(\xi,\eta,\phi;R)$ in which (N_ξ, N_η, n_ϕ) count the number of nodes and excitations in their respective degrees of freedom and open domains.

3. THE HYDROGEN H_2^+ MOLECULAR ION

In this section, we report numerical and graphical results on the properties of the homonuclear $H_2{}^+$ molecular ion confined in dihedral angles in its (001) and (021) gerade binding states and (011) ungerade antibinding state. Section 3.1 includes the results on the electronic energy as functions of the nuclear separation for the successive angles of confinement, including discussions about the united atom and dissociation limits. Section 3.2 analyzes the complete energy curves distinguishing between the binding and antibinding states, and determining, for the first ones, the equilibrium internuclear separation, the corresponding minimum of the molecular potential energy, and the asymptotic energy for large internuclear separations; correspondingly, we evaluate the Dunham parameters[19] to characterize the molecular well in its harmonic, anharmonic, and asymmetry properties, which are the starting point to study the molecular vibrations. Section 3.3 presents a comparison of the hyperboloidal coordinate wavefunctions for the free ion and the ion confined in the largest dihedral angle, in order to illustrate their common and different characteristics for the successive states. Section 3.4 includes the evaluation of the electric dipole moment acquired by the molecule due to the confinement in the dihedral angle and also as a manifestation of the breaking of the rotational symmetry around the internuclear axis.

3.1 Electronic Energies Including the United Atom and Dissociation Limits

For the homonuclear $H_2{}^+$ molecular ion, $Z_1 = Z_2 = 1$, the evaluation of the hyperboloidal coordinate function is based on the matrix equation (12) in its tridiagonal form, and that of the spheroidal coordinate in Eq. (14). For a chosen value of the confining angle ϕ_0, the states with $n_\phi = 1$ in Eq. (8) take the value $\mu = \pi/\phi_0$; the input parameters are the nuclear separation and the electronic energy W. The eigenstates are found when the matching of the electronic energy and the separation constants from both equations occurs, as discussed in Section 2.4.

Figure 1A and B contain the illustrative electronic energy $W(e^2/2a_0)$ curves as functions of the internuclear separation R, changing by three orders of magnitude, for angles of confinement $\phi_0 = 2\pi$, $3\pi/2$, π, and $\pi/2$ from bottom to top for the states (001) gerade and (021) gerade, respectively.

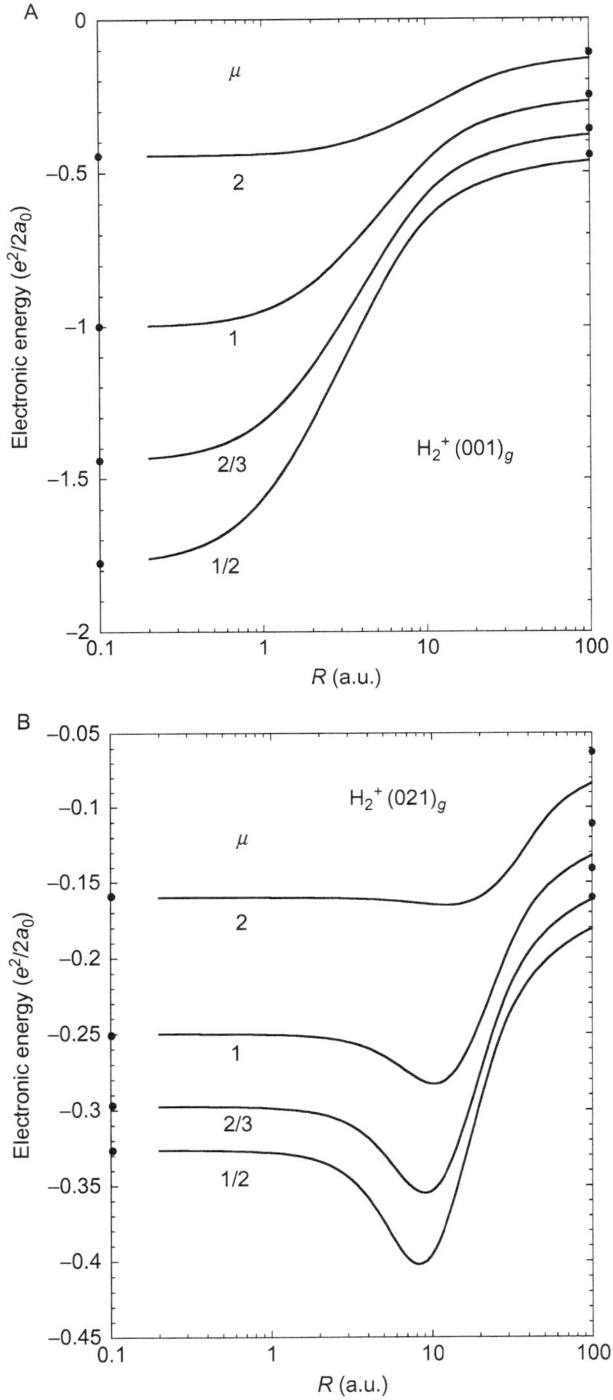

Figure 1 Electronic energy versus internuclear separation for $H_2{}^+$ confined in different dihedral angles: (A) ground state $(001)_g$; (B) second excited state $(021)_g$.

Let us first discuss Fig. 1A for the ground state with no nodes in the three degrees of freedom, apart from the two in the azimuthal angle at the boundary meridian planes. The behavior of the electronic energy curves for the different angles of confinement is qualitatively similar, exhibiting each a monotonic increase as the nuclear separation increases. On the vertical axis on the left, the black points indicate the energies of the united atom in the respective situations of confinement in the limit when R becomes vanishingly small; they correspond to a nuclear charge $Z_1 + Z_2 = 2$ binding the electron in a hydrogen-like atom with energy $-2^2/(\mu + 1)^2(e^2/2a_0)$, where in general $\nu = \eta_\xi + n_\eta + \mu + 1$ is the principal quantum label.[13] Notice that the extrapolation of the $W(R)$ energy curves to the left leads then to the respective values (in units of $e^2/2a_0$) $-16/9$, $-36/25$, -1, and $-4/9$ from bottom to top. On the other hand, the dissociation limits for very large values of the nuclear separation correspond to the points on the right vertical axis with the energy of an equally confined dissociated hydrogen atom with nuclear charge 1 in the ground state, with the respective energies $-4/9$, $-9/25$, $-1/4$, and $-1/9$. Figure 1B shows the analogous behavior of the electronic energy curves $W(\rho)$ for the state (021), with minima forming in the vicinity of $R \approx 10$. For smaller distances, they flatten out to the left tending to the hydrogen-like energy levels with nuclear charge 2 and $\nu = \mu + 3$: $-16/49$, $-36/121$, $-1/4$, and $-4/25$ as marked on the left vertical axis. On the right side, the corresponding values for a nuclear charge of 1 and $\nu = \mu + 2$: $-4/25$, $-9/64$, $-1/9$ and $-1/16$ are all above the energy curves up to $R \approx 100$. Closer convergence to these limiting values was verified for $R \approx 300$. Note that, for the state (021) the energy curves merge to their respective dissociation limits using $\nu = \mu + 2$ instead of $\nu = \mu + 3$, as anticipated for an equally confined H-atom in its excited state. This is in contrast with the correct trend of the ground-state energy curves toward the dissociation limit using the latter value of ν for an equally confined H-atom as shown in Fig. 1A. This ambiguity reflects the limitations of the BO approximation to properly describe the dissociation limit in general, as discussed in Ref. 25.

It is also important to compare the familiar results for the free ion with those of the confined ion in the dihedral angle $\phi_0 = 2\pi$, recognizing their differences at the level of the boundary conditions of their azimuthal eigenfunctions: periodic for the free ion, and vanishing at the meridian planes of the confining angle for the latter. Specifically, for the largest angle, the electron is excluded from the meridian half-plane $(x = 0, y, z)$.

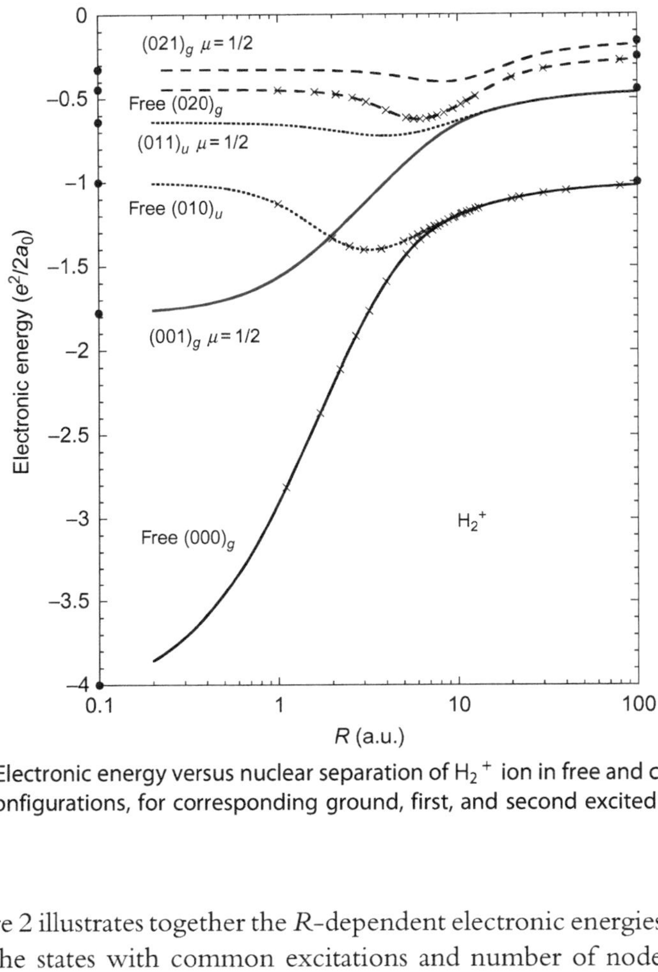

Figure 2 Electronic energy versus nuclear separation of $H_2{}^+$ ion in free and confined in $\phi_0 = 2\pi$ configurations, for corresponding ground, first, and second excited states.

Figure 2 illustrates together the R-dependent electronic energies for both ions in the states with common excitations and number of nodes, in the interval (−1, 1) of η. The curves for the confined ion are the ones from Fig. 1A and B, with $\mu = 1/2$; below each one are their counterparts for the free ion in the σ states with $m = 0$ and $N_\eta = 0, 1, 2$, respectively. Notice the united atom limits of their electronic energies on the left-hand axis at the values −4, −1, and −4/9 characteristic of the lower He, hydrogen-like atom eigenstates. Notice the similar behavior of the corresponding pairs of electronic energy curves for the respective ground, first, and second excited states. Additionally, for large nuclear separations, the ground state energy curve of the confined ion and the first excited energy curve of the free ion, with different η parities, cross each other at $R \approx 2$; and asymptotically, the successive pairs of ground and first excited states approach together their respective common dissociation limits, at −1 for the free ion and at −4/9 for the confined ion on the right side. The dissociation limits share the same energies, but the respective atomic states have different parities and excitations.

3.2 Molecular Potential Energy Curves and Dunham Parametrization

The total molecular energy combines the electronic energy and the Coulomb repulsion internuclear energy. The latter prevents the nuclei from coming close together and, combined with the electronic energy, determines the binding or unbinding nature of the different electronic states.

Figure 3A and B illustrates the potential energy curves $E(R)$ for the $(001)_g$ ground and $(021)_g$ excited electronic states, respectively. Both include the confining angles of the figures in the previous section, as well as some intermediate ones $\phi_0 = 11\pi/6$, $5\pi/3$, $8\pi/9$, $7\pi/9$, and $2\pi/3$, with their respective μ values. For short nuclear separations, the Coulomb repulsion energy dominates; in the intermediate region of nuclear separations $4 < R < 15$ in Fig. 3A and $10 < R < 30$ in Fig. 3B, the potential energy wells resulting from the combinations of the nuclear repulsion and the electronic energy can be identified; for very large separations, the nuclear repulsion becomes negligible and the electronic contribution dominates: in Fig. 3A, the dissociation limits of the previous section and their counterparts for the other angles can be identified on the right-hand side; in Fig. 3B, the dissociation limits are not reached yet, as already recognized in Fig. 1B. Apart from the differences in the energy scales of Fig. 3A and B, the confinement effect of the dihedral angles diminishing from bottom to top in the figures $\phi_0 = 2\pi$ to π is that the position of the minimum in the potential energy curves is displaced for larger nuclear separations, and the depth of the potential well is reduced.

Table 1 reports the total energies for the gerade (001) ground and (021) second excited states, discussed in the previous paragraph, and also for the ungerade (011) first excited state, for nuclear separations R from 0.2 to 100 and for chosen values of μ determined by the confining dihedral angle. The reader may compare the numerical values from this table for the gerade states with the corresponding points in the respective curves of Fig. 3A and B. Here, we concentrate on the energies for the ungerade (011) first excited state: Its four rows of entries in the middle, for successive values of μ, can be followed from small to large nuclear separations, exhibiting a monotonic variation from positive to negative values, fairly rapidly up to the vicinity of $R \approx 10$, and then very slow for larger separations and approaching the same dissociation limits as the ground state in the four rows above for $R \approx 100$. This monotonic behavior makes this first excited state to be antibinding, in contrast with the binding character of its two neighboring gerade

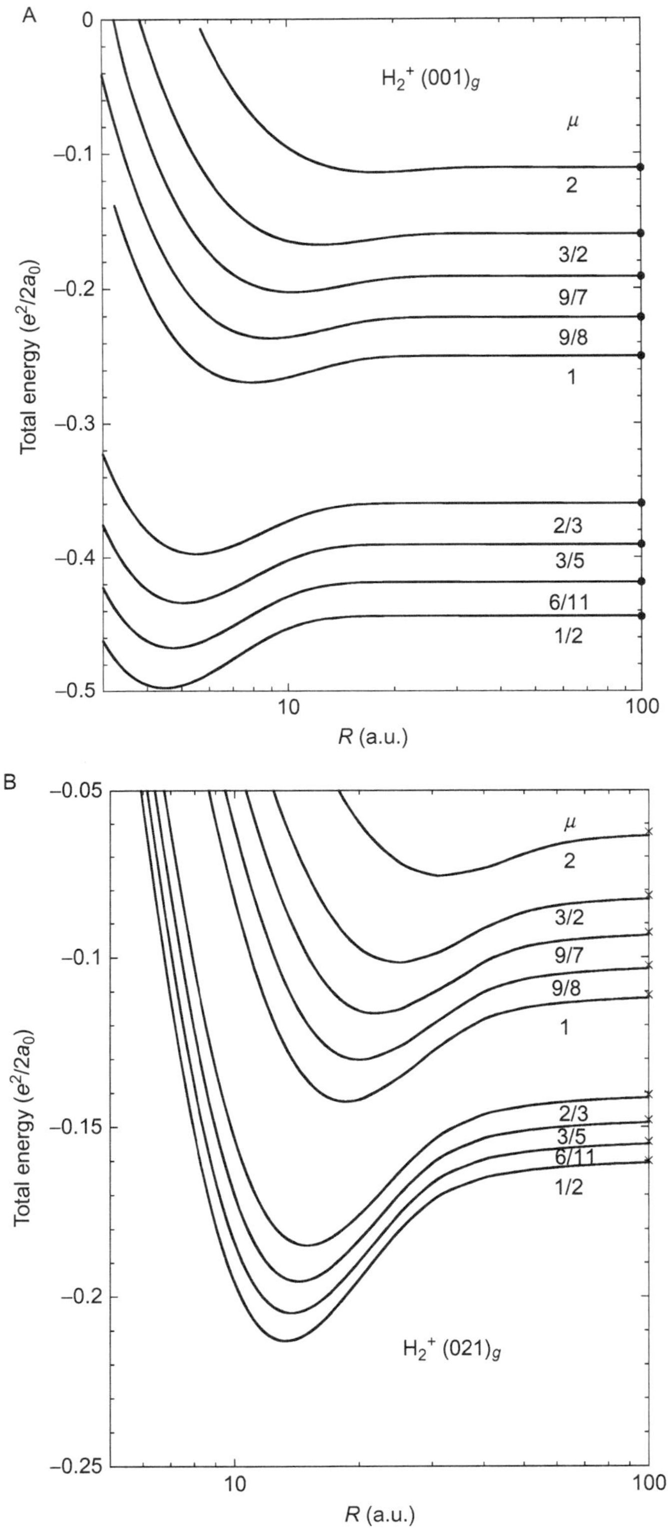

Figure 3 Potential energy curves for $H_2{}^+$ confined in a set of selected dihedral angles defined by $\mu = \pi/\phi_0$. (A) $(001)_g$ state; (B) $(021)_g$ state.

Table 1 Total Energies for the Ground State (001), First Excited State (011), and Second Excited State (021) of the Confined $H_2{}^+$ Ion for a Set of Internuclear Distances and Different Dihedral Angles Defined by $\mu = \pi/\phi_0$

	R **(a.u.)**	**0.2**	**0.5**	**1.0**	**3.0**	**5.0**	**10.0**	**20.0**	**30.0**	**50.0**	**100.0**
μ											
2		9.55578	3.55695	1.56092	0.25890	0.02967	−0.09509	−0.11376	−0.11096	−0.11076	−0.11105
1	(001)	9.00262	3.01531	1.05178	−0.10622	−0.24277	−0.26543	−0.25017	−0.24976	−0.24993	−0.24999
2/3		8.56786	2.60282	0.68971	−0.32281	−0.39612	−0.37306	−0.35988	−0.35989	−0.35997	−0.35999
1/2		8.23924	2.29998	0.43635	−0.46281	−0.49603	−0.45337	−0.44375	−0.44379	−0.44384	−0.44385
2		9.75000	3.75000	1.75001	0.41702	0.15211	−0.03517	−0.10062	−0.10912	−0.11075	−0.11105
1	(011)	9.55544	3.55486	1.55289	0.20730	−0.06023	−0.21983	−0.24875	−0.24974	−0.24993	−0.24999
2/3		9.43713	3.43523	1.42881	0.05942	−0.20926	−0.34459	−0.35963	−0.35989	−0.35997	−0.35999
1/2		9.35983	3.35632	1.34451	−0.04947	−0.31648	−0.43444	−0.44368	−0.44379	−0.44384	−0.44385
2		9.83999	3.83998	1.83993	0.50602	0.23831	0.03533	−0.05971	−0.07558	−0.06923	−0.06362
1	(021)	9.74998	3.74985	1.74940	0.41126	0.13547	−0.08274	−0.14196	−0.12789	−0.11496	−0.11201
2/3		9.70243	3.70217	1.70124	0.35754	0.07092	−0.15343	−0.17579	−0.15422	−0.14399	−0.14143
1/2		9.67358	3.67321	1.67186	0.32294	0.02627	−0.19584	−0.19488	−0.17163	−0.16302	−0.16063

All energies are in units of $e^2/2a_0$.

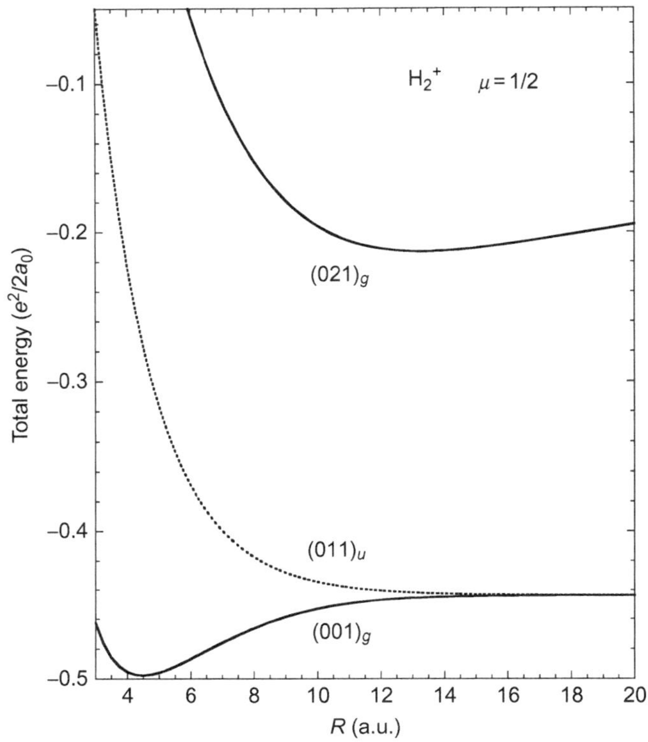

Figure 4 Potential energy curves for the ground $(001)_g$, first $(011)_u$, and second $(021)_g$ excited states of $H_2{}^+$ confined in a dihedral angle $\phi_0 = 2\pi$.

ground and second excited states with well-defined potential wells. Figure 4, for $\mu = 1/2$ ($\phi_0 = 2\pi$), shows the energy curves of the three states together, with the continuous lines the ground and second excited states below and above, respectively, and the first excited state with the broken line coming down from the left and going asymptotically to the right from above to meet the ground state also asymptotically to the same limit from below.

The determination of the minimum of the potential energy curves is important for the study of the molecular vibrations. It requires accurate evaluation of the total energies for the successive values of the nuclear separation. Tables 2 and 3 display the equilibrium total energies E_{eq} and corresponding internuclear distances R_{eq} for a set of confining dihedral angles ranging from $\pi/2$ to 2π for the ground (001) and second excited states (021) of $H_2{}^+$, respectively. Also shown are the corresponding Dunham[19] parameters (A_0, A_1, A_2), obtained after adjustment of the potential energy curves around the equilibrium configuration through a fourth-degree polynomial: $E(R) - E(R_{eq}) = A_0 z^2(1 + A_1 z + A_2 z^2)$ with $z = (R - R_{eq})/R_{eq}$. These parameters allow us to estimate the rotational constant B_e, the vibrational

Table 2 Numerical Values of the Equilibrium Internuclear Distance R_{eq}, Total Energy E_{eq}, Dunham Parameters (A_0, A_1, A_2), Rotational Constant B_e, Harmonic Frequency ω_e, Anharmonicity Constant $\omega_e x_e$, and Raman Frequency ω_R, for a Set of Dihedral Confining Angles ϕ_0 for the H_2^+ Ion in Its Ground (001) State

ϕ_0(deg)	μ	$R_{eq}(a_0)$	$E_{tot}^{eq}(e^2/2a_0)$	A_0	A_1	A_2	B_e (cm^{-1})	ω_e (cm^{-1})	$\omega_e x_e$ (cm^{-1})	ω_R (cm^{-1})
90	2	17.954	−0.11407	0.0314	−2.4828	2.8455	0.4	101.1	2.7	95.7
120	3/2	12.420	−0.16751	0.0497	−2.1406	2.6327	0.8	183.9	3.6	176.7
140	9/7	10.367	−0.20251	0.0618	−2.0991	2.6659	1.1	245.6	4.7	236.1
160	9/8	8.957	−0.23644	0.0741	−1.9706	1.7359	1.5	311.4	6.7	297.5
180	1	7.930	−0.26903	0.0849	−1.9758	2.4631	1.9	376.4	6.9	362.6
270	2/3	5.505	−0.39743	0.1297	−1.9566	2.4377	3.9	670.2	13.9	642.4
300	3/5	5.081	−0.43389	0.1418	−1.6996	2.4107	4.6	759.3	8.3	742.6
330	6/11	4.736	−0.46754	0.1544	−1.8946	2.1986	5.3	849.9	18.3	813.3
360	1/2	4.470	−0.49763	0.1654	−1.9148	1.9139	6.0	932.1	23.9	884.2
Free[a]	0	1.997	−1.20527	0.2053	−1.7363	2.1329	30.0	2324.3	73.5	2177.2

[a]Reference 9.
The values of the azimuthal quantum number, μ, are shown. Also, the corresponding quantities for the free molecular system are displayed for comparison.

Table 3 Numerical Values of the Equilibrium Internuclear Distance R_{eq}, Total Energy E_{eq}, Dunham Parameters (A_0, A_1, A_2), Rotational Constant B_e, Harmonic Frequency ω_e, Anharmonicity Constant $\omega_e x_e$, and Raman Frequency ω_R, for a Set of Dihedral Confining Angles ϕ_0 for the H_2^+ Ion in Its Second Excited (021) State

ϕ_0(deg)	μ	$R_{eq}(a_0)$	$E_{tot}^{eq}(e^2/2a_0)$	A_0	A_1	A_2	B_e (cm^{-1})	ω_e(cm^{-1})	$\omega_e x_e$(cm^{-1})	ω_R(cm^{-1})
90	2	31.881	−0.07580	0.0566	−1.9424	1.9510	0.1	76.5	0.5	75.5
120	3/2	24.823	−0.10153	0.0769	−1.8682	2.1016	0.2	114.4	0.7	113.1
140	9/7	22.056	−0.11664	0.0891	−1.8777	2.1162	0.2	138.7	0.8	137.0
160	9/8	20.069	−0.13024	0.1006	−1.9927	2.0508	0.3	161.9	1.3	159.3
180	1	18.601	−0.14249	0.1104	−2.0033	2.2170	0.3	183.0	1.5	180.1
270	2/3	14.960	−0.18481	0.1452	−1.9753	2.3337	0.5	260.9	2.0	256.8
300	3/5	14.278	−0.19546	0.1540	−1.9733	2.3466	0.6	281.6	2.2	277.1
330	6/11	13.730	−0.20486	0.1619	−1.9826	2.3539	0.6	300.2	2.4	295.3
360	1/2	13.292	−0.21301	0.1687	−1.9825	2.3497	0.7	316.6	2.6	311.4
Free	0	8.833	−0.35010	0.2848	−2.0322	2.6047	1.5	618.9	5.9	607.1

The values of the azimuthal quantum number, μ, are shown. Also, the corresponding quantities for the free molecular system are displayed for comparison.

frequency ω_e, anharmonicity constant $\omega_e x_e$, and Raman frequency ω_R, which are reported in Tables 2 and 3 for the different confinement conditions.

In all cases, the Dunham parameters were obtained using a fine grid of 20 equally spaced energy values around the equilibrium position. For completeness, the corresponding values for the free system in its ground state (000)[9] are also shown in Table 2 (last row with $\mu = 0$). As may be gathered from this table, the characteristic values for the confined ion in its ground state (001) differ considerably from its free counterpart, even for the largest confining dihedral angle $\phi_0 = 2\pi$. Indeed, the equilibrium bond length and energy change from $R_{eq} = 1.997$, $E_{eq} = -1.20527e^2/2a_0$ for the free system to $R_{eq} = 4.470$, $E_{eq} = -0.49763e^2/2a_0$ for the most favorable confining situation with $\phi_0 = 2\pi$. This abrupt change is also observed for the respective values of B_e, ω_e, $\omega_e x_e$, and ω_R, pointing to the dramatic effect produced by dihedral confinement on the system properties. For smaller dihedral confining angles $\pi/2 \leq \phi_0 \leq 2\pi$, a monotonic increase in the equilibrium bond length and energy is observed, with corresponding decrease in the rotational and vibrational parameters. A similar behavior for the previously mentioned characteristic quantities is observed from Table 3 for the second excited state (021) of the confined ion, with larger values in the equilibrium bond lengths and energies and lower values in the rotational and vibrational parameters.

From the above analysis, we may conclude that dihedral confinement of the H_2^+ molecular ion leads to rapidly changing properties in its rotational and vibrational properties as the confining angle ϕ_0 varies.

3.3 Hyperboloidal Coordinate Wavefunctions

This section is motivated by the need to show convincingly that the identifications of eigenstates in the previous subsections are correct. A comparison of the nodes in the hyperboloidal coordinate wavefunctions for the familiar free-ion configuration and for the ion confined in the largest dihedral angle $\phi_0 = 2\pi$ is the appropriate tool for the analysis of the situation.

The difference between the free and confined configurations of the ion resides, obviously, in the different boundary conditions of their azimuthal wavefunctions: periodic for the first with the consequent integer magnetic quantum numbers $m = 0, 1, 2, \ldots$; and the vanishing of the wavefunctions at the meridian planes of the confining angle ϕ_0 leading to the noninteger eigenvalues $\mu = n_\phi \pi/\phi_0$ for $n_\phi = 1, 2, 3, \ldots$. Consequently, the usual characterization of the states $\sigma, \pi, \delta, \ldots$ for the free ion is not applicable in the second case. The number of nodes, in the respective degrees of freedom in the wavefunctions, is the relevant element to identify the corresponding excitations. For the states under consideration $n_\phi = 1$, the only nodes in their common wavefunction are those associated with the boundaries $\phi = 0$ and $\phi = \phi_0$; there are no nodes in the open interval $0 < \phi < \phi_0$, in correspondence with the nodeless σ-states of the free ion.

Next, we consider the number of hyperbolic coordinate nodes in the states (000), (010), and (020) for the free ion and the corresponding states (001), (011), and (021) for the confined ion with $\mu = 1/2$, including their variations with the nuclear separation.

Figure 5A–C corresponds to the R-dependence of the respective eigenstates of the free ion in $H_{N_\eta}(\eta)$ versus η plots, evaluated from the diagonalization of Eq. (12) for $\mu = m = 0$. In Fig. 5A, there are no nodes in the interval [−1, 1], congruent with the identification of the ground state; its symmetry under the transformation $\eta \to -\eta$ is appreciated, including their maximum values at the equatorial plane $\eta = 0$; the variations with the

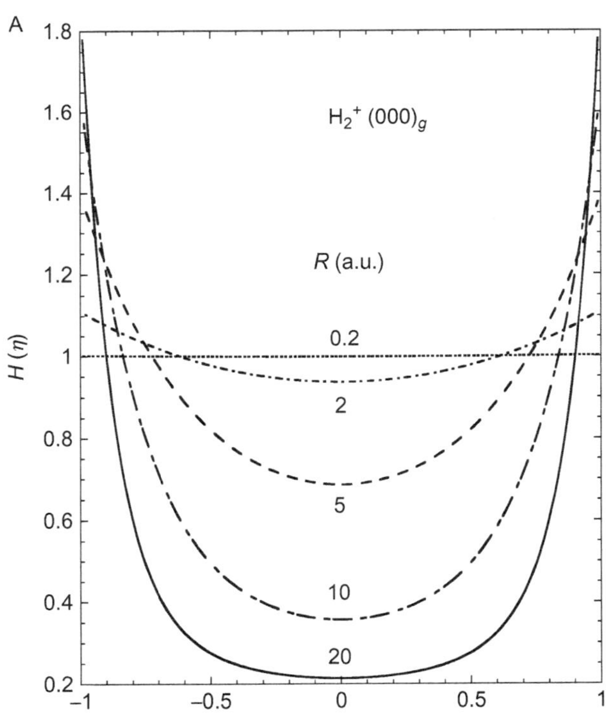

Figure 5—Cont'd

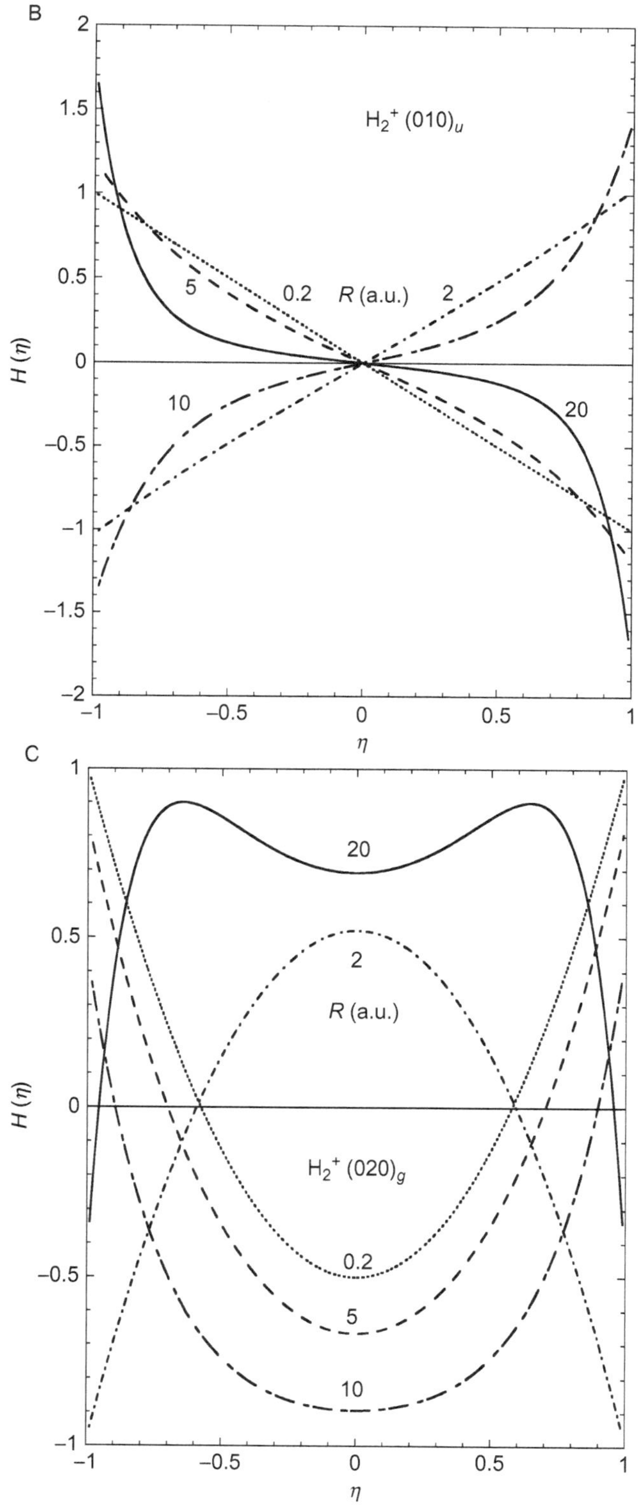

Figure 5 Varying characteristics of the hyperboloidal coordinate eigenfunctions for the free H_2^+ ion as the internuclear separation changes: (A) ground state $(000)_g$; (B) first excited state $(010)_u$; and (C) second excited state $(020)_g$.

successive values of R can also be recognized. In Fig. 5B, all the plots share the single and common node at $\eta = 0$, with the identification of $N_\eta = 1$; the lines are antisymmetric, with their larger ordinate values at the ends of the interval of η; their shapes are quasi-linear for small values of the nuclear separations and their curvatures increase as the separations increase. Each of the lines plotted in Fig. 5C is symmetric and exhibits two nodes, with the identification of $N_\eta = 2$, with minima and maxima at $\eta = 0$, and finite values at the ends of the physical interval; for $R = 0.2$, 2, 5, and 10, the concavities remain the same in the whole interval, and for $R = 20$, the minimum at the middle and the equidistant maxima can be appreciated.

Now let us consider the corresponding plots of Fig. 6A–C for the confined ions. Their respective number of nodes $N_\eta = 0, 1, 2$ coincide with those of their counterparts for the free ion, providing confidence in the identification of their orders of excitation. Their respective symmetries or antisymmetries also hold. The noticeable differences between Figs. 5 and 6 occur near their left and right ends manifested as maxima or minima,

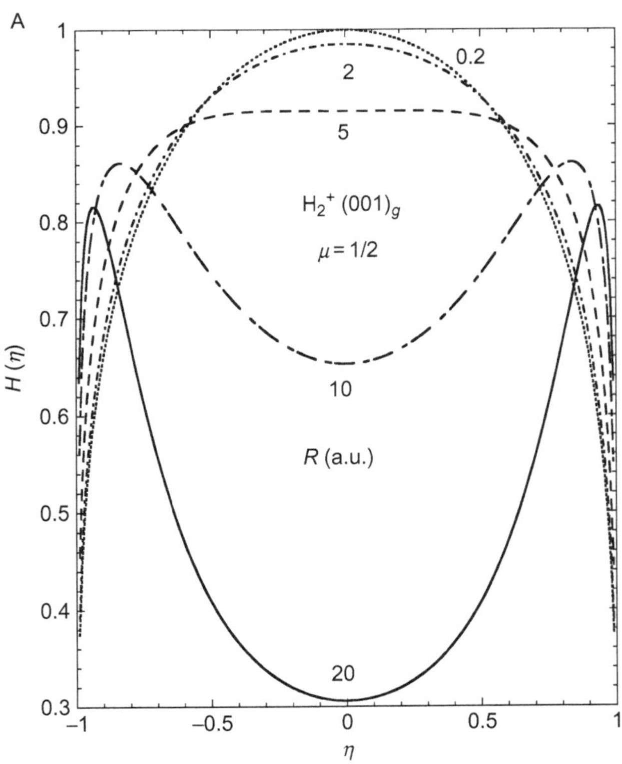

Figure 6—Cont'd

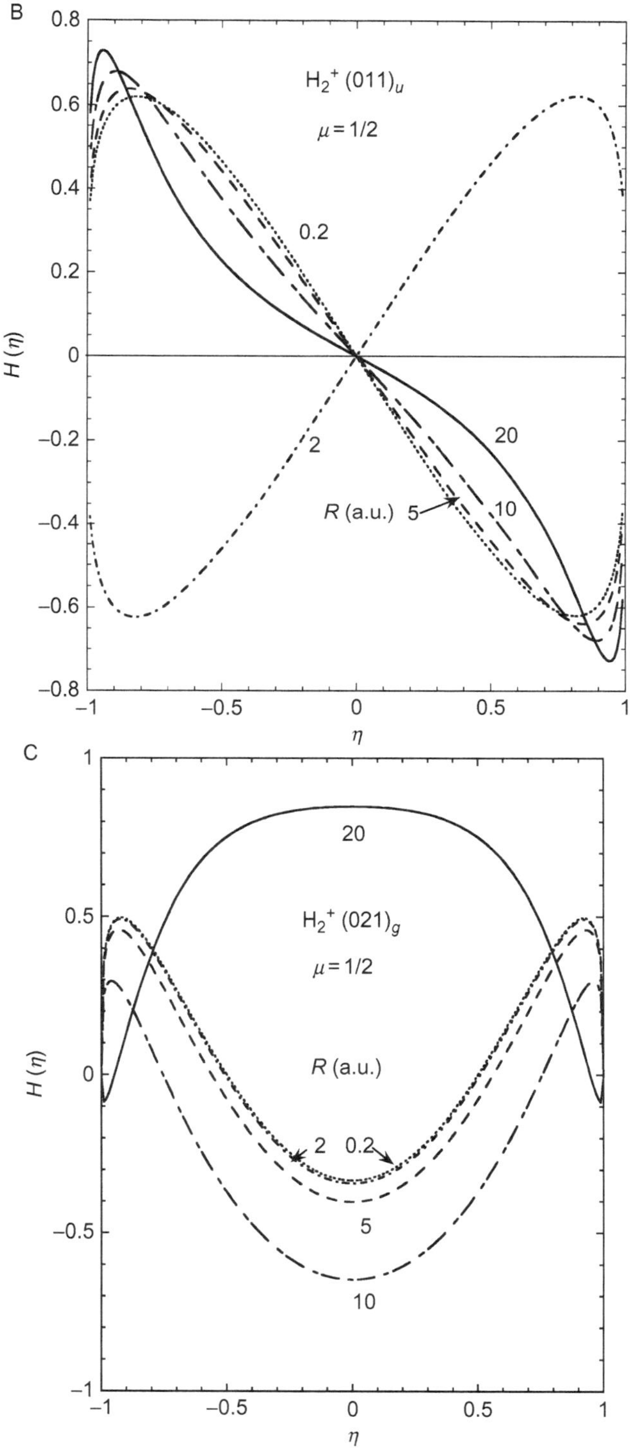

Figure 6 Same as in Fig. 5 for the $H_2{}^+$ ion dihedrally confined in an angle $\phi_0 = 2\pi$: (A) ground state $(001)_g$; (B) first excited state $(011)_u$; and (C) second excited state $(021)_g$.

and with variations which under extrapolation lead to zero values for $\eta = -1$ and $\eta = 1$. The tendencies can be recognized in Fig. 6A, and better appreciated in Fig. 6B and C according to the respective coordinate scales, which were chosen to appreciate the variations for the different nuclear separations in the respective domains. The observed behavior is explained by the presence of the removing singularity factor $(1-\eta^2)^{\mu/2}$ in Eq. (9); it does not occur in Fig. 5 because $\mu = 0$ there; it also occurs for the excited states of the free ion with $\mu = 1, 2, \ldots$ and for any state of the confined atom with $\mu \geq 1/2$.

3.4 Electric Dipole Moment

In general, free homonuclear diatomic molecules cannot have an electric dipole moment because of their invariance under rotations around the internuclear axis and reflection symmetry in the plane bisecting the internuclear axis. However, under confinement in dihedral angles, the rotational invariance is broken and the molecule acquires such an electric dipole moment. We have investigated this effect in the two-dimensional hydrogen atom confined in angles[20] and also in the hydrogen atom confined in dihedral angles.[13] Here we do it for the $H_2{}^+$ molecular ion confined in dihedral angles in its ground state. The corresponding calculation involves the electron average position in prolate spheroidal coordinates:

$$\begin{aligned}\mathbf{d} &= -e\langle \mathbf{r}\rangle = -e\langle \hat{\mathbf{i}}x + \hat{\mathbf{j}}y + \hat{\mathbf{k}}z\rangle \\ &= -e(R/2)\langle h_{\xi,\eta}(\hat{\mathbf{i}}\cos\phi + \hat{\mathbf{j}}\sin\phi) + \hat{\mathbf{k}}\xi\eta\rangle\end{aligned} \tag{16}$$

with

$$h_{\xi,\eta} = (\xi^2-1)^{1/2}(1-\eta^2)^{1/2}.$$

The average value in Eq. (16) is calculated using the molecular wavefunction given by Eq. (3), which satisfies the boundary condition for dihedral confinement. Following the procedure of Ref. 13, the final expression for the dipole moment becomes

$$\mathbf{d} = -\frac{R^4 e}{16}\left\{\frac{2}{\phi_0}\cdot\frac{4\mu^2}{4\mu^2-1}\sin\left(\frac{\phi_0}{2}\right)\left[\hat{\mathbf{i}}\cos\left(\frac{\phi_0}{2}\right) + \hat{\mathbf{j}}\sin\left(\frac{\phi_0}{2}\right)\right]I_\perp + \hat{\mathbf{k}}I_z\right\}, \tag{17}$$

where the transverse and longitudinal terms become

$$I_\perp = \iint \Xi^2(\xi)H^2(\eta)(\xi^2-1)^{1/2}(1-\eta^2)^{1/2}(\xi^2-\eta^2)\,\mathrm{d}\xi\,\mathrm{d}\eta, \tag{17a}$$

Table 4 Numerical Values of the Transverse Dipole Moment $d_\perp$ Relative to the Origin for a Set of Dihedral Confining Angles ϕ_0 for the H_2^+ Molecule in Its Ground (001) State

ϕ_0 (deg)	$d_\perp$ (ea_0)
90	9.158
120	6.087
140	5.168
160	4.110
180	3.502
270	1.943
300	1.692
330	1.362
360	1.145

All quantities are calculated for the equilibrium internuclear distances shown in Table 2.

$$I_z = \iint \Xi^2(\xi) H^2(\eta) \xi\eta \left(\xi^2 - \eta^2\right) \mathrm{d}\xi \mathrm{d}\eta, \tag{17b}$$

which are evaluated in the domains $-1 \leq \eta \leq 1$ and $1 \leq \xi < \infty$, respectively, and considering the normalized wavefunctions $\Xi(\xi)$ and $H(\eta)$. Clearly, the longitudinal component I_z [Eq. (17b)] becomes zero and $I_\perp$ is the only remaining nonzero integral. Table 4 displays the values of the transverse dipole moment at the equilibrium internuclear position for different values of the dihedral angle ϕ_0. An increasing value of this property is observed as the confining angle decreases, indicating the readjustment of the electronic distribution as the meridian planes approach each other. We note that the center of electronic charge is at the equatorial plane bisecting the internuclear axis, and also at the meridian plane bisecting the confining dihedral angle, so that its electric dipole moment has no component along the axis and is contained in that plane as well as in the equatorial plane, because the center of nuclear charge is at the origin, where the axis and both planes meet. This is a particular feature only appearing due to the dihedral confinement.

4. THE HeH^{2+} MOLECULAR ION

The calculation for the (001) and (011) eigenstates of the heteronuclear HeH^{2+} molecular ion confined in dihedral angles and the analysis of their electronic energies, molecular potential energies, wavefunctions, and metastability of the second state are presented in numerical and graphical forms in the successive subsections. The heteronuclear nature of the system requires the use of the pentagonal matrix in Eq. (12) for the hyperboloidal coordinate eigenfunctions.

4.1 Electronic Energies Including the United Atom and Dissociation Limits

The evaluation of the electronic energies and eigenfunctions of the ground state (001) and the first excited state (011) of the ion confined in dihedral angles was implemented for different nuclear separations.

Figure 7 illustrates the variations of the electronic energies for nuclear separations in the range $0.1 < R < 100$, and the specific values of $\phi_0 = 2\pi$ ($\mu = 1/2$), for both states. The calculations for other confining angles lead to similar variations for the respective states. While the variations in the ground state are monotonically increasing in the entire range, those of the excited state include a dip in the intermediate region of a few units of nuclear separations. In this case, the united atom limits of lithium ($Z = 3$) hydrogen-like atom in the same situation of confinement at -4 and $-36/25$ are

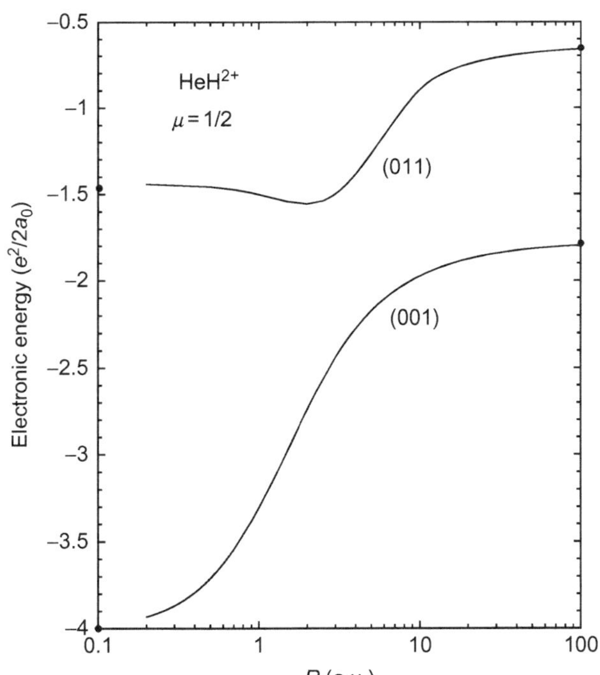

Figure 7 Electronic energy versus nuclear separation for the ground (001) and first excited states (011) of the HeH^{2+} ion dihedrally confined in an angle $\phi_0 = 2\pi$.

identified on the left for $R \rightarrow 0$; and also the dissociation limits of He ($Z=2$) hydrogen-like atom at $-16/9$ and $-16/25$, respectively.

4.2 Molecular Potential Energy Curves

The addition of the Coulomb repulsion energy between the helium and hydrogen nuclei, $2/R$, to the electronic energy leads to the total molecular energy $E(R)$ as a function of the nuclear separation.

Figure 8 is constructed from such an addition to the respective points of Fig. 7. It shows the potential energy curves for the respective ground and excited states, using a normal horizontal scale instead of a logarithmic one for the nuclear separation axis; notice also the change of the vertical scale in order to include the fast variation of the Coulomb energy for small separations. Now, the ground state is monotonically decreasing, approaching the asymptotic limit of molecular dissociation in He^+ from above. This allows us to identify its antibinding nature. On the other hand, for the first excited state, the original dip in the electronic energy is converted into a corresponding dip in the potential energy curve, preceding the

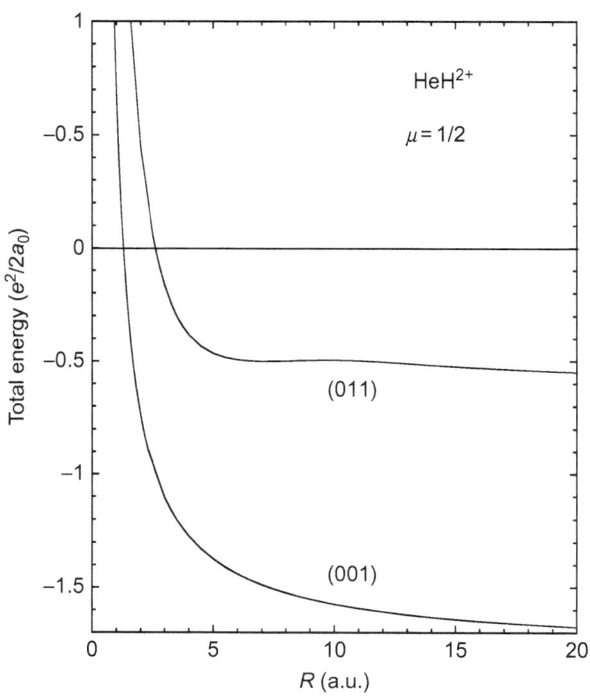

Figure 8 Potential energy curves for the ground (001) and first excited states (011) of the HeH^{2+} ion dihedrally confined in an angle $\phi_0 = 2\pi$.

corresponding approach from above to the asymptotic limit of dissociation. We go on to a more detailed analysis of the dip region, not only for $\phi_0 = 2\pi$ but also for other angles.

Figure 9 exhibits the potential energy curves for the same excited state and for the confining angles $\phi_0 = 2\pi$, $11\pi/6$, $3\pi/5$, $3\pi/2$, and π: with a logarithmic scale for the nuclear separation and a normal scale for the total energy. This choice of scales amplifies the region of the dips, and the corresponding analysis is simplified. It can be recognized that the dip is more noticeable for the original largest angle, and its characteristics become reduced as the confining angle is reduced, practically disappearing in the upper curve with $\phi_0 = \pi$, in which the meridian half-planes defining the angle become a plane. Dynamically, the important element of that upper potential energy curve is its monotonically decreasing behavior, which identifies it as antibinding; this also holds for the two curves below with μ values of 2/3 and 3/5, as the reader can ascertain by following them down from left to right. The same exercise allows the identification of small-height potential barriers to the right of the dip, forming a local well

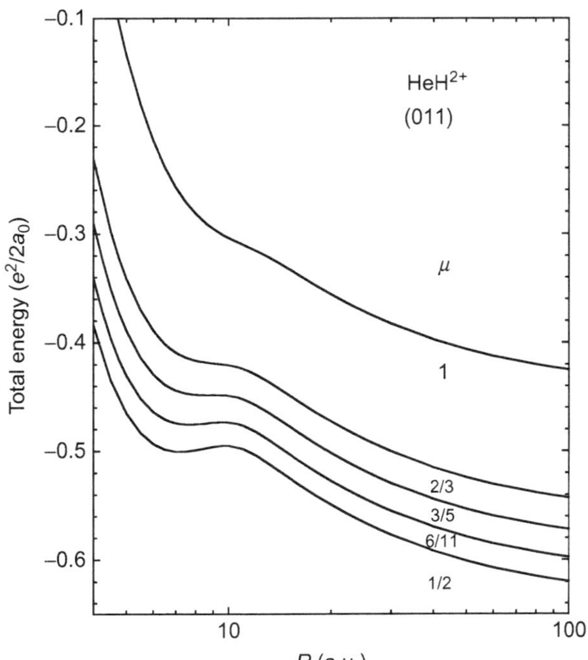

Figure 9 Behavior of the potential energy curve for the first excited state of HeH^{2+} as the confinement dihedral angle changes from $\phi_0 = 2\pi$ to $\phi_0 = \pi$.

Table 5 Total Energies for the Ground (001) and First Excited State (011) of the Confined HeH^{2+} Ion for a Set of Internuclear Distances and Different Dihedral Angles Defined by $\mu = \pi/\phi_0$

μ	R (a.u.)	0.2	0.5	1.0	3.0	5.0	10.0	20.0	30.0	50.0	100.0
2/3		16.79402	4.92836	1.20455	−0.75716	−1.10357	−1.23914	−1.33986	−1.37329	−1.39999	−1.41999
3/5	(001)	16.52749	4.69153	1.01357	−0.88519	−1.15916	−1.36180	−1.46239	−1.49580	−1.52249	−1.54250
6/11		16.28446	4.47813	0.84339	−1.00120	−1.27204	−1.47416	−1.57465	−1.60804	−1.63473	−1.65474
1/2		16.06750	4.28957	0.69436	−1.10447	−1.37318	−1.57492	−1.67533	−1.70872	−1.73540	−1.75541
2/3		18.73273	6.72444	2.69995	0.02531	−0.34119	−0.42064	−0.47275	−0.50031	−0.52410	−0.54290
3/5[c]	(011)	18.66654	6.65597	2.62508	−0.04687	−0.38907	−0.44889	−0.50177	−0.52944	−0.55328	−0.57211
6/11[b]		18.60840	6.59548	2.55805	−0.11000	−0.43023	−0.47365	−0.52723	−0.55499	−0.57888	−0.59773
1/2[a]		18.55811	6.54287	2.49904	−0.16434	−0.46525	−0.49509	−0.54923	−−0.57707	−0.44384	−0.44385

[a] $R_{eq} = 7.200$, $E_{eq} = -0.50049$.
[b] $R_{eq} = 7.684$, $E_{eq} = -0.47558$.
[c] $R_{eq} = 8.682$, $E_{eq} = -0.44825$.
Also, the equilibrium internuclear distance R_{eq} and energy E_{eq} for the excited states with $\mu = 1/2$, 6/11, and 3/5 are given at the bottom. All energies are in units of $e^2/2a_0$.

for the larger angles of confinement. For $\phi_0 = 2\pi$, the minimum of the potential well is at $R_m \approx 7$ and the maximum at $R_m \approx 10$. Such a well may bind the molecule, but the molecule may also tunnel across the potential barrier. The state is metastable.

For the free molecular ion, such a metastable state was first predicted theoretically—within the Born–Oppenheimer approximation—by Bates and Carson[21] six decades ago and first confirmed experimentally only two decades ago by Ben-Itzhak et al.[22,23] Interestingly enough, this metastable state appears also under dihedral confinement conditions, as discussed before. In this case, a shallower minimum dip in the potential energy appears as the dihedral confining angle decreases from $\phi_0 = 2\pi$ to $\phi_0 = 3\pi/2$ (see Fig. 9), as may be observed from Table 5, where the total HeH^{2+} energies are shown for varying internuclear separations and dihedral confining angles, including the corresponding equilibrium position. It is therefore expected to find the presence of metastable states too for the dihedrally confined HeH^{2+} molecular ion within the range of dihedral angles $3\pi/2 \leq \phi_0 \leq 2\pi$.

Notice that the solutions in this work for $\phi_0 = 2\pi$ do not connect with those of the free system because they satisfy different boundary conditions. This type of situation has been well known for confinement in semi-infinite spaces with different boundaries.[24] The important conclusion is that the metastability in the free system is still present for confinement in large enough dihedral angles. This is also a motivation to examine the electronic eigenfunctions of such state, as we do in the next section.

4.3 Hyperboloidal and Spheroidal Coordinate Eigenfunctions

Here, we present the plots of the hyperboloidal coordinate eigenfunctions for the ground (001) and first excited (011) states, in order to illustrate their common and differentiating characteristics.

Figure 10 corresponds to the ground state with parameters $\mu = 1/2$ and $R = 7$. Its wavefunction $H_0(\eta)$ has zero nodes in the domain $(-1, 1)$ and nodes at $\eta = \pm 1$ associated with its singularity removing factor [Eq. (9)]. The proton and the He nucleus are assumed to be at $\eta = -1$ and $\eta = 1$, respectively. The variations of the wavefunction express the preference of the electron to be closer to the He nucleus.

The corresponding Fig. 11, for the first excited state, with $\mu = 1/2$ and $R = 7$, illustrates the eigenfunction $H_1(\eta)$. It also vanishes at the ends of the domain and has one node inside the domain. In this case, the electron is

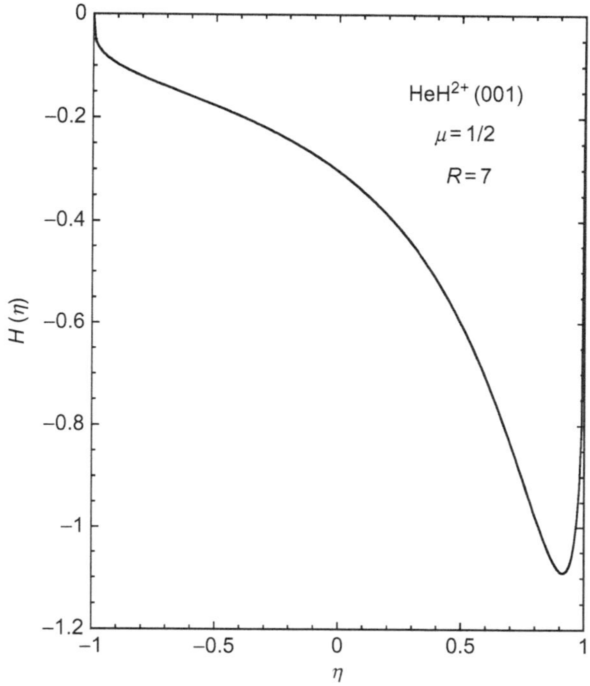

Figure 10 Hyperboloidal coordinate eigenfunction for the ground state of the HeH^{2+} ion confined in a dihedral angle $\phi_0 = 2\pi$ and for a nuclear separation $R = 7$.

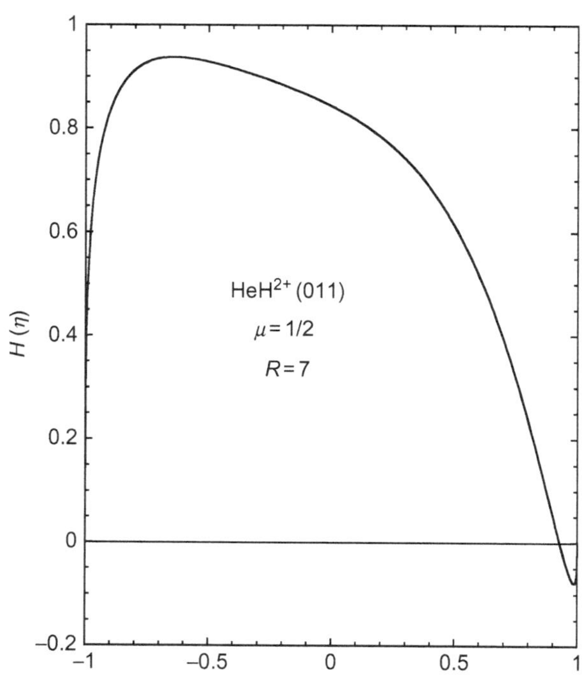

Figure 11 Same as in Fig. 10 for the first excited state of the HeH^{2+} ion.

more likely to be found below the equatorial plane on the side where the proton is located.

Figure 12 refers to the same state and parameters of the previous paragraph, illustrating the matching spheroidal coordinate eigenfunction $\Xi_0(\xi)$, in the domain $1 \leq \xi < \infty$. Notice its zero value at the beginning of the domain, associated with the singularity removing factor in the Jaffé solution [Eq. (13)], and no additional nodes in the rest of the domain. The curve can be described as a localized peak at the left, with a very steep rising until its extreme value is reached at $\xi \approx 1.1$, after which a monotonic lowering occurs with diminishing steepness and is continued on towards zero asymptotically. We can interpret it as a localization of the electron in the interval $1 \leq \xi \leq 1.5$, which is equivalent to the distance interval of $3.5 < z < 4.6$ for the equilibrium nuclear separation $R = 7$ in Fig. 10.

Here, we also present the comparison of the wavefunctions in the hyperboloidal coordinate for the ion confined in the largest dihedral angle $\phi_0 = 2\pi$ in the ground (001) and first excited (011) states, and their counterparts (000) and (010) σ-states for the free ion, following their changes for the same internuclear separations.

Figure 13A and B for the ground state of the free and confined ions, respectively, is compared first. The corresponding wavefunctions for the nuclear separations $R = 2$, 10, and 20 exhibit very similar behaviors in the range $-0.9 < \eta < 0.9$, corresponding to the respective practically horizontal dotted lines, and the other two in the segmented and continuous lines rising slowly between $-0.9 < \eta < 0.5$ and more rapidly beyond. Their differences reside in their behaviors in the vicinities of $\eta = -1$ and $\eta = 1$: sudden drops to zero are observed for the three lines on the left and also on the right for the dotted line; for the other two, pronounced peaks in the vicinity of $\eta \approx 0.9$ with their right steep drops to zero at $\eta = 1$, for the confined ion. The lines for the free ion remain finite at both ends. All of them do not have nodes in the interval $(-1, 1)$ of η, justifying their characterization as $N_\eta = 0$.

Now we go on to Fig. 14A and B for the first excited state of the confined (011) and free (010) ions, describing them consecutively in that order. For the confined ion, the hyperboloidal coordinate wavefunction is shown for the nuclear separations $R = 0.2$, 3, 4, 7, 10, and 20. All of the lines have the common features of one node in the interval $(-1, 1)$ of η, at the position $\eta \approx 0.1$ for the almost straight dotted line, and 0.6, 0.75, 0.9, 0.93, and 0.95 for the successive ones; and the nodes at the ends of the interval $[-1, 1]$ of η. The corresponding wavefunctions in Fig. 14B for the free ion correspond to the

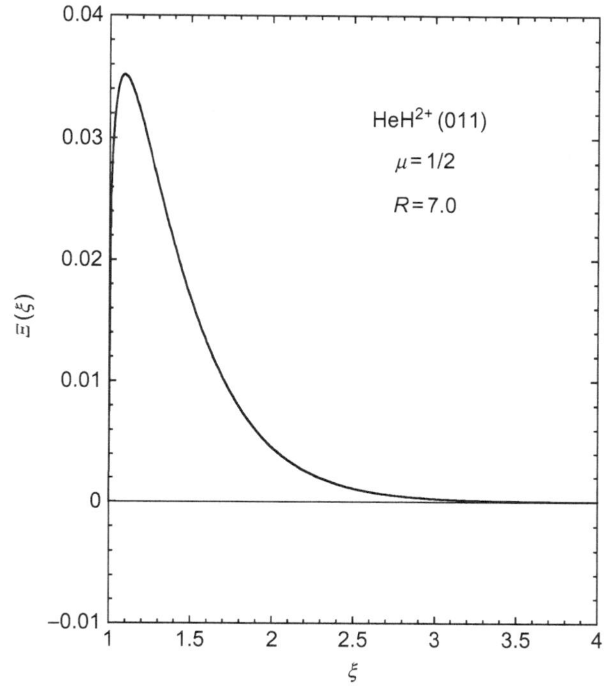

Figure 12 Spheroidal coordinate eigenfunction for the first excited state of the HeH^{2+} ion confined in a dihedral angle $\phi_0 = 2\pi$ and for a nuclear separation $R = 7$.

nuclear separations of 0.2, 3, 4, 6, 10, and 20. Only the first three show one node at $\eta \approx 0.5$, 0.85, and 0.93; the other three remain above the zero ordinate. The quantum number $N_\eta = 1$ can be confidently assigned only to the wavefunctions with one node.

5. DISCUSSION AND CONCLUSIONS

This section discusses the important concepts and results about exact solutions of the Schrödinger equation—within the Born–Oppenheimer approximation—for one-electron diatomic molecules, in general, and for the homonuclear $H_2{}^+$ and the heteronuclear HeH^{2+} molecular ions, confined in dihedral angles in particular.

The key distinction between the free ions and the ions confined in dihedral angles is the difference in the boundary conditions of their azimuthal eigenfunctions, leading to their respective integer m and noninteger μ eigenvalues of the component of the angular momentum along the internuclear axis. This breaking of the rotational symmetry around the internuclear axis

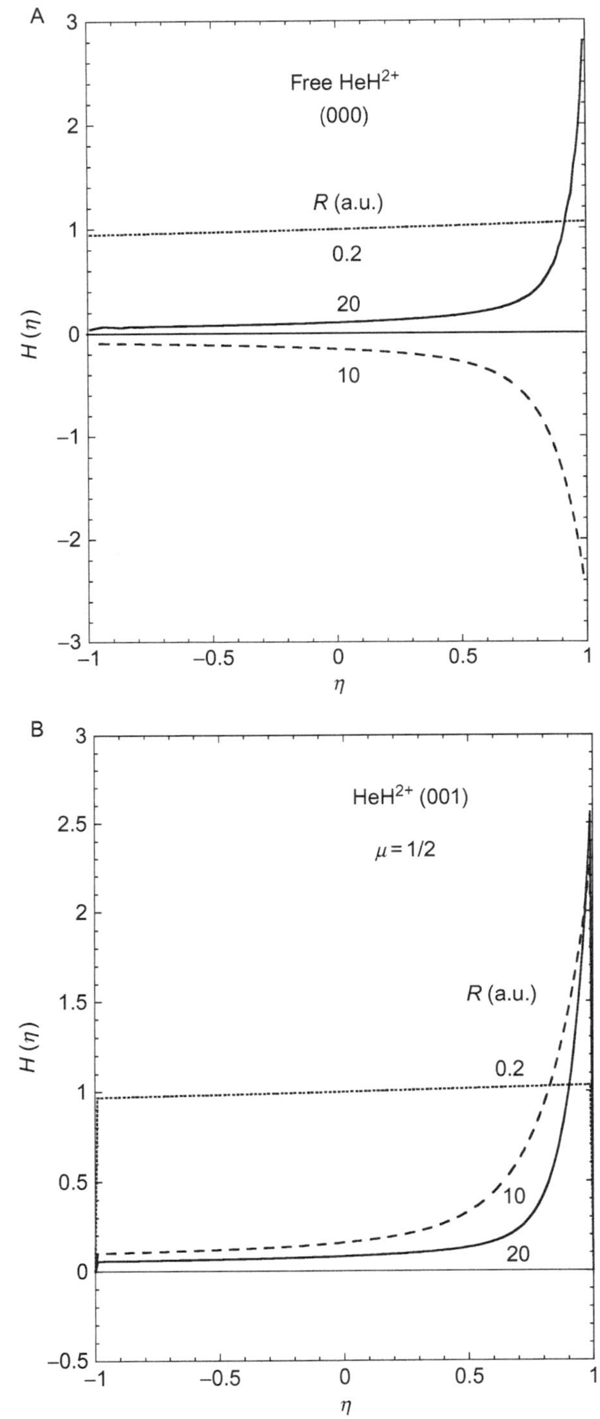

Figure 13 Varying characteristics of the hyperboloidal coordinate eigenfunctions for the (A) free and (B) confined in $\phi_0 = 2\pi$, HeH^{2+} ion in their respective ground states, for a set of internuclear separations.

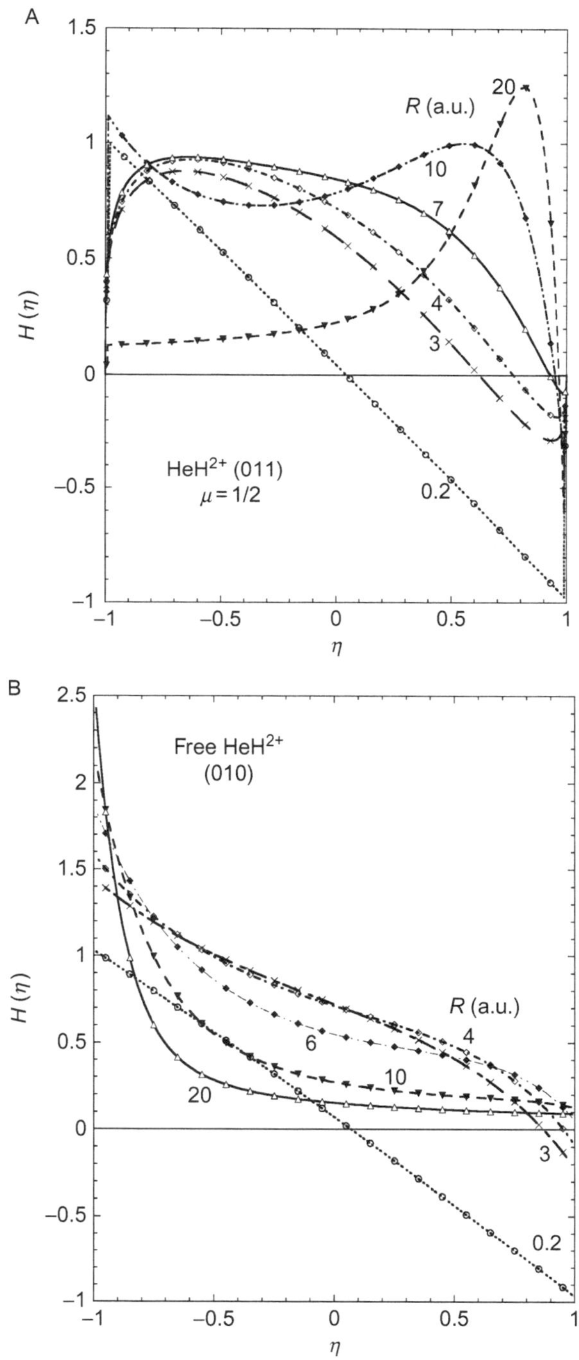

Figure 14 Same as Fig. 13 for the (A) confined in $\phi_0 = 2\pi$ and (B) free HeH^{2+} ion in its respective first excited states, for a wider span of internuclear separations. The different symbols are included to guide the eye.

produces the consequent breaking of the symmetry of reflection in the plane bisecting that axis for homonuclear molecules. The restoration of the reflection symmetry for the spheroidal harmonics in the hyperboloidal coordinate plays a key role for the construction of the solutions of Eq. (5) via its matrix form of Eq. (12), as detailed in the Appendices.

In the case of the homonuclear H_2^+ molecular ion, the results about its electronic energies, potential energy curves, hyperboloidal coordinate wavefunctions, and electric dipole moment as functions of the internuclear separation and the confining angle are original and reported for the first time.

The electronic energies of the $(001)_g$, $(011)_u$, and $(021)_g$ eigenstates are systematically above their counterparts for the free ion, σ-states $(000)_g$, $(010)_u$, and $(020)_g$, due to the differences between their azimuthal excitations $\mu = \pi/\phi_0$ and $m = 0$, respectively. Consequently, the respective potential energy curves, for the binding states of even parities $(001)_g$ and $(021)_g$, have minima with increasing equilibrium separations and decreasing depths, as the confining angle diminishes from its largest value of $\phi_0 = 2\pi$. The comparative analysis of the nodes of the hyperboloidal coordinate wavefunctions for the free and confined ions illustrates their similarities and differences and, most important, provides confidence in the consistent assignment of their N_η quantum number.

The transverse electric dipole moment is a manifestation of the rotational symmetry breaking; the absence of a longitudinal component attests the well-defined parity of the ground state. We are also in the position to evaluate other properties of the system, like the pressure distribution on the confining meridian planes, the hyperfine structure in its dipole–dipole interaction form, because the Fermi-contact term is suppressed. Time limitations did not allow us to do it for this contribution.

For the heteronuclear HeH^{2+} molecular ion, the study was implemented, also for the first time, and reported for the electronic energies, potential energy curves, and the analysis of the hyperboloidal coordinate wavefunctions. The results on the potential energy curves show consistency for the confinement in dihedral angles starting from the largest $\phi_0 = 2\pi$. The ground state (001) is antibinding as exhibited by its monotonically decreasing potential energy curve. The first excited state (011) exhibits potential energy curves with a local minimum followed by a maximum and with a monotonic decrease for the larger angles of confinement, providing the elements to identify it as a metastable state; for smaller angles of confinement, the monotonic decrease occurs for all increasing nuclear separations, making them antibinding. The analysis of the nodes in the hyperboloidal coordinate wavefunction is consistent for the ion confined in the angle $\phi_0 = 2\pi$, for both ground and excited states with the confident identification of their η-excitations; however, for the free ion in the excited state, the identification is reliable for the smaller values of the nuclear separation, and no node appears for the larger ones. The reasons for this behavior are still to be clarified after further analysis.

APPENDICES

A Breaking of η-Parity Symmetry and Its Restoration

In Section 2.3, the symmetry-breaking effect of the confinement in dihedral angles is manifest via the no longer integer magnetic quantum number μ. Here, we analyze its consequences on the parity of the corresponding hyperboloidal coordinate-associated Legendre polynomials $(1-\eta^2)^{\mu/2} P^{\mu}_{n_\eta+\mu}(\eta)$.

In fact, $\lambda = n_\eta + \mu$ plays the role of the label for the magnitude $\lambda(\lambda+1)$ of the square of the angular momentum operator. The corresponding Legendre polynomial has alternative hypergeometric function representations:

$$P^{\mu}_{n_\eta+\mu}(\eta) = {}_2F_1\left(-n_\eta, n_\eta + 2\mu + 1; \mu + 1; \frac{1-\eta}{2}\right)$$
$$P^{\mu}_{n_\eta+\mu}(\eta) = {}_2F_1\left(-n_\eta, n_\eta + 2\mu + 1; \mu + 1; \frac{1+\eta}{2}\right). \tag{A1}$$

For integer values of μ, both representations are the same involving only even or odd powers of η for the corresponding even or odd values of n_η; consequently, they have a definite parity $(-1)^{n_\eta}$ under the exchange $\eta \to -\eta$. On the other hand, for noninteger values of μ, each representation contains both even and odd powers of η; hence their η-parity is broken.

In the case of homonuclear molecules, η-parity is conserved, and here, we must restore it. This is readily accomplished using the superposition of both representations

$$F^{\pm}_{n_\eta,\mu}(\eta) = \frac{1}{2}\left[{}_2F_1\left(-n_\eta, n_\eta + 2\mu + 1; \mu + 1; \frac{1-\eta}{2}\right)\right.$$
$$\left. + (-1)^{n_\eta}{}_2F_1\left(-n_\eta, n_\eta + 2\mu + 1; \mu + 1; \frac{1+\eta}{2}\right)\right] \tag{A2}$$

which have $\pm$ parities for n_η even or odd, respectively.

This appendix is an adaptation and extension of the corresponding discussion for spherical harmonics in Refs. 13,14. The first polynomials are contained in Table 4 of Ref. 14 which we adapt here:

$$\begin{aligned} F^+_{0,\mu} &= 1 \\ F^-_{1,\mu} &= \eta \\ F^+_{2,\mu} &= [-1 + (2\mu+3)\eta^2]/2(\mu+1) \\ F^-_{2,\mu} &= [-3\eta + (2\mu+5)\eta^3]/2(\mu+1). \end{aligned} \tag{A3}$$

B Recurrence Relations and Normalization of μ-Associated Legendre Polynomials with Definite Parity

Equations (29)–(32) of Ref. 14 contain the lowering and raising of the polar-angle quantum number operators for the μ-associated Legendre polynomials with definite parity. Their adaptations for the respective hyperboloidal coordinate polynomials lead to

$$\left[n_\eta \eta - (\eta^2 - 1)\frac{d}{d\eta}\right] F^\pm_{n_\eta,\mu} = n_\eta F^\mp_{n_\eta - 1,\mu} \tag{B.1}$$

$$\left[(n_\eta + 2\mu + 1)\eta + (\eta^2 - 1)\frac{d}{d\eta}\right] F^\pm_{n_\eta,\mu} = (n_\eta + 2\mu + 1) F^\mp_{n_\eta + 1,\mu}. \tag{B.2}$$

The sum of these two operators cancels the derivative terms and leads to the simple recurrence relation:

$$\eta F^\pm_{n_\eta,\mu} = \frac{n_\eta}{2n_\eta + 2\mu + 1} F^\mp_{n_\eta - 1,\mu} + \frac{n_\eta + 2\mu + 1}{2n_\eta + 2\mu + 1} F^\mp_{n_\eta + 1,\mu}. \tag{B.3}$$

This is the starting point to construct the matrix elements of η in the orthonormal basis of spheroidal harmonics with definite parity,

$$\langle \eta | n_\eta \mu \rangle = N_{n_\eta,\mu} (1 - \eta^2)^{\mu/2} F^\pi_{n_\eta,\mu}(\eta), \tag{B.4}$$

as formulated in Section 2.4. Naturally, the normalization constants have to be evaluated first. Here, we evaluate them explicitly for the lower states:

$$\langle 0\mu|\,0\mu\rangle = N_{0,\mu}^2\int_{-1}^{1}(1-\eta^2)^{\mu}\mathrm{d}\eta = N_{0,\mu}^2\int_0^1 t^{\mu}(1-t)^{-1/2}\mathrm{d}t = N_{0,\mu}^2 B\left(\mu+1,\frac{1}{2}\right)$$
$$= N_{0,\mu}^2\frac{\Gamma(\mu+1)\Gamma(\frac{1}{2})}{\Gamma(\mu+3/2)}$$

$$\langle 1\mu|\,1\mu\rangle = N_{1,\mu}^2\int_{-1}^{1}(1-\eta^2)^{\mu}\eta^2\mathrm{d}\eta = N_{1,\mu}^2\int_0^1 t^{\mu}(1-t)(1-t)^{-1/2}\mathrm{d}t$$
$$= N_{1,\mu}^2\left[B\left(\mu+1,\frac{1}{2}\right) - B\left(\mu+2,\frac{1}{2}\right)\right]$$
$$= N_{1,\mu}^2 B\left(\mu+1,\frac{1}{2}\right)\left[1-\frac{\mu+1}{\mu+3/2}\right] = N_{1,\mu}^2\frac{\Gamma(\mu+2)\Gamma(\frac{1}{2})}{\Gamma(\mu+5/2)}$$

$$\langle 2\mu|\,2\mu\rangle = N_{2,\mu}^2\int_{-1}^{1}(1-\eta^2)^{\mu}\left[\frac{-1+(2\mu+3)\eta^2}{2(\mu+1)}\right]^2\mathrm{d}\eta$$
$$= N_{2,\mu}^2\left[\frac{B(\mu+1,\frac{1}{2}) - 2(2\mu+3)B(\mu+2,\frac{1}{2}) + (2\mu+3)^2 B(\mu+3,\frac{1}{2})}{4(\mu+1)^2}\right]$$
$$= \frac{N_{2,\mu}^2}{4(\mu+1)^2}B\left(\mu+1,\frac{1}{2}\right)\left[1-\frac{2(2\mu+3)(\mu+1)}{\mu+3/2}+\frac{(2\mu+3)^2(\mu+1)(\mu+2)}{(\mu+3/2)(\mu+5/2)}\right]. \tag{B.5}$$

The integrations over η in the interval $[-1,\ 1]$ equal twice the integrations in the interval $[0,\ 1]$. Then we make the change of variable $1-\eta^2=t$ equivalent to $\eta^2=1-t$ with $2\eta\mathrm{d}\eta=-\mathrm{d}t$. The resulting integrals have the general form of the Beta functions:

$$B(x,y) = \int_0^1 t^{x-1}(1-t)^{y-1}\mathrm{d}t \tag{B.6}$$

with the identifications of $x=n_\eta+1$ and $y=1/2$ as indicated in each case.

The recurrence relation (B.3) becomes, in terms of the normalized functions, in the ket notation:

$$\frac{\eta|n_\eta\mu\rangle}{N_{n_\eta,\mu}} = \frac{n_\eta}{2n_\eta+2\mu+1}\frac{|n_\eta-1\mu\rangle}{N_{n_\eta-1,\mu}} + \frac{n_\eta+2\mu+1}{2n_\eta+2\mu+1}\frac{|n_\eta+1\mu\rangle}{N_{n_\eta+1,\mu}}. \tag{B.7}$$

Their explicit forms for the successive values of $n_\eta = 0,1,2,3$ become:

$$
\begin{aligned}
\eta|0\mu\rangle &= \frac{N_{0,\mu}}{N_{1,\mu}}|1\mu\rangle = \frac{1}{\sqrt{2\mu+3}}|1\mu\rangle \\
\eta|1\mu\rangle &= \frac{1}{2\mu+3}\frac{N_{1,\mu}}{N_{0,\mu}}|0\mu\rangle + \frac{2\mu+2}{2\mu+3}\frac{N_{1,\mu}}{N_{2,\mu}}|2\mu\rangle \\
\eta|2\mu\rangle &= \frac{2}{2\mu+5}\frac{N_{2,\mu}}{N_{1,\mu}}|1\mu\rangle + \frac{2\mu+3}{2\mu+5}\frac{N_{2,\mu}}{N_{3,\mu}}|3\mu\rangle \\
\eta|3\mu\rangle &= \frac{3}{2\mu+7}\frac{N_{3,\mu}}{N_{2,\mu}}|2\mu\rangle + \frac{2\mu+4}{2\mu+7}\frac{N_{3,\mu}}{N_{4,\mu}}|4\mu\rangle.
\end{aligned}
\tag{B.8}
$$

C Matrix Elements of η and η^2 on the Basis of Orthonormalized Spheroidal Harmonics with Definite Parity

The matrix elements of our interest follow from the above recurrence relations projected with the successive bra-eigenstates:

$$
\begin{aligned}
\langle 1\mu|\eta|0\mu\rangle &= \langle 0\mu|\eta|1\mu\rangle = \frac{1}{\sqrt{2\mu+3}} \\
\langle 1\mu|\eta|2\mu\rangle &= \langle 2\mu|\eta|1\mu\rangle = \sqrt{\frac{2(2\mu+2)}{(2\mu+3)(2\mu+5)}} \\
\langle 2\mu|\eta|3\mu\rangle &= \langle 3\mu|\eta|2\mu\rangle = \sqrt{\frac{3(2\mu+3)}{(2\mu+5)(2\mu+7)}} \\
\langle 3\mu|\eta|4\mu\rangle &= \langle 4\mu|\eta|3\mu\rangle = \sqrt{\frac{4(2\mu+4)}{(2\mu+7)(2\mu+9)}}
\end{aligned}
\tag{C.1}
$$

The generalization can be made using mathematical induction, with the results:

$$
\begin{aligned}
\langle n_\eta - 1\mu|\eta|n_\eta\mu\rangle &= \sqrt{\frac{(n_\eta+2\mu)n_\eta}{(2n_\eta+2\mu+1)(2n_\eta+2\mu-1)}} \\
\langle n_\eta + 1\mu|\eta|n_\eta\mu\rangle &= \sqrt{\frac{(n_\eta+2\mu+1)(n_\eta+1)}{(2n_\eta+2\mu+1)(2n_\eta+2\mu+3)}}
\end{aligned}
\tag{C.2}
$$

recognizing the respective selection rules $n'_\eta = n_\eta - 1$ and $n'_\eta = n_\eta + 1$ for the quantum number n'_η.

The matrix elements of η^2 follow from the multiplication by η of the recurrence relation [Eq. (B.3)] and the successive application of the same relation:

$$\begin{aligned}\frac{\eta^2|n_\eta\mu\rangle}{N_{n_\eta,\mu}} &= \frac{n_\eta}{2n_\eta+2\mu+1}\frac{\eta|n_\eta-1\mu\rangle}{N_{n_\eta-1,\mu}} + \frac{n_\eta+2\mu+1}{2n_\eta+2\mu+1}\frac{\eta|n_\eta+1\mu\rangle}{N_{n_\eta+1,\mu}} \\ &= \frac{n_\eta}{2n_\eta+2\mu+1}\left[\frac{n_\eta-1}{2n_\eta+2\mu-1}\frac{|n_\eta-2\mu\rangle}{N_{n_\eta-2,\mu}} + \frac{n_\eta+2\mu}{2n_\eta+2\mu-1}\frac{|n_\eta\mu\rangle}{N_{n_\eta,\mu}}\right] \\ &+ \frac{n_\eta+2\mu+1}{2n_\eta+2\mu+1}\left[\frac{n_\eta+1}{2n_\eta+2\mu+3}\frac{|n_\eta\mu\rangle}{N_{n_\eta,\mu}} + \frac{n_\eta+2\mu+2}{2n_\eta+2\mu+3}\frac{|n_\eta+2\mu\rangle}{N_{n_\eta+2,\mu}}\right]\end{aligned} \tag{C.3}$$

the starting nonvanishing matrix elements are:

$$\begin{aligned}
\langle 0\mu|\eta^2|0\mu\rangle &= \frac{1}{2\mu+3} \\
\langle 2\mu|\eta^2|0\mu\rangle &= \frac{2\mu+2}{2\mu+3}\frac{N_{0,\mu}}{N_{2,\mu}} = \frac{2\mu+2}{2\mu+3} \\
\langle 1\mu|\eta^2|1\mu\rangle &= \frac{1}{2\mu+3} + \frac{2\mu+2}{2\mu+3}\cdot\frac{2}{2\mu+5} \\
\langle 3\mu|\eta^2|1\mu\rangle &= \frac{1}{2\mu+7}\frac{N_{1.\mu}}{N_{3,\mu}} \\
\langle 0\mu|\eta^2|2\mu\rangle &= \frac{2}{2\mu+5}\cdot\frac{1}{2\mu+3}\cdot\frac{N_{2,\mu}}{N_{0,\mu}} \\
\langle 2\mu|\eta^2|2\mu\rangle &= \frac{2}{2\mu+5}\cdot\frac{2\mu+2}{2\mu+3} + \frac{2\mu+3}{2\mu+5}\cdot\frac{3}{2\mu+7} \\
\langle 4\mu|\eta^2|2\mu\rangle &= \frac{2\mu+3}{2\mu+5}\cdot\frac{2\mu+4}{2\mu+7}\cdot\frac{N_{2,\mu}}{N_{4,\mu}} \\
\langle 1\mu|\eta^2|3\mu\rangle &= \frac{3}{2\mu+7}\cdot\frac{2}{2\mu+5}\cdot\frac{N_{1,\mu}}{N_{3,\mu}} \\
\langle 3\mu|\eta^2|3\mu\rangle &= \frac{3}{2\mu+7}\cdot\frac{2\mu+3}{2\mu+5} + \frac{2\mu+4}{2\mu+7}\cdot\frac{4}{2\mu+9} \\
\langle 5\mu|\eta^2|3\mu\rangle &= \frac{2\mu+4}{2\mu+7}\cdot\frac{2\mu+5}{2\mu+9}\cdot\frac{N_{3,\mu}}{N_{5,\mu}} \\
\langle 2\mu|\eta^2|4\mu\rangle &= \frac{4}{2\mu+9}\cdot\frac{3}{2\mu+7}\cdot\frac{N_{2,\mu}}{N_{4,\mu}} \\
\langle 4\mu|\eta^2|4\mu\rangle &= \frac{4}{2\mu+9}\cdot\frac{2\mu+4}{2\mu+7} + \frac{2\mu+5}{2\mu+9}\cdot\frac{5}{2\mu+11} \\
\langle 3\mu|\eta^2|5\mu\rangle &= \frac{5}{2\mu+11}\cdot\frac{4}{2\mu+9}\cdot\frac{N_{3,\mu}}{N_{5,\mu}}.
\end{aligned} \tag{C.4}$$

Their generalization from mathematical induction is incorporated in Eq. (12).

REFERENCES

1. Cottrell, T. L. Molecular Energy at High Pressure. *Trans. Faraday Soc.* **1951**, *47*, 337.
2. Le Sar, R.; Herschbach, D. R. Electronic and Vibrational Properties of Molecules at High Pressures. Hydrogen Molecule in a Rigid Spheroidal Box. *J. Phys. Chem.* **1981**, *85*, 2798.
3. Le Sar, R.; Herschbach, D. R. Polarizability and Quadrupole Moment of a Hydrogen Molecule in a Spheroidal Box. *J. Phys. Chem.* **1983**, *87*, 5202.
4. Cruz, S. A.; Colín-Rodríguez, R. Spheroidal Confinement of a Single Electron and of the Hydrogen Atom, the H_2^+ and HeH^{++} Molecular Ions with Arbitrary Nuclear Positions Along the Major Axis. *Int. J. Quantum Chem.* **2009**, *109*, 3041.
5. Colín-Rodríguez, R.; Díaz-García, C.; Cruz, S. A. The Hydrogen Molecule and the H_2^+ Molecular Ion Inside Padded Prolate Spheroidal Cavities with Arbitrary Nuclear Positions. *J. Phys. B: At. Mol. Opt. Phys.* **2011**, *44*, 241001.
6. Singh, K. K. Theory of Boundary Perturbation + Compressed Hydrogen Molecular Ion. *Physica* **1964**, *30*, 211.
7. Gorecki, J.; Byers-Brown, W. On the Ground State of the Hydrogen Molecule-Ion H_2^+ Enclosed in Hard and Soft Spherical Boxes. *J. Chem. Phys.* **1988**, *89*, 2138.
8. Ley-Koo, E.; Cruz, S. A. The Hydrogen Atom and the H_2^+ and HeH^{++} Molecular Ions Inside Prolate Spheroidal Boxes. *J. Chem. Phys.* **1981**, *74*, 4603.
9. Mateos-Cortés, S.; Ley-Koo, E.; Cruz, S. A. Hydrogen Molecular Ion Inside Penetrable Prolate Spheroidal Boxes: Electronic and Vibrational Properties. *Int. J. Quantum Chem.* **2002**, *86*, 376.
10. Sarsa, A.; Le Sech, C. Study of Quantum Confinement of H_2^+ Ion and H_2 Molecule with Monte Carlo. Respective Role of the Electron and Nuclei Confinement. In: *Electronic Structure of Quantum Confined Atoms and Molecules*; Sen, K. D. Ed.; Springer: Switzerland, 2014; (Chapter 9, p 227, and references therein).
11. Ndengué, S. A.; Motapon, O. Spatial and Shell-Confined One Electron Atomic and Molecular Systems: Structure and Dipole Polarizability. In: *Electronic Structure of Quantum Confined Atoms and Molecules*; Sen, K. D. Ed.; Springer: Switzerland, 2014; (Chapter 7, p 169, and references therein).
12. Lo, J. M. H.; Klobukowski, M.; Diercksen, G. H. F. Low-Lying Excited States of the Hydrogen Molecule in Cylindrical Harmonic Confinement. *Adv. Quantum Chem.* **2005**, *48*, 59.
13. Ley-Koo, E.; Sun, G. H. Surface Effects in the Hydrogen Atom Confined by Dihedral Angles. In: *Electronic Structure of Quantum Confined Atoms and Molecules*; Sen, K. D. Ed.; Springer: Switzerland, 2014; (Chapter 1, p 1).
14. Ley-Koo, E.; Sun, G. H. Ladder Operators for Quantum Systems Confined by Dihedral Angles. *SIGMA* **2012**, *8*, 060.
15. Jaffé, G. Z. On the Theory of Hydrogen Molecular Ion. *Z. Phys.* **1934**, *87*, 535.
16. Abramowitz, M.; Stegun, I. A. *Handbook of Mathematical Functions*. Dover: New York, 1965.
17. Erikson, H. A.; Hill, E. L. A Note on the One-Electron States of Diatomic Molecules. *Phys. Rev.* **1949**, *75*, 29.
18. Hunter, G.; Pritchard, H. O. Born-Oppenheimer Separation for Three-Particle Systems. II. Two-Center Wavefunctions. *J. Chem. Phys.* **1967**, *46*, 2146.
19. Dunham, J. L. The Energy Levels of a Rotating Vibrator. *Phys. Rev.* **1932**, *41*, 721.
20. Chaos-Cador, L.; Ley-Koo, E. Two-Dimensional Hydrogen Atom Confined in Circles, Angles, and Circular Sections. *Int. J. Quantum Chem.* **2005**, *103*, 369.
21. Bates, D. R.; Carson, T. R. Exact Wave Functions of HeH^{2+}. *Proc. R. Soc. Lond.A* **1956**, *234*, 207.

22. Ben-Itzhak, I.; Gertner, I.; Heber, O.; Rosner, B. Experimental Evidence for the Existence of the 2pσ Bound State of HeH^{2+} and Its Decay Mechanism. *Phys. Rev. Lett.* **1993**, *71*, 1347.
23. Ben-Itzhak, I.; Chen, Z.; Esry, B. D.; Gertner, I.; Heber, O.; Lin, C. D.; Rosner, B. Mean Lifetime of the Bound 2pσ State of HeH^{2+}. *Phys. Rev. A* **1994**, *49*, 1774.
24. Ley-Koo, E. The Hydrogen Atom Confined in Semi-infinite Spaces Limited by Conoidal Boundaries. *Adv. Quantum Chem.* **2009**, *57*, 79.
25. Esry, B. D.; Sadeghpour, H. R. Adiabatic Formulation of Heteronuclear Hydrogen Molecular Ion. *Phys. Rev. A* **1999**, *60*, 3604.

CHAPTER SIX

Angular Momentum Theory in Bases of Lamé Spheroconal Harmonics

Ricardo Méndez-Fragoso*[,1], Eugenio Ley-Koo†
*Facultad de Ciencias, Universidad Nacional Autónoma de México, Circuito Exterior S/N Delegación Coyoacán, Ciudad Universitaria, México
†Instituto de Física, Universidad Nacional Autónoma de México, México D.F., México
[1]Corresponding author: e-mail address: rich@ciencias.unam.mx

Contents

Abstract

This contribution reviews and illustrates some of the key concepts of angular momentum theory in bases of Lamé spheroconal harmonics, which has been developed in connection with research about the free rotations of asymmetric molecules. The review starts from the reparametrization of the Hamiltonian of the physical system, separating it into its spherical rotor and asymmetry distribution Hamiltonian, including their symmetries and commutativities; the symmetries of their common eigenfunctions and

Advances in Quantum Chemistry, Volume 71
ISSN 0065-3276
http://dx.doi.org/10.1016/bs.aiq.2015.02.003

eigenenergies are also identified and explicitly illustrated, as functions of the asymmetry parameters of the molecule. The review also covers the identification of three sets of ladder operators for the rotational eigenstates of molecules with any asymmetry, and of four-step ladder operators for the most symmetric eigenstates in four different species of the most asymmetric molecules. Our most recent work illustrated the combined use of the four-step ladder operators and the cartesian components of the angular momentum operators to connect all the eigenstates of the most asymmetric molecules within each of the four different species. The reviews of the last two paragraphs include also pending matters in our previous works, for which the present contribution contains the answers: the matrix elements of the angular momentum operators between the rotational normalized eigenstates (1) with the same eigenvalue of angular momentum and different species; (2) with neighboring values of angular momentum $\ell' = \ell \pm 1$, different parities and species, connected by the linear momentum operators, for molecules of any asymmetry including the most asymmetric.

1. INTRODUCTION

This is the written version of our contribution to the Festschrift in honor of Professor Frank Harris, preceded by the Workshop Concepts of Mathematical Physics in Chemistry, held in Playa del Carmen, Quintana Roo, México in the second week of December 2014. The relevant references go back to the pioneering work of Kramers and Ittmann,[1] who showed that the Schrödinger equation for the rotations of free asymmetric molecules involve exact solutions in terms of Lamé spheroconal harmonic polynomials; and to the work of Patera and Winternitz on the representations of the rotational group using Lamé and Heun polynomials.[2] The field of rotational molecular spectroscopy did not develop within the framework of Kramers and Ittmann,[1] because at that time the numerical evaluation of Lamé polynomials was not sufficiently developed; instead, the development took place along the perturbation theory approach from the starting points of the prolate and oblate symmetric molecules.[3] More recently, Piña[4] and Valdés & Piña[5] made use of the separation of the Hamiltonian of asymmetric molecules into its spherical rotor and asymmetric Hamiltonian contributions, in order to study some properties of the spectra of rotational spectra of asymmetric molecules and, in particular, the most asymmetric molecules; for the latter, Piña[4] identified that the eigenstates of species [1], [y], [xz] and [xyz] with vanishing asymmetry distribution energies involve Lamé polynomials that become Jacobi polynomials, for which another previous work of the same author[6] had constructed ladder operators allowing the four-step changes of angular momentum among the states with $\ell = 4n$, $4n + 1$, $4n + 2$ and $4n + 3$, with $n = 0, 1, 2, \ldots$ respectively.

Our own works[7,8] were initially motivated by reference[4] and implemented within its framework of separating the Hamiltonian into its spherical rotor and asymmetry distribution contributions, with the objective of developing and implementing methods of reliable evaluations of the Lamé spheroconal polynomials for the characterization of the rotational states of any asymmetry. The first one[7] introduced the matrix evaluation of the common eigenfunctions and eigenenergies of the three Hamiltonians by constructing the matrix of the asymmetry-distribution Hamiltonian in the alternative bases of spherical harmonics with rotational symmetries around the axes x, y or z; such a matrix for each value of the angular momentum label ℓ is of finite size $(2\ell \times +1) \times (2\ell + 1)$ and tridiagonal, due to the quadratic dependence of the Hamiltonian on the components of the angular momentum. Under the diagonalization of such matrices the eigenvectors and eigenenergies are obtained for the eigenstates of the successive species: [1], [xy], [xz], [yz] with positive parities for ℓ even in numbers of $(\ell/2) + 1$ for the first, $\ell/2$ for the last three, and a total $2\ell + 1$; [x], [y], [z], [xyz] with negative parity for ℓ odd in numbers of $(\ell + 1)/2$ for the first three, $(\ell - 1)/2$ for the last one, and again a total of $2\ell + 1$. The second reference[8] was implemented by using spheroconal coordinates, showing the separation of the Schrödinger equation for the asymmetry-distribution Hamiltonian, provided the geometric parameters of the coordinates have the proper connection with the dynamical asymmetry-distribution parameters, into Lamé ordinary differential equations in the respective elliptical-cone coordinates. The construction of the solutions of the latter involves the identification and separation of their singularities with the complementary polynomial factors. The coefficients in the polynomials satisfy three-term recurrence relations involving the separation constants; such relations can be written in the matrix form of an eigenvalue problem, which after diagonalization yields the respective eigenvalues of the separation constant and eigenvectors as the sets of coefficients of each Lamé polynomial. The relevant equations of this reference are included in Appendix of this manuscript for the readers interested in the details. In the comparison of the two methods of evaluating the rotational states, using spherical harmonics and spheroconal harmonics, the second reference pointed out "it may be worthwhile exploring the extension of the theory of angular momentum based on spheroconal harmonics."

Our subsequent set of investigations dealt successively with the identification and construction of a common generating function for the hydrogen atom wave functions in spherical, spheroconal, parabolic and prolate spheroidal coordinates[9], the hydrogen atom confined in a semi-infinite space with elliptical-cone boundary[10], our contribution to the 50th Sanibel Symposium on "Lamé spheroconal harmonics in atoms and molecules,"[11] and on the rotations of asymmetric molecules and the hydrogen atom in free and confined configurations.[12] The latter was motivated by the previous one, as a review of the set of investigations including the original results for the molecules confined by elliptical-cone boundaries. Its section 4 "On developing the theory of angular momentum in bases of Lamé spheroconal harmonics" identified complete radial and angular momentum raising and lowering operators for a free particle in three dimensions in 4.1) spherical and 4.2.2) spheroconal coordinates, 4.2.1) angular momentum transformations of Lamé spheroconal harmonic polynomials and 4.3) a spheroconal toolbox to continue the constructions of the theory.

Our participation in the International Symposium on Superintegrability, Exact Solvability and Special Functions, motivated us to identify and construct three sets of ladder operators for spheroconal harmonic polynomials.[13] The first set connects rotational eigenstates with the same magnitude of the angular momentum ℓ and the same species $[AB]$, and different and complementary angular excitations $n_1' = n_1 \pm 2$, $n_2' = n_2 \mp 2$, subject to $n_1 + n_2 = n_1' + n_2' = \ell + n^{[AB]}$, $h_{n_1}^A + h_{n_2}^B = h_{n_1'}^A + h_{n_2'}^B = \ell(\ell + 1)$, where $n^{[AB]} = 0, 1, 2, 3$ according to the number of cartesian coordinate factors in AB. The second set connects eigenstates with the same ℓ and different species, corresponding to the cartesian components of the angular momentum operator $\hat{L}_x$, $\hat{L}_y$, $\hat{L}_z$. The third set connects eigenstates with neighboring values of angular momentum, $\ell' = \ell \pm 1$, different parities, kinds and species corresponding to the linear momentum operators $\hat{p}_x$, $\hat{p}_y$, $\hat{p}_z$.

Our recent participation in the "30th International Symposium on Methods of Group Theory in Physics" allowed us to extend the connections of the most symmetric eigenstates of the most asymmetric molecules with Piña's four-step ladder operators, by applying the angular momentum operators $\hat{L}_x$, $\hat{L}_y$, $\hat{L}_z$ to those eigenstates in order to connect them with their companions sharing the same eigenvalues ℓ and changing to the appropriate species, as illustrated in figs. 2 and 3 in Ley-Koo.[14]. Due to space restrictions these illustrations are schematic, and restricted to the connections leading to the 2ℓ companion eigenstates for each values of ℓ. The problem contains more connections deserving a complete analysis, which we have implemented for the present contribution.

The main body of the paper is structured as follows: Section 2 reviews the symmetries in the rotations of the asymmetric molecules at the levels of the complete spherical-rotor and asymmetry-distribution Hamiltonians, their common eigenfunctions, and their eigenenergies, for any asymmetry, including the particular case of the most asymmetric molecules. Section 3 reviews briefly the successive ladder and shift operators, illustrating their actions between rotational eigenstates for molecules of any asymmetry, including the most asymmetric. The actions of the angular momentum operators already discussed in the previous paragraph can be represented in a compact way presenting their matrices in the bases of spheroconal harmonics. The actions of the linear momentum operators on the successive eigenstates with $\ell = 0, 1, 2, 3, 4$ illustrate the connections between eigenstates of neighboring columns in the fig. 3 of Ley-Koo,[14], including the equivalent connections à la Piña. Section 4 discusses the new results presented in this contribution in connection with the development of the theory of angular momentum. Appendix includes the explicit spheroconal descriptions of the positions, linear and angular momentum, and Hamiltonian operators. The main text makes reference to them where appropriate.

2. SYMMETRIES IN THE ROTATIONS OF ASYMMETRIC MOLECULES

This section reviews successively the symmetries in the rotations of asymmetric molecules at the levels of their Hamiltonians, their eigenfunctions and their eigenenergies as functions of their asymmetry distribution parameters, including their manifestations for the most asymmetric molecules.

2.1 The Complete Hamiltonian and Its Separation in the Spherical-Rotor and Asymmetry-Distribution Contributions

The Hamiltonian for a molecule in the body-fixed frame of reference, with moments of inertia $I_1 < I_2 < I_3$ relative to the principal axes, has the form[4]:

$$\hat{H} = \frac{1}{2}\left(\frac{\hat{L}_x^2}{I_1} + \frac{\hat{L}_y^2}{I_2} + \frac{\hat{L}_z^2}{I_3}\right) \tag{1}$$

Piña chose the reparametrization

$$\frac{1}{I_i} = Q + Pe_i \tag{2}$$

in which,

$$Q = \frac{1}{3}\left(\frac{1}{I_1} + \frac{1}{I_2} + \frac{1}{I_3}\right) \tag{3}$$

is the average of the trace of the matrix of inverses of the moments of inertia,

$$P^2 = \frac{2}{9}\left[\left(\frac{1}{I_1} - \frac{1}{I_2}\right)^2 + \left(\frac{1}{I_1} - \frac{1}{I_3}\right)^2 + \left(\frac{1}{I_2} - \frac{1}{I_3}\right)^2\right] \tag{4}$$

defines the magnitude of the asymmetry P of the molecule; and the asymmetry distribution parameters e_i are such that $e_1 > e_2 > e_3$, following the ordering of the moments of inertia,

$$e_1 + e_2 + e_3 = 0 \text{ and } e_1^2 + e_2^2 + e_3^2 = 3/2. \tag{5}$$

Notice that the last two equations indicate that only one of the e_is can be chosen independently; the reparametrization goes from three moments of inertia to Q, P and one of the e_is.

The Hamiltonian takes the new form

$$\hat{H} = \frac{1}{2}Q\hat{L}^2 + \frac{1}{2}P\left(e_1L_x^2 + e_2L_y^2 + e_3L_z^2\right) = \frac{1}{2}Q\hat{L}^2 + P\hat{H}^* \tag{6}$$

showing its separation into its first term, as the spherical-rotor Hamiltonian, and its second term the asymmetry-distribution Hamiltonian involving the product of the magnitude of the asymmetry and

$$\hat{H}^* = \frac{1}{2}\left(e_1L_x^2 + e_2L_y^2 + e_3L_z^2\right) \tag{7}$$

with only one independent parameter e_i. This can be contrasted with the dependence on three independent parameters of the complete Hamiltonian, as used by Kramers and Ittmann.[1]

The square of the angular momentum operator commutes with any of its cartesian components $\hat{L}_i$, $i = x, y, z$, as well as with their squares. Correspondingly, it also commutes with $\hat{H}$ and $\hat{H}^*$:

$$\left[\hat{L}^2, \hat{H}\right] = 0, \ \left[\hat{L}^2, \hat{H}^*\right] = 0, \tag{8}$$

The complete, spherical-rotor and asymmetry-distribution Hamiltonians also commute by pairs:

$$\left[\hat{H}^Q, \hat{H}\right] = 0, \ \left[\hat{H}^Q, \hat{H}^*\right] = 0, \ \left[\hat{H}^*, \hat{H}\right] = 0. \tag{9}$$

Consequently, the three Hamiltonians share common eigenfunctions. The eigenfunctions of $\hat{H}^Q$ are eigenfunctions of $\hat{L}^2$ and have the spectrum $\frac{\hbar^2}{2} Q\ell(\ell+1)$, $\ell = 0, 1, 2, \ldots$ The common eigenfunctions of $\hat{L}^2$ and $\hat{H}^*$ are analyzed in the next section. Since the Hamiltonians are quadratic in the components of the angular momentum, they are invariant under the parity transformations $x \rightarrow -x$, $y \rightarrow -y$, $z \rightarrow -z$ and also commute with each of those operations. As a consequence, they also share eigenfunctions with the respective parities.

The restrictions on the asymmetry-distribution parameters in Eq. (5) suggest an alternative parametrization in terms of an angle $0^\circ < \sigma < 60^\circ$:

$$e_1 = \cos\sigma, \ e_2 = \cos(\sigma - 120^\circ), \ e_3 = \cos(\sigma + 120^\circ) \tag{10}$$

A Mercedes-Benz logo, Fig. 1, symbolic of one of our birthday presents for Frank, illustrates the values of the asymmetry distribution parameters for symmetric prolate molecules $e_1 = 1$, $e_2 = e_3 = -1/2$, $\sigma = 0^\circ$; symmetric oblate molecules $e_1 = e_2 = 1/2$, $e_3 = -1$, $\sigma = 60^\circ$; and for the most asymmetric molecules $e_1 = \sqrt{3}/2, e_2 = 0, e_3 = -\sqrt{3}/2, \sigma = 30^\circ$, half way angularly between the two previous extreme positions.

When the asymmetry distribution Hamiltonian in Eq. (7) is considered for the asymmetry parameters σ and $60^\circ - \sigma$, it may be established that

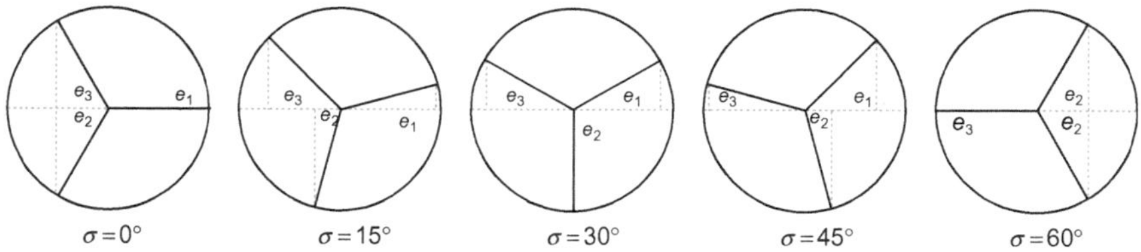

Figure 1 Angular positions of asymmetry distribution circle.

$e_1(60° - \sigma) = -e_3(\sigma)$, $e_2(60° - \sigma) = -e_2(\sigma)$, $e_3(60° - \sigma) = -e_1(\sigma)$, i. e., the parameters e_i for "complementary" angles are the negatives of the values of the original ones with the exchange between $1 \rightleftarrows 3$ and $2 \to 2$. If these transformations are accompanied with the exchange of $\hat{L}_x^2$ and $\hat{L}_{x'}^2$, and $\hat{L}_y^2 \to \hat{L}_y^2$ it follows that $\hat{H}^*(60° - \sigma) = -\hat{H}^*(60°)$. Some of the consequences of this symmetry of the asymmetry distribution Hamiltonian on its eigenfunctions and eigenenergies are described and illustrated in the following sections.

2.2 Common Eigenfunctions of the Square of Angular Momentum and Asymmetry-Distribution Hamiltonian

Our second symbolic gift for Frank consisted of spheroconal harmonic balloons and flowers, as depicted in Fig. 2.[8] The interested readers may find in the reference and in Appendix of this manuscript the mathematical details about the construction and classification of the common eigenfunctions of $\hat{L}^2$ and $\hat{H}^*$.

Here we limit ourselves to point out that under the exchange of the elliptical cone coordinates and their parameters χ_i, k_i^2 with $i = 1, 2$: $1 \rightleftarrows 2$, the cartesian coordinates, Eq. (A.1), the linear momentum operators, Eq. (A.8), and the angular momentum operators, Eq. (A.12), behave in the respective ways $x \rightleftarrows z$, $y \to y$; $\hat{p}_x \rightleftarrows \hat{p}_z$, $\hat{p}_y \to \hat{p}_y$; $\hat{L}_x \rightleftarrows -\hat{L}_z$, $\hat{L}_y \to -\hat{L}_y$.

The separability of the two eigenvalue equations requires the condition that the geometric parameters of the spheroconal coordinates and the asymmetric distribution parameters of the molecule are related by Eq. (A.16). This means that the geometry of the coordinates is tailored to fit the asymmetry of the molecule. The geometric parameters for the prolate and oblate symmetric molecules are $(k_1^2 = 0, k_2^2 = 1)$ and $(k_1^2 = 1, k_2^2 = 0)$, respectively; and for the most asymmetric molecules $(k_1^2 = k_2^2 = 1/2)$, also halfway between the others. The parameters for any given molecule are restricted by the condition $k_1^2 + k_2^2 = 1$, needed for the sum of the squares of x, y, z in Eq. (A.1) to add up to the square of the radial coordinate. The exchange $k_1^2 \rightleftarrows k_2^2$ goes together with the changes $e_1 \rightleftarrows -e_3$, $e_2 \to -e_2$.

Now, we are in the position to describe Fig. 2, which contains the eleven spheroconal harmonics for $\ell = 5$: three of each species u_x, u_y and u_z, and two of species $u_x u_y u_z$ displayed from bottom to top in that order in succession; it also corresponds to the order of increasing values of their eigenenergies E^*. Their variations from left to right illustrate the effects of the change in the asymmetry distribution for different molecules. Starting from the

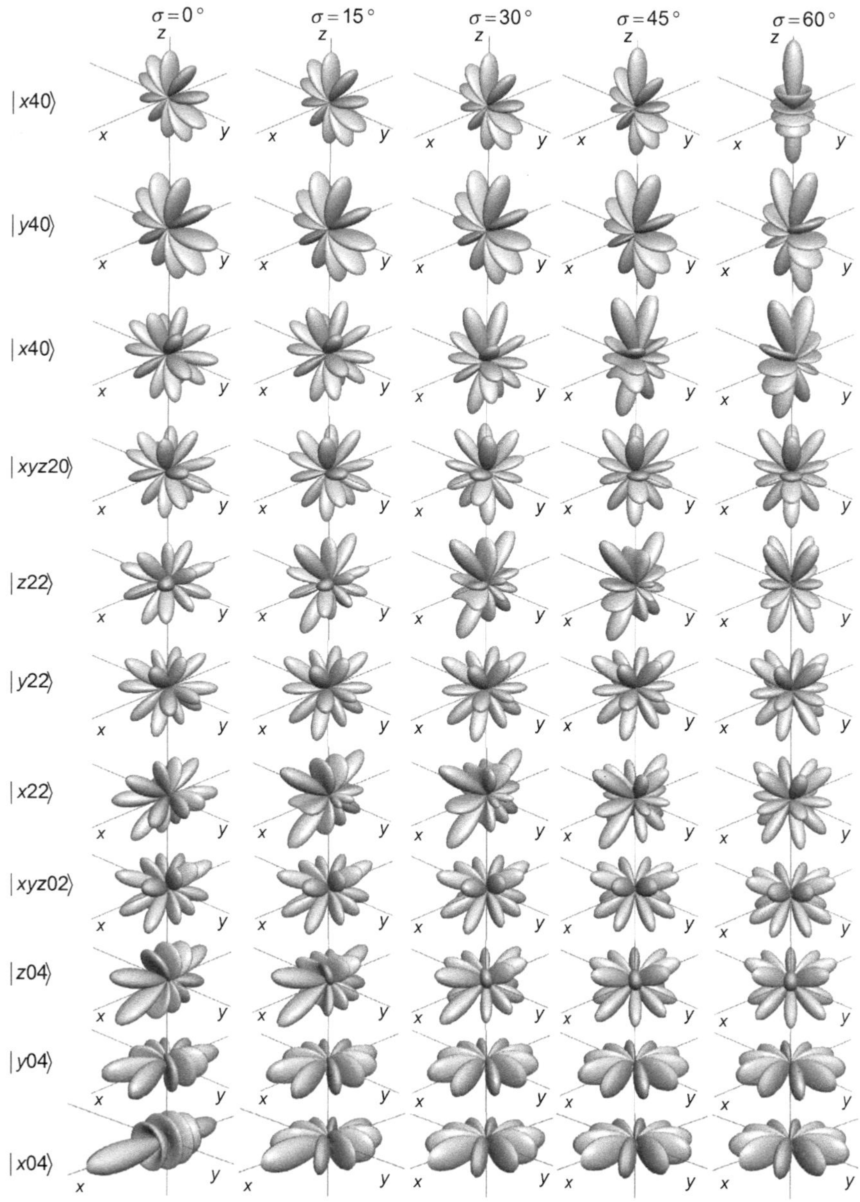

Figure 2 Polar graphs of rotational eigenfunctions with $\ell = 5$.

outer columns on the left and on the right, corresponding to the symmetric prolate and oblate molecules, the reader may recognize the familiar spherical harmonics shapes: from balloon shape with the x-axis symmetry at the bottom on the left, the successive changes along the way going up, until the flower shape at the top is reached. Jumping to the extreme right at the top, the balloon shape is met again with the z-axis of symmetry, and going down the same shapes as before are found successively, including the flower shape around the xy-plane. The most asymmetric molecule belongs in the middle column; starting from the bottom and top flowers with the same shape and different orientations, the reader may go up and down to identify the same characteristics in each successive pair of eigenfunctions, until the center of the figure is reached for the most symmetric eigenstate of the most asymmetric molecule. The reader may also compare the same correspondences in shape and orientations for the "complementary" asymmetries for molecules going up the second column and down for the fourth column, with $\sigma = 15°$ and $45°$, respectively. All the spheroconal harmonics exhibit five nodal surfaces: for the symmetric molecules there are combinations of circular cones, including equatorial planes, and meridian planes; for the asymmetric molecules they are elliptic cylindrical cones, including cartesian coordinate planes determinated by their respective parities, behind their classification in species.

2.3 Asymmetry-Distribution Energies as Functions of the Asymmetry Distribution Parameter

This section illustrates graphically the symmetries of the asymmetry-distribution eigenenergies $E^{*\ell[AB]}_{n_1 n_2}$ as functions of the asymmetry-distribution parameter. We do it also borrowing some data from our previous work,[7] appearing here in Table 1 and Fig. 3. The symmetries are naturally connected with those of the Hamiltonian and the associated eigenfunctions.

The analysis of the data in Table 1 follows the same order as the one of the previous section. In fact, the left and right side columns correspond to the symmetric prolate and oblate molecules. For the prolate case, the lowest energy state at the bottom is non degenerate at -7.5, and the five following pairs, going up, are degenerate. For the oblate case, the top energy state is non degenerate at 7.5, and the five following pairs, going down, are degenerate sharing the eigenvalues with same magnitudes and opposite signs as their counterparts of the prolate molecules. It is interesting to follow the changes of the eigenenergies as the asymmetry parameter changes from $\sigma = 0°$ to $\sigma = 60°$. For $\sigma = 15°$, notice the removal of the degeneracies and change of the eigenenergies, going up, compared to the prolate case, and similarly, for $\sigma = 45°$ going down compared with the oblate case; also notice the same number in both columns with their opposite signs, on the

Table 1 Energy Eigenvalues for $\ell = 5$ of Molecules with Different Values of the Parameter σ

σ	0°	15°	30°	45°	60°
$\|ABn_1n_2\rangle$					
$\|z40\rangle$	11.25	10.91578693906	9.954222983497	8.536090033911	7.5
$\|y40\rangle$	11.25	10.91568701090	9.949874371066	8.474382041736	6.75
$\|x40\rangle$	4.5	4.508041166928	4.644434067912	5.303300858899	6.75
$\|xyz20\rangle$	4.5	4.5	4.5	4.5	4.5
$\|z22\rangle$	−0.75	−0.30918522126	1.185401612798	3.216394830184	4.5
$\|y22\rangle$	−0.75	−0.50016213090	0	0.50016213090	0.75
$\|x22\rangle$	−4.5	−3.216394830184	−1.185401612798	0.30918522126	0.75
$\|xyz02\rangle$	−4.5	−4.5	−4.5	−4.5	−4.5
$\|z04\rangle$	−6.75	−5.303300858899	−4.644434067912	−4.508041166928	−4.5
$\|y04\rangle$	−6.75	−8.474382041736	−9.949874371066	−10.91568701090	−11.25
$\|x04\rangle$	−7.5	−8.536090033911	−9.954222983497	−10.91578693906	−11.25

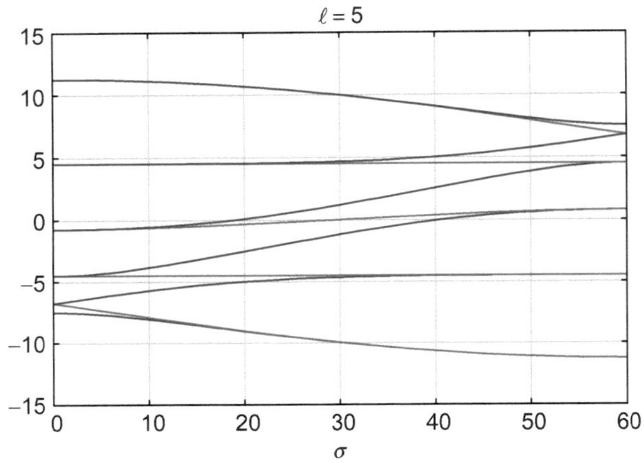

Figure 3 Eigenenergy lines as functions of the asymmetry parameter σ.

way up on the left and down on the right. For $\sigma = 30^\circ$, the most asymmetric molecules, notice the vanishing eigenenergy of the most symmetric state at the middle, and its equidistant pairs of energy levels below and above with energies of the same magnitude and opposite sign [*ds*]22 & [*sd*]22, [*dcsscd*] 02 & [*dcsscd*]20, [*ds*]04 & [*sd*]40, [*cc*]04 & [*cc*]40, and [*sd*]04 & [*ds*]40, respectively. When the five columns and eleven rows of eigenenergies are considered together their antisymmetry relative to $E^*_{22}(\sigma = 30^\circ) = 0$ is recognized. Of course, it is an illustration of the connection of the symmetry of the Hamiltonians $\hat{H}^*_{x \rightleftarrows z}(60^\circ - \sigma) = -\hat{H}^*(\sigma)$. Going back from the middle to the right and to the left, we can follow also the restoration of the double degeneracies in the oblate and prolate molecules, respectively.

Figure 3 displays the data of the table in graphical form. The description of the symmetry exhibited by the eigenenergy lines can be stated very concisely: They are antisymmetric under inversion relative to the point $\sigma = 0^\circ$, $E^* = 0$, in agreement with the analysis of the previous paragraph. Equivalently, rotation of the figure itself by 180° around the same point is self-mapping.

Here the illustration has been implemented for $\ell = 5$, but its conclusion is general. The interested reader may find the figures for other values of ℓ in Ref. 7.

2.4 The Most Symmetric Eigenstates of the Most Asymmetric Molecules

The last section of Piña's reference,[4] with the title "The most asymmetric molecule" was focused in the most symmetric eigenstates of such a molecule with vanishing asymmetry distribution energy. In fact, the solutions of the Lamé ordinary differential equation for such cases after removal of the respective singularity factors become Jacobi polynomials, represented in that reference as $f_n(\alpha, \gamma, z^2)$ in its Eqs. (102)–(106). The subsequent Eqs. (107)–(108) describe the four-step ladder operators[6] connecting the successive eigenstates $\Lambda_{\ell+4,0}$ and $\Lambda_{\ell,0}$ where ℓ is the angular momentum quantum number and $E^* = 0$ refers to the vanishing value of the asymmetry-distribution energy.

Our comments about such eigenstates is motivated by our need to use them explicitly in our recent work,[14] and in the present one. First, we prefer to use the hypergeometric function representation for the Jacobi polynomials, ${}_2F_1(-n, n+\alpha; \gamma; z^2)$, which allows its direct evaluation[15] as illustrated in Table 2.

Second, for the most asymmetric molecules, with parameters $e_1 = -e_3 = \sqrt{3}/2$ and $e_2 = 0$, Eqs. (98)–(100) in Piña[4] involve the variables $x = e_3 + (e_2 - e_3)\text{sn}^2(\chi_1|k_1^2) = \sqrt{3}z/2$, and parameters $2E^*/\hbar^2 = -\sqrt{3}b/2$. It must be recognized that for the most asymmetric molecules the relevant variable is $z^2 = 1 - \text{sn}^2(\chi_1|k_1^2) = \text{cn}^2(\chi_1|k_1^2)$, and $b = 0$ for the most symmetric eigenstates.

We can also add our own version of such eigenstates. In fact, the vanishing of the asymmetry-distribution energy

$$E^* = e_1 h_{n_1}^A + e_3 h_{n_2}^B = \frac{\sqrt{3}}{2}\left(h_{n_1}^A - h_{n_2}^B\right) \tag{11}$$

for the most asymmetric molecules demands the equality of the complementary eigenvalues of the Lamé equation $h_{n_1}^A = h_{n_2}^B$, which in turn requires the

Table 2 Jacobi Polynomials from Hypergeometric Function Representation

n	$\ell = 4n$, $\alpha = \frac{1}{4}$ $\gamma = \frac{3}{4}$	$\ell = 4n+1$, $\alpha = \frac{3}{4}$ $\gamma = \frac{5}{4}$	$\ell = 4n+2$, $\alpha = \frac{5}{4}$ $\gamma = \frac{3}{4}$	$\ell = 4n+3$, $\alpha = \frac{7}{4}$ $\gamma = \frac{5}{4}$
0	1	1	1	1
1	$1 - \frac{5}{3}z^4$	$1 - \frac{7}{5}z^4$	$1 - 3z^4$	$1 - \frac{11}{5}z^4$
2	$1 - 6z^4 + \frac{39}{7}z^8$	$1 - \frac{22}{5}z^4 + \frac{11}{3}z^8$	$1 - \frac{26}{3}z^4 + \frac{13\cdot 17}{3\cdot 7}z^8$	$1 - 6z^4 + \frac{19}{3}z^8$
3	$1 - 13z^4 + \frac{13\cdot 17}{7}z^8 - \frac{13\cdot 17}{11}z^{12}$	$1 - 9z^4 + 19z^8 - \frac{3\cdot 19\cdot 23}{9\cdot 13}z^{12}$	$1 - 17z^4 + 17z^8 - 17\cdot 25z^{12}$	$1 - \frac{3\cdot 19}{5}z^4 + \frac{19\cdot 23}{15}z^8 - \frac{3\cdot 19\cdot 23}{5\cdot 13}z^{12}$

equalities of the species $[A] = [B]$ and the degrees of excitations $n_1 = n_2$. The corresponding eigenvalues, satisfying the restriction $h_{n_1}^A + h_{n_2}^B = \ell(\ell+1)$, are simply $h_n^A = \ell(\ell+1)/2$. For the species [1], $[y = cc]$, $[xz = dssd]$ and $[xyz = dcsscd]$ and the respective values of $\ell = 4n, 4n+1, 4n+2, 4n+3$ for a chosen value of n, the number of nodes is also $\ell = n_1 + n_2 + n^{AB} = 4n + n^{AB}$; therefore, the common degree of excitation is $n_1 = n_2 = 2n$, as identified in figs. 2 and 3 of Ley-Koo.[14] The eigenfunctions themselves can be generated from the recurrence relations of Eqs. (32), (34), (37), and (39) of Ley-Koo and Méndez-Fragoso.[8] Table 3 illustrates such Lamé polynomials as the counterparts of those in Table 2. The reader may ascertain their equivalences except for a normalization constant.

Here we also cite Eqs. (107) and (108) from Ref. 4 for the four-step ladder operators

$$\left[2z(z^2-1)\frac{\mathrm{d}}{\mathrm{d}z} - (\ell+4)\left(z^2 - \frac{\ell+2}{2\ell+5}\right)\right]\Lambda_{\ell+4,0} = -\frac{\sqrt{(\ell+1)(\ell+2)(\ell+3)(\ell+4)}}{2\ell+5}\Lambda_{\ell,0}, \tag{12}$$

Table 3 Lamé Polynomials for the Most Symmetric Eigenstates of the Most Asymmetric Molecules

n	$\ell = 4n[1]$	$\ell = 4n+1[y]$	$\ell = 4n+2[xz]$	$\ell = 4n+3[xyz]$
0	1	1	1	1
1	$1-5\mathrm{sn}^2(\chi_i)$ $+\frac{5}{2}\mathrm{sn}^4(\chi_i)$	$1-7\mathrm{sn}^2(\chi_i)$ $+\frac{7}{2}\mathrm{sn}^4(\chi_i)$	$1-3\mathrm{sn}^2(\chi_i)$ $+\frac{3}{2}\mathrm{sn}^4(\chi_i)$	$1-\frac{11}{3}\mathrm{sn}^2(\chi_i)$ $+\frac{11}{6}\mathrm{sn}^4(\chi_i)$
2	$1-18\mathrm{sn}^2(\chi_i)$ $+48\mathrm{sn}^4(\chi_i)$ $-39\mathrm{sn}^6(\chi_i)$ $+\frac{39}{4}\mathrm{sn}^8(\chi_i)$	$1-22\mathrm{sn}^2(\chi_i)$ $+66\mathrm{sn}^4(\chi_i)$ $-55\mathrm{sn}^6(\chi_i)$ $+\frac{55}{4}\mathrm{sn}^8(\chi_i)$	$1-\frac{26}{3}\mathrm{sn}^2(\chi_i)$ $+\frac{11\cdot 26}{3\cdot 5}\mathrm{sn}^4(\chi_i)$ $-\frac{13\cdot 17}{3\cdot 5}\mathrm{sn}^6(\chi_i)$ $+\frac{13\cdot 17}{3\cdot 4\cdot 5}\mathrm{sn}^8(\chi_i)$	$1-10\mathrm{sn}^2(\chi_i)$ $+24\mathrm{sn}^4(\chi_i)$ $-19\mathrm{sn}^6(\chi_i)$ $+\frac{19}{4}\mathrm{sn}^8(\chi_i)$
3	$1-39\mathrm{sn}^2(\chi_i)$ $+\frac{13\cdot 37}{2}\mathrm{sn}^4(\chi_i)$ $-\frac{12\cdot 13\cdot 17}{5}\mathrm{sn}^6(\chi_i)$ $+\frac{13\cdot 17\cdot 47}{20}\mathrm{sn}^8(\chi_i)$ $-\frac{3\cdot 7\cdot 13\cdot 17}{20}\mathrm{sn}^{10}(\chi_i)$ $+\frac{7\cdot 13\cdot 17}{40}\mathrm{sn}^{12}(\chi_i)$	$1-45\mathrm{sn}^2(\chi_i)$ $+\frac{15\cdot 41}{2}\mathrm{sn}^4(\chi_i)$ $-7\cdot 41\cdot 19\mathrm{sn}^6(\chi_i)$ $+3\cdot 16\cdot 19\mathrm{sn}^8(\chi_i)$ $-\frac{3\cdot 19\cdot 23}{4}\mathrm{sn}^{10}(\chi_i)$ $+\frac{19\cdot 23}{8}\mathrm{sn}^{12}(\chi_i)$	$1-17\mathrm{sn}^2(\chi_i)$ $+\frac{17\cdot 47}{10}\mathrm{sn}^4(\chi_i)$ $-\frac{2\cdot 17\cdot 23}{5}\mathrm{sn}^6(\chi_i)$ $+\frac{3^2\cdot 17\cdot 19}{4\cdot 5}\mathrm{sn}^8(\chi_i)$ $-\frac{3\cdot 15\cdot 17}{4}\mathrm{sn}^{10}(\chi_i)$ $+\frac{5\cdot 17}{8}\mathrm{sn}^{12}(\chi_i)$	$1-19\mathrm{sn}^2(\chi_i)$ $+\frac{3\cdot 17\cdot 19}{10}\mathrm{sn}^4(\chi_i)$ $-\frac{2^4\cdot 19\cdot 23}{5\cdot 7\cdot 8}\mathrm{sn}^6(\chi_i)$ $+\frac{3\cdot 19\cdot 23\cdot 61}{5\cdot 7\cdot 8}\mathrm{sn}^8(\chi_i)$ $-\frac{3^3\cdot 19\cdot 23}{2^2\cdot 5\cdot 7}\mathrm{sn}^{10}(\chi_i)$ $+\frac{3^2\cdot 19\cdot 23}{2^3\cdot 7\cdot 5}\mathrm{sn}^{12}(\chi_i)$

$$\left[2z(z^2-1)\frac{\mathrm{d}}{\mathrm{d}z}+(\ell+1)\left(z^2-\frac{\ell+3}{2\ell+5}\right)\right]\Lambda_{\ell,0}=\frac{\sqrt{(\ell+1)(\ell+2)(\ell+3)(\ell+4)}}{2\ell+5}\Lambda_{\ell+4,0}, \tag{13}$$

useful for the readers interested in their lowering and raising actions connecting the most symmetric eigenstates in the entries of the Tables 2 and 3.

3. ACTIONS OF OPERATORS OF ANGULAR MOMENTUM AND LINEAR MOMENTUM ON ROTATIONAL EIGENSTATES

This section is motivated by some questions left open in our most recent work,[14] and for which we have constructed some illustrative and quantitative answers. In fact, Fig. 3 in that work were intended to illustrate how angular momentum operators, applied to the most symmetric eigenstates of the most asymmetric molecules, lead to their 2ℓ companion eigenstates of the same and other species and excitations with the common value ℓ. The calculations themselves, as already performed in Ref. 13 for the lower values of ℓ and any asymmetry, indicate that the connections involve not only the individual eigenstate of our choice, but also all its companions of the same species. Sections 3.2 and 3.3 address this question, starting from an analysis of the angular momentum operators and their actions on an eigenstate of a specific species leading to all the eigenstates of another well defined species, guiding the reader along the successive steps of the calculations, and using the matrix representation of the operators in the spheroconal harmonic basis to exhibit quantitatively the connected eigenstates, including their proportions. On the other hand, the penultimate sentence in Ref. 14 suggested how four successive applications of the linear momentum operators on the most symmetric eigenstates of the most asymmetric molecules should have an action equivalent to Piña's four-step ladder operators. Here we illustrate the schematic connections for the first tier of eigenstates and on to the next tier.

Of course, the connections can be continued for the successive values of ℓ. Again, the connections occur not only for the specific eigenstate of our choice but for all companion eigenstates of the same species. Sections 3.1 and 3.4 are the counterparts of 3.2 and 3.3 for the actions and matrix representations of the linear momentum operators.

While the $\hat{L}_i$ ladder operators produce vertical displacements within each column and tier in Fig. 3 of Ley-Koo,[14] the $\hat{p}_i$ shift operators produce horizontal displacements connecting states of neighboring columns, including when they are in the next tier, above or below. It is also appropriate to dedicate this "QUANTUM LEGO" set of ladder and shift operators to Frank.

3.1 Structure and Actions of Linear Momentum Operators

The cartesian components of the linear momentum operators of Eq. (A.8) in spheroconal coordinates contain the following common elements: the angular part of the square of the scale factor in the denominator of the common factor; the terms in the first order partial derivatives with respect to the radial coordinate with the director cosines of the radial unit vector as coefficients, and with respect to χ_1 and χ_2, with coefficients corresponding to the components of the unit vectors $\hat{\chi}_1$ and $\hat{\chi}_2$ along the respective cartesian unit vectors Eq. (A.6), including their scale factors which have already been factored out. The rotational eigenfunctions are the products of the singularity removing factors $A(\chi_1)B(\chi_2)$ and their associated Lamé polynomials $\Lambda^A_{n_1}(\chi_1)\Lambda^B_{n_2}(\chi_2)$ with even power[s] of sn $^2(\chi_i)$.

Now we proceed to describe the actions of the successive terms of the operators on the successive factors of the eigenfunctions, with special attention on the changes in their angular parts in order to identify the species of the resulting eigenfunctions, as well as their orders of excitations.

The terms with the radial derivative are easier to analyze, since they require a simple multiplication of dn (χ_1)sn (χ_2), cn (χ_1)cn (χ_2) and sn (χ_1)dn (χ_2) by the respective $A(\chi_1)$ and $B(\chi_2)$. Actually, the multiplication can be carried out directly in the cartesian representation with the results of Table 4.

The entries with squares of the cartesian coordinates are equivalent to [1] as species, since they become polynomials in sn $^2(\chi_i)$. The derivatives with

Table 4 Multiplication Table of Cartesian Species

AB	1	x	y	z	xy	xz	yz	xyz
x	x	x^2	xy	xz	x^2y	x^2z	xyz	x^2yz
y	y	xy	y^2	yz	xy^2	xyz	y^2z	xy^2z
z	z	xz	yz	z^2	xyz	xz^2	yz^2	xyz^2

respect to χ_i of the even powers $\mathrm{sn}^2(\chi_1)$ contain the common factor $2\mathrm{sn}(\chi_i)\,\mathrm{cn}(\chi_i)\mathrm{dn}(\chi_i)$. Upon multiplication by the coefficients of the respective derivatives and the original species factors, the respective Tables 5 and 6 can be constructed with the corresponding identifications of the connected species. The entries for $AB = 1$ involve only the product of the common derivative and the components of $\hat{\chi}_1$ and $\hat{\chi}_2$.

Comparison of the Tables 4–6 lead us to the conclusion that the three terms in $\hat{p}_x$, $\hat{p}_y$ and $\hat{p}_z$ lead to the same connections of eigenstates according to their species.

Table 5 Multiplication Table for Contributions of Derivatives with Respect to χ_1 in $\hat{p}_x, \hat{p}_y$, and $\hat{p}_z$ for the Successive Species

AB		**1**	***x***	***y***	***z***	***xy***	***xz***	***yz***	***xyz***
$\hat{p}_x$	χ_1	s^2dc^2	$s^2c^2d^2$	s^2c^2cd	s^2dc^2s	$s^2d^2c^2c$	$s^2d^2c^2s$	s^2dc^2cs	$s^2s^2c^2cs$
	χ_2	s	s^2	sc	sd	s^2c	s^2d	scd	s^2cd
		x	1	xy	xz	y	z	xyz	yz
$\hat{p}_y$	χ_1	s^2d^2c	s^2d^2cd	$s^2d^2c^2$	s^2d^2cs	$s^2d^2c^2d$	s^2d^2cds	$s^2d^2c^2s$	$s^2d^2c^2ds$
	χ_2	c	cs	c^2	cd	s	csd	c^2d	c^2sd
		y	xy	1	yz	x	xyz	z	xz
$\hat{p}_z$	χ_1	c^2d^2s	c^2d^2sd	c^2d^2sc	$c^2d^2s^2$	c^2d^2sdc	$c^2d^2ds^2$	$c^2d^2s^2c$	$c^2d^2s^2cd$
	χ_2	d	ds	dc	d^2	dsc	d^2s	d^2c	d^2cs
		z	xz	yz	1	xyz	x	y	xy

Table 6 Multiplication Table for Contributions of Derivatives with Respect to χ_2 in $\hat{p}_x, \hat{p}_y$, and $\hat{p}_z$ for the Successive Species

AB		**1**	***x***	***y***	***z***	***xy***	***xz***	***yz***	***xyz***
$\hat{p}_x$	χ_1	d	d^2	dc	ds	d^2c	d^2s	dcs	d^2cs
	χ_2	sc^2d^2	$s^2c^2d^2$	sc^2d^2c	sc^2d^2d	$s^2c^2d^2c$	$s^2c^2d^2d$	sc^2d^2cd	$s^2c^2d^2cd$
		x	1	xy	xz	y	z	xyz	yz
$\hat{p}_y$	χ_1	c	cd	c^2	cs	c^2d	cds	c^2s	c^2ds
	χ_2	s^2cd^2	s^2cd^2s	$s^2c^2d^2$	s^2cd^2d	s^2c^2s	s^2csd	$s^2c^2d^2d$	s^2c^2sd
		y	xy	1	yz	x	xyz	z	xz
$\hat{p}_z$	χ_1	s	sd	sc	s^2	sdc	s^2d	s^2c	s^2dc
	χ_2	dc^2s^2	dc^2s^2s	dc^2s^2c	$d^2c^2s^2$	dc^2s^2sc	$d^2c^2s^2s$	$d^2c^2s^2c$	$d^2c^2s^2sc$
		z	xz	yz	1	xyz	x	y	xy

After the common species factors are factorized, the tables also contain the even power factors of the elliptical integral functions in the respective χ_1 and χ_2 coordinates associated with each of the three terms. They still have to be multiplied, including the necessary coefficients and signs, with the two polynomials in the radial term, and the derivative polynomial that was factorized and the other polynomial in the two angular terms.

Now we have to analyze how to identify the target Lamé polynomials of each species and different excitations. The presence of the square of the scale factor in the denominator is the key to find the answer: The polynomial obtained from the combination of the three terms must be divisible by the square of the scale factor, so that the quotient leads in general to a superposition of the Lamé functions of the chosen species with coefficients to be determined. The product of the factors must be equal to the polynomial and the coefficients can be determined by comparing the coefficients of the linearly independent powers of $\mathrm{sn}^{2s}(\chi_1)\mathrm{sn}^{2t}(\chi_2)$. Our experience is that the task can be accomplished by considering only the angular terms; except when starting from the $\ell = 0$ state, which requires the multiplication by x, y and z to reach the eigenstates with $\ell = 1$.

3.2 Structure and Actions of the Angular Momentum Operators

The cartesian components of angular momentum operators in spheroconal coordinates, Eq. (A.12), contain the same elements appearing in the linear momentum ones, but in different combinations. This is easy to understand via Eq. (A.11), which ensures that the radial component does not exist, and the $\hat{\chi}_1$ and $\hat{\chi}_2$ components are exchanged including a change of sign. The square of the scale factor is also a common factor.

Now, the action of the $\hat{L}_i$ operators on the rotational eigenstates leads only to two angular terms; in the first one there is the common factor $2\mathrm{sn}(\chi_1)\mathrm{cn}(\chi_1)\mathrm{dn}(\chi_1)$ from deriving the polynomials in powers of $\mathrm{sn}^2(\chi_1)$; and likewise for the second with $1 \to 2$. Let us proceed to identify the changes in species of the eigenstates in the respective terms following the same methodology of the previous section and leading to Tables 7 and 8.

Comparison between the species of rotational eigenstates connected by the derivative terms for χ_1 and χ_2 elliptical cone coordinates in the

Table 7 Changes of Species Due to Angular Momentum Operators on $A(\chi_1)$

A		**1**	**x**	**y**	**z**	**xy**	**xz**	**yz**	**xyz**
$\hat{L}_x$	χ_1	sd^2c	sd^2cd	sd^2c^2	s^2d^2c	sd^2c^2d	s^2d^2cd	$s^2d^2c^2$	$s^2d^2c^2d$
	χ_2	cd	cds	c^2d	cd^2	sc^2d	cd^2s	c^2d^2	c^2sd^2
		yz	xyz	z	y	xz	xy	1	x
$\hat{L}_y$	χ_1	sc^2d	sc^2d^2	sc^2dc	s^2c^2d	sc^2d^2c	$s^2c^2d^2$	s^2c^2dc	$s^2c^2d^2c$
	χ_2	sd	s^2d	sdc	sd^2	s^2dc	s^2d^2	sd^2c	s^2d^2c
		xz	z	xyz	x	yz	1	xy	y
$\hat{L}_z$	χ_1	s^2cd	s^2cd^2	s^2c^2d	s^2cds	$s^2c^2d^2$	s^2cd^2s	s^2c^2ds	$s^2c^2d^2s$
	χ_2	sc	s^2c	sc^2	scd	s^2c^2	s^2cd	sc^2d	s^2c^2d
		xy	y	x	xyz	1	yz	xz	z

Table 8 Changes of Species Due to Angular Momentum Operators on $B(\chi_2)$

B		**1**	**x**	**y**	**z**	**xy**	**xz**	**yz**	**xyz**
$\hat{L}_x$	χ_1	sc	scd	sc^2	s^2c	sc^2d	s^2cd	s^2c^2	s^2c^2d
	χ_2	s^2cd	$s^{c}ds$	s^2c^2d	s^2cd^2	s^2c^2sd	s^2csd^2	$s^2c^2d^2$	$s^2c^2d^2s$
		yz	xyz	z	y	xz	xy	1	x
$\hat{L}_y$	χ_1	sd	sd^2	scd	s^2d	sd^2c	s^2d^2	s^2dc	s^2d^2c
	χ_2	sc^2d	s^2c^2d	sc^2dc	sc^2d^2	s^2c^2dc	$s^2c^2d^2$	sc^2d^2c	$s^2c^2d^2c$
		xz	z	xyz	x	yz	1	xy	y
$\hat{L}_z$	χ_1	cd	cd^2	c^2d	dcs	c^2d^2	cd^2s	c^2ds	c^2d^2s
	χ_2	scd^2	s^2cd^2	sc^2d^2	scd^2d	$s^2c^2d^2$	s^2cd^2d	sc^2d^2d	$s^2c^2d^2d$
		xy	y	x	xyz	1	yz	xz	z

successive components of the angular momentum are found to be the same, respectively. They also coincide with the connections using the cartesian coordinate representation of the operators and harmonic functions. The information can be presented in a concise graphical manner as in Fig. 4.

From here on, the discussion about completing the evaluation of the actions of the $\hat{L}_i$ operators on the rotational eigenfunctions of a given species is an adaptation of the same methodology at the end of section 3.1. It involves the identification of the polynomial obtained after the factorization of the common species removing singularity factor. Such a polynomial can be written as the product of the square of the scale factor and a linear superposition of the Lamé polynomials of the same species and all the

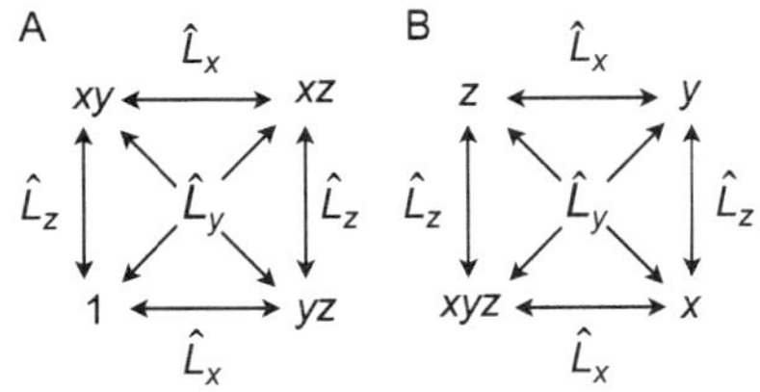

Figure 4 Part (A) involves the even parity states in their four species and part (B) holds for the corresponding species with odd parities.

different excitations. The coefficients in the linear superposition become determined by comparing the polynomial with the product of its factors via the linear independence of the products of even powers of sn (χ_1) and sn (χ_2).

3.3 Angular Momentum Matrices in Spheroconal Harmonic Bases

In our work on ladder operators for Lamé harmonic polynomials,[13] we evaluated explicitly the actions of the angular momentum operators on the $\ell = 1, 2, 3$ eigenstates for any value of the asymmetry distribution parameter. In writing the present contribution we realize that the most precise manner to present that information is in a matrix form, specially in relation with the eigenstates connected by each of the three operators and in what proportions.

First, we illustrate the structure of the successive matrices of $\hat{L}_x$, $\hat{L}_y$ and $\hat{L}_z$ divided by $(-i\hbar)$, indicating with a dark circle, $\bullet$, the position of the connected eigenstates and zeros in the others, complemented by the Tables 9 and 10 of the numerical values of the non zero entries. Some of the circles are replaced with an asterisk, *, for the connections in Fig. 3 of Ley-Koo.[14]

For $\ell = 1$

$$\begin{pmatrix} 0 & 0 & 0 \\ 0 & 0 & -1 \\ 0 & 1 & 0 \end{pmatrix}, \begin{pmatrix} 0 & 0 & -1 \\ 0 & 0 & 0 \\ 1 & 0 & 0 \end{pmatrix}, \begin{pmatrix} 0 & -1 & 0 \\ 1 & 0 & 0 \\ 0 & 0 & 0 \end{pmatrix}$$

For $\ell = 2$.

$$\begin{pmatrix} 0 & 0 & 0 & \bullet & 0 \\ 0 & 0 & -1 & 0 & 0 \\ 0 & 1 & 0 & 0 & 0 \\ \bullet & 0 & 0 & 0 & \bullet \\ 0 & 0 & 0 & \bullet & 0 \end{pmatrix}, \begin{pmatrix} 0 & 0 & * & 0 & 0 \\ 0 & 0 & 0 & -1 & 0 \\ \bullet & 0 & 0 & 0 & \bullet \\ 0 & 1 & 0 & 0 & 0 \\ 0 & 0 & * & 0 & 0 \end{pmatrix}, \begin{pmatrix} 0 & \bullet & 0 & 0 & 0 \\ \bullet & 0 & 0 & 0 & \bullet \\ 0 & 0 & 0 & 1 & 0 \\ 0 & 0 & -1 & 0 & 0 \\ 0 & \bullet & 0 & 0 & 0 \end{pmatrix}$$

For $\ell = 3$

$$\begin{pmatrix} 0 & 0 & 0 & * & 0 & 0 & 0 \\ 0 & 0 & \bullet & 0 & 0 & 0 & \bullet \\ 0 & \bullet & 0 & 0 & 0 & \bullet & 0 \\ \bullet & 0 & 0 & 0 & \bullet & 0 & 0 \\ 0 & 0 & 0 & * & 0 & 0 & 0 \\ 0 & 0 & \bullet & 0 & 0 & 0 & \bullet \\ 0 & \bullet & 0 & 0 & 0 & \bullet & 0 \end{pmatrix}, \begin{pmatrix} 0 & 0 & \bullet & 0 & 0 & 0 & \bullet \\ 0 & 0 & 0 & * & 0 & 0 & 0 \\ \bullet & 0 & 0 & 0 & \bullet & 0 & 0 \\ 0 & \bullet & 0 & 0 & 0 & \bullet & 0 \\ 0 & 0 & \bullet & 0 & 0 & 0 & \bullet \\ 0 & 0 & 0 & * & 0 & 0 & 0 \\ \bullet & 0 & 0 & 0 & \bullet & 0 & 0 \end{pmatrix}, \begin{pmatrix} 0 & \bullet & 0 & 0 & 0 & \bullet & 0 \\ \bullet & 0 & 0 & 0 & \bullet & 0 & 0 \\ 0 & 0 & 0 & * & 0 & 0 & 0 \\ 0 & 0 & \bullet & 0 & 0 & 0 & \bullet \\ 0 & \bullet & 0 & 0 & 0 & \bullet & 0 \\ \bullet & 0 & 0 & 0 & \bullet & 0 & 0 \\ 0 & 0 & 0 & * & 0 & 0 & 0 \end{pmatrix}$$

Table 9 Matrix Elements for $\ell = 2$

(i,j)	$\langle i \vert \hat{L}_x \vert j \rangle$	(i,j)	$\langle i \vert \hat{L}_y \vert j \rangle$	(i,j)	$\langle i \vert \hat{L}_z \vert j \rangle$
1,4	$C^1(2k_2^2- -2\sqrt{1-k_1^2k_2^2})$	1,3	$-2C^1$	1,2	$C^1(2k_1^2+ +2\sqrt{1-k_1^2k_2^2})$
4,1	$-2(1+k_1^2)+ +2\sqrt{1-k_1^2k_2^2}$	3,1	$2(k_1^2-k_2^2)- -4\sqrt{1-k_1^2k_2^2}$	2,1	$2(1+k_2^2)+ +2\sqrt{1-k_1^2k_2^2}$
4,5	$-2(1+k_1^2)- -2\sqrt{1-k_1^2k_2^2}$	3,5	$2(k_1^2-k_2^2)+ +4\sqrt{1-k_1^2k_2^2}$	2,5	$2(1+k_2^2)- -2\sqrt{1-k_1^2k_2^2}$
5,4	$C^1(-2k_1^2- -2\sqrt{1-k_1^2k_2^2})$	5,3	$2C^1$	5,2	$C^1(-2k_1^2+ +2\sqrt{1-k_1^2k_2^2})$

The common factor is $C^1 = -4\sqrt{1-k_1^2k_2^2}$.

Table 10 Matrix Elements of Angular Momentum Operators for $\ell = 3$

(i,j)	$\langle i\|\hat{L}_x\|j\rangle$	(i,j)	$\langle i\|\hat{L}_y\|j\rangle$	(i,j)	$\langle i\|\hat{L}_z\|j\rangle$
1,4	$C^1_{4,1}$	1,3	$C^1_{3,2}$	1,2	$C^1_{2,3}$
2,3	$C^1_{3,1}$	1,7	$C^1_{7,2}$	1,7	$C^1_{6,3}$
2,7	$C^1_{7,1}$	2,4	$C^1_{4,2}$	2,1	$C^1_{1,2}$
3,2	$C^1_{2,1}$	3,1	$C^1_{1,2}$	2,6	$C^1_{5,3}$
3,6	$C^1_{6,1}$	3,5	$C^1_{5,2}$	3,4	$C^1_{4,3}$
4,1	$2\bar{a}^d_1(h^d_0(k^2_1))$	4,2	$2(k^2_1 - k^2_2) - 4\sqrt{4 - k^2_1 k^2_2}$	4,3	$-2\bar{a}^d_1(h^d_2(k^2_2))$
4,5	$2\bar{a}^d_1(h^d_2(k^2_1))$	4,6	$2(k^2_1 - k^2_2) + 4\sqrt{4 - k^2_1 k^2_2}$	4,7	$2\bar{a}^d_1(h^d_0(k^2_2))$
5,4	$C^2_{4,1}$	5,3	$C^2_{3,2}$	5,2	$C^2_{2,3}$
6,3	$C^2_{3,1}$	5,7	$C^2_{7,2}$	5,6	$C^2_{6,3}$
6,7	$C^2_{7,1}$	6,4	$C^2_{4,2}$	6,1	$C^2_{1,3}$
7,2	$C^2_{2,1}$	7,1	$C^2_{1,2}$	6,5	$C^2_{5,3}$
7,6	$C^2_{6,1}$	7,5	$C^2_{5,2}$	7,4	$C^2_{4,3}$

The explicit expressions for the entries in this table are found for any asymmetry in Ref. 13.

Notice that the number of significant entries in the successive matrices are 2, 6, 12 with the generalization to $\ell(\ell+1)$. Notice the symmetrical positions of the entries relative to the main diagonal, associated with the double arrows in Fig. 4, including the zeros in the main diagonal. Notice that under reflection in the other diagonal: the positions in $\hat{L}_y$ are symmetric, but for $\hat{L}_x$ and $\hat{L}_z$ they get exchanged. For those rows or columns with a single entry the connection is unique, which occurs only between the monomial solutions for $\ell = 1$ and 2; and for several entries their number counts also the number of connections with the companion states of the connected species. For the latter, the relative values of the entries measure the probability amplitudes of the connections.

Next, we also present the matrices for $\ell = 4$ and the numerical Table 11 of their respective twenty entries, only for the most asymmetric molecules, evaluated after the Workshop. Here for the first time, there appear three entries in the yz, xz and xy columns of the respective matrices connecting them with the three $|[1]\,04\rangle$, $|[1]\,22\rangle$, $|[1]\,40\rangle$ eigenstates.

Table 11 Matrix Elements of Angular Momentum Operators for $\ell = 4$

(i,j)	$\langle i \vert \hat{L}_x \vert j \rangle$	(i,j)	$\langle i \vert \hat{L}_y \vert j \rangle$	(i,j)	$\langle i \vert \hat{L}_y \vert j \rangle$
1,4	$\frac{-2\sqrt{13}-\sqrt{22}+\sqrt{13}\sqrt{22}+20}{52}$	1,3	$\frac{\sqrt{7}+\sqrt{13}}{26}$	1,2	$\frac{-2\sqrt{13}-\sqrt{22}-\sqrt{13}\sqrt{22}-20}{52}$
1,8	$\frac{-2\sqrt{13}+\sqrt{22}-\sqrt{13}\sqrt{22}+20}{52}$	1,7	$\frac{-\sqrt{7}+\sqrt{13}}{26}$	1,6	$\frac{-2\sqrt{13}+\sqrt{22}+\sqrt{13}\sqrt{22}-20}{52}$
2,3	$\frac{1}{2}\left(1-\frac{1-\sqrt{7}}{\sqrt{22}}\right)$	2,4	$-2-\frac{5}{\sqrt{22}}$	2,1	$5+\sqrt{13}+\frac{2}{11}(8+3\sqrt{13})$
2,7	$\frac{1}{2}\left(1-\frac{1+\sqrt{7}}{\sqrt{22}}\right)$	2,8	$1-\frac{5}{\sqrt{22}}$	2,5	$5\left(1-\frac{2}{11}\right)$
3,2	$\frac{1}{2}\left(-7+\sqrt{22}+\frac{5-2\sqrt{22}}{\sqrt{7}}\right)$	3,1	$2(-\sqrt{7}-\sqrt{13})$	2,9	$5-\sqrt{13}+\sqrt{\frac{2}{11}}(8-3\sqrt{13})$
3,6	$\frac{1}{2}\left(-7-\sqrt{22}+\frac{5+2\sqrt{22}}{\sqrt{7}}\right)$	3,5	$\frac{90}{\sqrt{7}}$	3,4	$\frac{5\sqrt{7}+7\sqrt{22}+2\sqrt{7}\sqrt{22}+49}{14}$
4,1	$-5+\sqrt{13}+\sqrt{\frac{2}{11}}(8-3\sqrt{13})$	3,9	$2(-\sqrt{7}+\sqrt{13})$	3,8	$\frac{5\sqrt{7}-7\sqrt{22}-2\sqrt{7}\sqrt{22}+49}{14}$
4,5	$-5\left(1+\sqrt{\frac{2}{11}}\right)$	4,2	$2-\frac{5}{\sqrt{22}}$	4,3	$-\frac{1}{2}\left(1+\frac{1+\sqrt{7}}{\sqrt{22}}\right)$
4,9	$-5-\sqrt{13}+\sqrt{\frac{2}{11}}(8+3\sqrt{13})$	4,6	$-1-\frac{5}{\sqrt{22}}$	4,7	$-\frac{1}{2}\left(1+\frac{1-\sqrt{7}}{\sqrt{22}}\right)$
5,4	$\frac{6+\sqrt{22}}{26}$	5,3	$-\frac{\sqrt{7}}{13}$	5,2	$\frac{-6+\sqrt{22}}{26}$
5,8	$\frac{6-\sqrt{22}}{26}$	5,7	$\frac{\sqrt{7}}{13}$	5,6	$\frac{-6-\sqrt{22}}{26}$
6,3	$\frac{1}{2}\left(1+\frac{1-\sqrt{7}}{\sqrt{22}}\right)$	6,4	$1+\frac{5}{\sqrt{22}}$	6,1	$5+\sqrt{13}-\sqrt{\frac{2}{11}}(8+3\sqrt{13})$

Continued

Table 11 Matrix Elements of Angular Momentum Operators for $\ell = 4$—cont'd

(i,j)	$\langle i\vert\hat{L}_x\vert j\rangle$	**(i,j)**	$\langle i\vert\hat{L}_y\vert j\rangle$	**(i,j)**	$\langle i\vert\hat{L}_y\vert j\rangle$
6,7	$\frac{1}{2}\left(1+\frac{1+\sqrt{7}}{\sqrt{22}}\right)$	6,8	$-2+\frac{5}{\sqrt{22}}$	6,5	$5\left(1+\sqrt{\frac{2}{11}}\right)$
7,2	$\frac{1}{2}\left(-7+\sqrt{22}+\frac{-5+2\sqrt{22}}{\sqrt{7}}\right)$	7,1	$2(\sqrt{7}-\sqrt{13})$	6,9	$5-\sqrt{13}-\sqrt{\frac{2}{11}}(8-3\sqrt{13})$
7,6	$\frac{1}{2}\left(-7-\sqrt{22}+\frac{-5-2\sqrt{22}}{\sqrt{7}}\right)$	7,5	$-\frac{90}{\sqrt{7}}$	7,4	$\frac{-5\sqrt{7}+7\sqrt{22}-2\sqrt{7}\sqrt{22}+49}{14}$
8,1	$-5+\sqrt{13}-\sqrt{\frac{2}{11}}(8-3\sqrt{13})$	7,9	$2(\sqrt{7}+\sqrt{13})$	7,8	$\frac{-5\sqrt{7}-7\sqrt{22}+2\sqrt{7}\sqrt{22}+49}{14}$
8,5	$-5\left(1-\sqrt{\frac{2}{11}}\right)$	8,2	$-1+\frac{5}{\sqrt{22}}$	8,3	$-\frac{1}{2}\left(1-\frac{1+\sqrt{7}}{\sqrt{22}}\right)$
8,9	$-5-\sqrt{13}-\sqrt{\frac{2}{11}}(8+3\sqrt{13})$	8,6	$2+\frac{5}{\sqrt{22}}$	8,7	$-\frac{1}{2}\left(1-\frac{1-\sqrt{7}}{\sqrt{22}}\right)$
9,4	$\frac{2\sqrt{13}-\sqrt{22}-\sqrt{13}\sqrt{22}+20}{52}$	9,3	$\frac{\sqrt{7}-\sqrt{13}}{26}$	9,2	$\frac{2\sqrt{13}-\sqrt{22}+\sqrt{13}\sqrt{22}-20}{52}$
9,8	$\frac{2\sqrt{13}+\sqrt{22}+\sqrt{13}\sqrt{22}+20}{52}$	9,7	$\frac{-\sqrt{7}-\sqrt{13}}{26}$	9,6	$\frac{2\sqrt{13}+\sqrt{22}-\sqrt{13}\sqrt{22}-20}{52}$

$$\begin{pmatrix} 0 & 0 & 0 & \bullet & 0 & 0 & 0 & \bullet & 0 \\ 0 & 0 & \bullet & 0 & 0 & 0 & \bullet & 0 & 0 \\ 0 & \bullet & 0 & 0 & 0 & \bullet & 0 & 0 & 0 \\ \bullet & 0 & 0 & 0 & * & 0 & 0 & 0 & \bullet \\ 0 & 0 & 0 & \bullet & 0 & 0 & 0 & \bullet & 0 \\ 0 & 0 & \bullet & 0 & 0 & 0 & \bullet & 0 & 0 \\ 0 & \bullet & 0 & 0 & 0 & \bullet & 0 & 0 & 0 \\ \bullet & 0 & 0 & 0 & * & 0 & 0 & 0 & \bullet \\ 0 & 0 & 0 & \bullet & 0 & 0 & 0 & * & 0 \end{pmatrix}, \begin{pmatrix} 0 & 0 & \bullet & 0 & 0 & 0 & \bullet & 0 & 0 \\ 0 & 0 & 0 & \bullet & 0 & 0 & 0 & \bullet & 0 \\ \bullet & 0 & 0 & 0 & * & 0 & 0 & 0 & \bullet \\ 0 & \bullet & 0 & 0 & 0 & \bullet & 0 & 0 & 0 \\ 0 & 0 & \bullet & 0 & 0 & 0 & \bullet & 0 & 0 \\ 0 & 0 & 0 & \bullet & 0 & 0 & 0 & \bullet & 0 \\ \bullet & 0 & 0 & 0 & * & 0 & 0 & 0 & \bullet \\ 0 & \bullet & 0 & 0 & 0 & \bullet & 0 & 0 & 0 \\ 0 & 0 & \bullet & 0 & 0 & 0 & \bullet & 0 & 0 \end{pmatrix},$$

$$\begin{pmatrix} 0 & * & 0 & 0 & 0 & \bullet & 0 & 0 & 0 \\ \bullet & 0 & 0 & 0 & * & 0 & 0 & 0 & \bullet \\ 0 & 0 & 0 & \bullet & 0 & 0 & 0 & \bullet & 0 \\ 0 & 0 & \bullet & 0 & 0 & 0 & \bullet & 0 & 0 \\ 0 & \bullet & 0 & 0 & 0 & \bullet & 0 & 0 & 0 \\ \bullet & 0 & 0 & 0 & * & 0 & 0 & 0 & \bullet \\ 0 & 0 & 0 & \bullet & 0 & 0 & 0 & \bullet & 0 \\ 0 & 0 & \bullet & 0 & 0 & 0 & \bullet & 0 & 0 \\ 0 & \bullet & 0 & 0 & 0 & \bullet & 0 & 0 & 0 \end{pmatrix}.$$

The squares of the matrices of angular momentum can be calculated and the entries for the three cartesian components include diagonal and off diagonal contributions in common positions. The addition of the three matrices leads to the matrix of the square of angular momentum: all its diagonal elements are found to be $\ell(\ell + 1)$, as it should be; and the off diagonal contributions from its components cancel out, ensuring that the matrix is $\ell(\ell + 1)$ times the unit matrix. The combinations $e_1\hat{L}_x^2 + e_2\hat{L}_y^2 + e_3\hat{L}_z^2$ lead to the diagonal matrix of $\hat{H}^*$ with the diagonal entries $E_{n_1 n_2}^{*\ell[AB]}$.

3.4 Linear Momentum Matrices in Spheroconal Harmonic Bases

In references 12 and 13, we also evaluated the actions of the linear momentum operators on the spheroconal eigenstates of the free particle and the rotations of symmetric molecules, respectively, and only for the lowest values of ℓ. Here, we also want to refer to the figure at the beginning of Section 3 sketching the connections between the eigenstates in neighboring columns in order to go from the $\ell = 0$ to the $\ell = 4$ most asymmetric eigenstates of species [1], in order to make some clarifying remarks based on the content of Section 3.1. Indeed, we must emphasize that the radial derivative coefficients in the operators, Eq. (A.8), are the director cosines of the radial unit vector, and also the spheroconal harmonics for $\ell = 1$. Their existence makes it possible to accomplish the shift from $\ell = 0$ to $\ell = 1$ in the three species x, y, z. The return from $\ell = 1$ states to the $\ell = 0$ state uses the derivative terms; or alternatively, via the multiplication by the respective director cosines of the unit radial vectors and adding them up to obtain 1 as the $\ell = 0$ eigenfunction. Naturally, the multiplication mentioned in the last sentence accomplishes at the same time the connection with the two eigenstates with $\ell = 2$ and species [1]. In the successive applications of the operators on any eigenstate $\ell[AB]$ the connections with the neighboring $\ell' = \ell \pm 1$ of the appropriate species $[A'B']$, described in the tables of Section 3.1, must be taken into consideration.

Next we illustrate the explicit connections for the successive lower ℓ eigenstates by showing the matrices of the cartesian components of the linear momentum operators in the bases of spheroconal harmonics. Due to the negative parity of the operators and the well-defined parities of the eigenfunctions the matrix contains blocks of diagonal vanishing matrices of dimensions $(2\ell + 1) \times (2\ell + 1)$ for the successive values of ℓ, bordered by rectangular matrices of dimensions $(2(\ell - 1) + 1) \times (2\ell + 1)$ and $(2\ell + 1) \times (2(\ell - 1) + 1)$ above and to the left, respectively, as well as matrices of dimension $(2(\ell + 1) + 1) \times (2\ell + 1)$ below and $(2\ell + 1) \times (2(\ell - 1) + 1)$ to the right associated with the selection rule $\ell' = \ell \pm 1$ for the angular momentum states connected by the operators, and additional blocks of vanishing matrices beyond the rectangular matrices. The rectangular matrices are illustrated for $\ell = 1,2,3,4$: Their non vanishing entries are identified from the tables in Section 3.1, and their numbers are counted to be 1, 4, 9, 16 and ℓ^2 in general. The symmetry of the p_y matrices stands out, while the complementarity of the p_x and p_z, matrices can be recognized also.

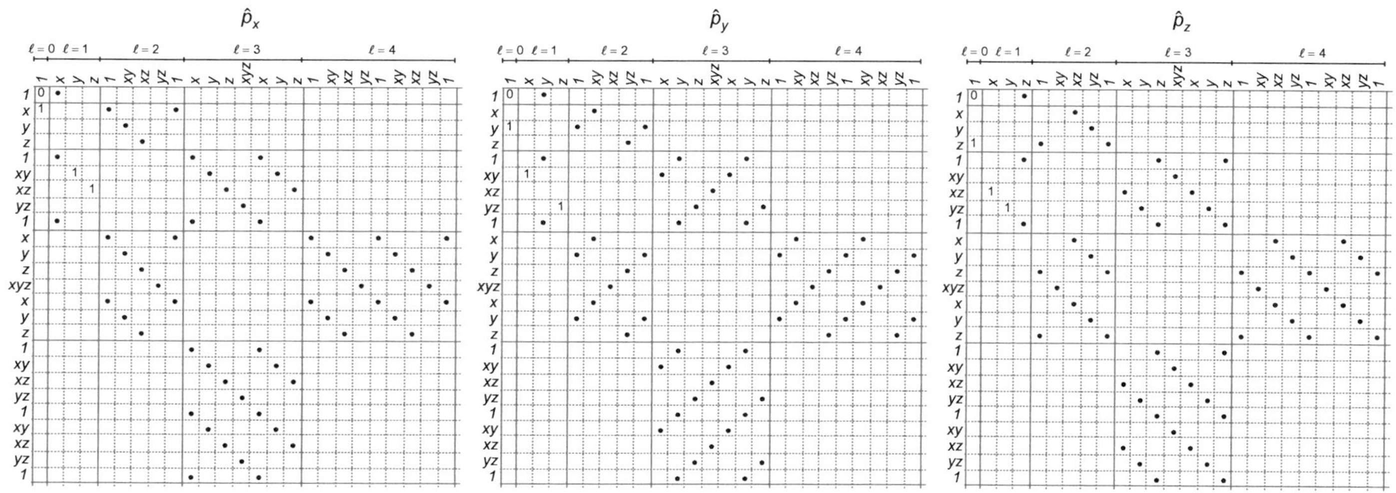

p̂x
p̂y
p̂z
ℓ = 0
ℓ = 1
ℓ = 2
ℓ = 3
ℓ = 4

Notice that the 1 × 3 and 3 × 1 matrices connecting the $\ell = 0$ to $\ell = 1$ states and viceversa, respectively, have only one entry making the connections unique. In turn, the two 3 × 5 and 5 × 3 matrices, connecting the $\ell = 1 \rightleftarrows \ell = 2$ eigenstates, show two rows or columns with single entries, with the third one with two entries associated with the two states of species [1]. For the two 5 × 7 and 7 × 5 matrices connecting the $\ell = 2 \rightleftarrows \ell = 3$ eigenstates, there is only one row or column with a single entry, and their companions contain two entries each associated with the xyz state, and the x, y, z species with excitations 02 and 20, respectively. For the 7 × 9 and 9 × 7 matrices connecting $\ell = 3 \rightleftarrows \ell = 4$ eigenstates, and from there on there are no single entries; their numbers correspond to the number of excited states for a given species $\ell - n^{AB} = n_1 + n_2$, Appendix A.4. When a linear momentum operator is applied to the eigenstates of a chosen value of ℓ and species $[AB]$, the connections to the eigenstates of both neighbors with $\ell' = \ell - 1$ and $\ell' = \ell + 1$ and the common matching species $[A'B']$, in the matrices to the left and right, or above and below, are obtained together. The graphic structure of matrices p_x, p_y, p_z, in bases of rotational eigenstates $\ell[AB]n_1n_2$, is illustrated below. The entries for the lowest values of ℓ for any asymmetry can be found in Ref. 13. Work is in progress to make the extensions contemplated in Fig. 5 and beyond. Table 12 contains a sample of the initial extensions for the most asymmetric molecules.

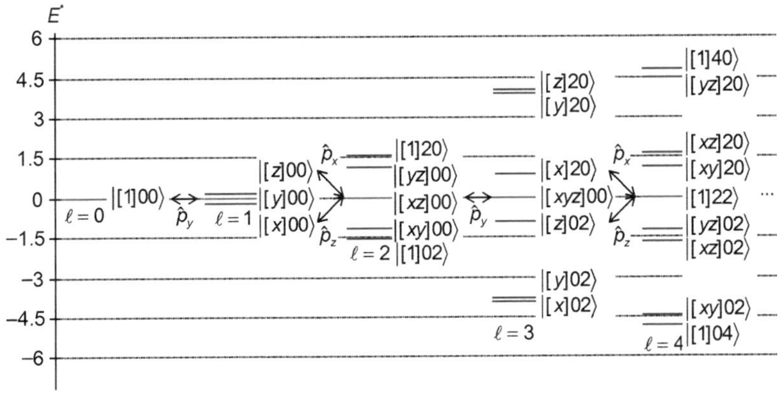

Figure 5 Actions of the linear momentum operators on the rotational eigenstates connecting them with eigenstates in neighboring columns.

Table 12 Linear momentum matrix elements connecting neighboring eigenstates starting from $\ell = 1$, $\ell = 2$, and arriving up to $\ell = 3$.

(i,j)	$p_x^{(i,j)}$	(i,j)	$p_y^{(i,j)}$	(i,j)	$p_z^{(i,j)}$
1,2	$\frac{2}{3}$	1,3	$\frac{2}{3}$	1,4	$\frac{2}{3}$
2,5	$-\frac{3}{5}(1+\sqrt{3})$	2,6	$\frac{2}{5}$	2,7	$\frac{3}{5}$
2,9	$-\frac{3}{5}(1-\sqrt{3})$	3,5	$\frac{6}{5}$	3,8	$\frac{3}{5}$
3,6	$\frac{2}{5}$	3,9	$\frac{6}{5}$	4,5	$\frac{3}{5}(1-\sqrt{3})$
4,7	$\frac{3}{5}$	4,8	$\frac{3}{5}$	4,9	$\frac{3}{5}(-1-\sqrt{3})$
5,2	$\frac{1+\sqrt{3}}{6}$	5,3	$-\frac{1}{3}$	5,4	$\frac{1-\sqrt{3}}{6}$
9,2	$\frac{1-\sqrt{3}}{6}$	9,3	$-\frac{1}{3}$	9,4	$\frac{1+\sqrt{3}}{6}$
10,5	$\frac{1}{5}\left(-6-\sqrt{3}-\frac{9}{\sqrt{6}}(1+\sqrt{3})\right)$	10,6	$\frac{-4-\sqrt{6}}{5}$	10,7	$\frac{1-\sqrt{3}}{5}$
10,9	$\frac{1}{5}\left(-6+\sqrt{3}+\frac{9}{\sqrt{6}}(1-\sqrt{3})\right)$	11,5	$\frac{-3-\sqrt{5}}{5}$	11,8	$\frac{3-\sqrt{15}}{15}$
11,6	$\frac{3+\sqrt{15}}{15}$	11,9	$\frac{-3+\sqrt{5}}{5}$	12,5	$\frac{1}{5}\left(-6+\sqrt{3}+\frac{9}{\sqrt{6}}(1-\sqrt{3})\right)$
12,7	$\frac{1+\sqrt{6}}{5}$	12,8	$\frac{-4+\sqrt{6}}{5}$	12,9	$\frac{1}{5}\left(-6-\sqrt{3}+\frac{9}{\sqrt{6}}(1+\sqrt{3})\right)$
13,8	-2	13,7	-2	13,6	-2
14,5	$\frac{1}{5}\left(-6-\sqrt{3}+\frac{9}{\sqrt{6}}(1+\sqrt{3})\right)$	14,6	$\frac{-4+\sqrt{6}}{5}$	14,7	$\frac{1+\sqrt{3}}{5}$
14,9	$\frac{1}{5}\left(-6+\sqrt{3}-\frac{9}{\sqrt{6}}(1-\sqrt{3})\right)$	15,5	$\frac{-3+\sqrt{5}}{5}$	15,8	$\frac{3+\sqrt{15}}{5}$
15,6	$\frac{3-\sqrt{15}}{15}$	15,9	$\frac{-3-\sqrt{5}}{5}$	16,5	$\frac{1}{5}\left(-6+\sqrt{3}+\frac{9}{\sqrt{6}}(-1+\sqrt{3})\right)$
16,7	$\frac{1-\sqrt{6}}{5}$	16,8	$\frac{-4-\sqrt{6}}{5}$	16,9	$\frac{1}{5}\left(-6-\sqrt{3}+\frac{9}{\sqrt{6}}(-1-\sqrt{3})\right)$

4. DISCUSSION

The oral presentation of our contribution at the Workshop was focused on the key concepts of the theory of angular momentum in the bases of Lamé spheroconal harmonics, with illustrations from the recent work on the most asymmetric molecules. This written version shares the same focus, with emphasis on the general validity and applicability of the concepts and methods for the molecules with any asymmetry. It is within this point of view that the symmetries of the Hamiltonians, their eigenfunctions and eigenenergies can be understood together, for molecules with a specific asymmetry distribution and also for all the asymmetry distributions, including their changes and connections as the asymmetry distribution parameter changes, as discussed and illustrated in Section 2. On the other hand, Section 3 has analysed in parallel the structures of the linear momentum and angular momentum operators, their connections as well as details on their actions on any rotational eigenstate $\ell[AB]$ leading to the complete sets of their matching companions $\ell'[A'B']$ with $\ell' = \ell \pm 1$, and $\ell[A'B']$, respectively, in the same calculation. The construction of their matrices in the bases of the rotational eigenstates is a significant step, providing quantitative answers to some of the problems recognized in Refs. 12–14. We intend to follow this line of inquiring to identify and calculate Clebsch-Gordan coefficients from the entries in those matrices.

The angular momentum theory reviewed in this contribution is a natural extension of the familiar theory using the basis of spherical harmonics. Figures 2 and 3 illustrate that the familiar cases of the prolate and oblate symmetric molecules, with the respective x and z axes of rotational rotational symmetry, involve such harmonics, eigenfunctions of the corresponding components of angular momentum. The traditional analysis of the rotational spectra of asymmetric molecules was developed using perturbation theory, taking the eigenfunctions and eigenenergies of those symmetric molecules as the starting points. The same figures also illustrate the exact solutions for molecules with all the possible distributions of asymmetry, including the most asymmetric, as well as the connections between the eigensolutions for molecules with "complementary" asymmetry distributions. The authors invite the interested readers to take the steps from circular to elliptical integral functions, in order to take some additional steps between the left and right sides of the figures, and further develop the theory as a tool for the analysis of the rotational spectra of asymmetric molecules and the processes in which they participate. The spheroconal harmonics also have applications in other asymmetric physical and chemical systems.

ACKNOWLEDGMENTS

The authors have the pleasure to congratulate Professor Frank Harris for his birthday, and to thank the organizers of the Workshop and the editors of the Festschrift in Advances in Quantum Chemistry, Professors John R. Sabin and Remigio Cabrera-Trujillo, for the invitations to participate in both.

Financial support from DGAPA PAPIIT IA 102414 and PAPIME PE 106615 are gratefully acknowledged.

APPENDIX

A.1 Cartesian and Spheroconal Coordinate Transformations

The transformation equations between the coordinates of our interest are

$$\begin{aligned} x &= r\mathrm{dn}(\chi_1|k_1^2)\mathrm{sn}(\chi_2|k_2^2),\\ y &= r\mathrm{cn}(\chi_1|k_1^2)\mathrm{cn}(\chi_2|k_2^2),\\ z &= r\mathrm{sn}(\chi_1|k_1^2)\mathrm{dn}(\chi_2|k_2^2), \end{aligned} \tag{A.1}$$

using Jacobi elliptical integral functions. The latter are connected by the relations:

$$\mathrm{cn}^2(\chi_i|k_i^2) = 1 - \mathrm{sn}^2(\chi_i|k_i^2),\ \mathrm{dn}^2(\chi_i|k_i^2) = 1 - k_i^2\mathrm{sn}^2(\chi_i|k_i^2), \tag{A.2}$$

and their derivatives are

$$\begin{aligned} \frac{\mathrm{d}}{\mathrm{d}\chi_i}\mathrm{sn}(\chi_i|k_i^2) &= \mathrm{cn}(\chi_i|k_i^2)\mathrm{dn}(\chi_i|k_i^2),\\ \frac{\mathrm{d}}{\mathrm{d}\chi_i}\mathrm{cn}(\chi_i|k_i^2) &= -\mathrm{sn}(\chi_i|k_i^2)\mathrm{dn}(\chi_i|k_i^2),\\ \frac{\mathrm{d}}{\mathrm{d}\chi_i}\mathrm{dn}(\chi_i|k_i^2) &= -k_i^2\mathrm{sn}(\chi_i|k_i^2)\mathrm{cn}(\chi_i|k_i^2). \end{aligned} \tag{A.3}$$

The parameters k_1^2 and k_2^2 are connected such that $k_1^2 + k_2^2 = 1$, in order to guarantee that the sum of the squares of x, y and z equals the square of the radial coordinate r. Fixed values of χ_1 and χ_2 define elliptical cones with axes along the z-axis and x-axis, respectively.[8] Notice that under the exchange of χ_1, k_i^2 and χ_2, k_2^2: $x \rightleftarrows z$ and $y \to y$.

The differential displacements in the three-dimensional space in the respective coordinates take the following forms:

$$
\begin{aligned}
& d\vec{r} = \hat{i}\mathrm{d}x + \hat{j}\mathrm{d}y + \hat{k}\mathrm{d}z = \\
& \left[\hat{i}\,\mathrm{dn}(\chi_1)\,\mathrm{sn}(\chi_2) + \hat{j}\,\mathrm{cn}(\chi_1)\,\mathrm{cn}(\chi_2) + \hat{k}\,\mathrm{sn}(\chi_1)\,\mathrm{dn}(\chi_2)\right]\mathrm{d}r \\
& + \left[-\hat{i}k_1^2\,\mathrm{sn}(\chi_1)\,\mathrm{cn}(\chi_1)\,\mathrm{sn}(\chi_2) - \hat{j}\,\mathrm{sn}(\chi_1)\,\mathrm{dn}(\chi_1)\,\mathrm{cn}(\chi_2)\right. \\
& \left. + \hat{k}\,\mathrm{cn}(\chi_1)\,\mathrm{dn}(\chi_1)\,\mathrm{dn}(\chi_2)\right]\mathrm{d}\chi_1 \\
& + \left[\hat{i}\,\mathrm{dn}(\chi_1)\,\mathrm{cn}(\chi_2)\,\mathrm{dn}(\chi_2) - \hat{j}\,\mathrm{cn}(\chi_1)\,\mathrm{sn}(\chi_2)\,\mathrm{dn}(\chi_2)\right. \\
& \left. - \hat{k}\,\mathrm{sn}(\chi_1)k_2^2\,\mathrm{sn}(\chi_2)\,\mathrm{cn}(\chi_2)\right] r\mathrm{d}\chi_2,
\end{aligned}
\tag{A.4}
$$

leading to the identification of the scale fators,

$$
h_r = 1, \quad h_\chi = h_{\chi_1} = h_{\chi_2} = r\sqrt{1 - k_1^2\mathrm{sn}^2(\chi_1) + k_2^2\mathrm{sn}^2(\chi_2)}
\tag{A.5}
$$

and the unit vectors,

$$
\begin{aligned}
\hat{r} &= \left[\hat{i}\,\mathrm{dn}(\chi_1)\,\mathrm{sn}(\chi_2) + \hat{j}\,\mathrm{cn}(\chi_1)\,\mathrm{cn}(\chi_2) + \hat{k}\,\mathrm{sn}(\chi_1)\,\mathrm{dn}(\chi_2)\right] \\
\hat{\chi}_1 &= \frac{r}{h_\chi}\left[-\hat{i}k_1^2\,\mathrm{sn}(\chi_1)\,\mathrm{cn}(\chi_1)\,\mathrm{sn}(\chi_2) - \hat{j}\,\mathrm{sn}(\chi_1)\,\mathrm{dn}(\chi_1)\,\mathrm{cn}(\chi_2)\right. \\
&\quad \left. + \hat{k}\,\mathrm{cn}(\chi_1)\,\mathrm{dn}(\chi_1)\,\mathrm{dn}(\chi_2)\right] \\
\hat{\chi}_2 &= \frac{r}{h_\chi}\left[\hat{i}\,\mathrm{dn}(\chi_1)\,\mathrm{cn}(\chi_2)\,\mathrm{dn}(\chi_2) - \hat{j}\,\mathrm{cn}(\chi_1)\,\mathrm{sn}(\chi_2)\,\mathrm{dn}(\chi_2)\right. \\
&\quad \left. - \hat{k}\,\mathrm{sn}(\chi_1)k_2^2\,\mathrm{sn}(\chi_2)\,\mathrm{cn}(\chi_2)\right].
\end{aligned}
\tag{A.6}
$$

Here we simplify the notation for the arguments of the Jacobi functions $(\chi_i|k_i^2) \to (\chi_i)$, $i = 1{,}2$. The director cosines of the unit radial vector $\hat{r}$ are identified in both Eq. (A.1) and the first of Eq. (A.6). The reader may ascertain that the set of unit vectors $\hat{r}$, $\hat{\chi}_1$, $\hat{\chi}_2$ is orthonormal and right handed.

A.2 Linear Momentum Operators

The gradient operator in spheroconal coordinates can be written immediately, by using the unit vectors and scale factors of Eqs. (A.5) and (A.6):

$$
\nabla = \hat{r}\frac{\partial}{\partial r} + \frac{1}{h_\chi}\left(\hat{\chi}_1\frac{\partial}{\partial \chi_1} + \hat{\chi}_2\frac{\partial}{\partial \chi_2}\right).
\tag{A.7}
$$

The linear momentum quantum operator is simply $\hat{\vec{p}} = -i\hbar\nabla$. Our interest here is in its cartesian components, which follow directly by taking the projections of $\hat{r}$, $\hat{\chi}_1$ and $\hat{\chi}_2$ along $\hat{i}$, $\hat{j}$, $\hat{k}$, respectively to obtain:

$$\begin{aligned}
\hat{p}_x &= -i\hbar\left[\mathrm{dn}(\chi_1)\mathrm{sn}(\chi_2)\frac{\partial}{\partial r} + \right.\\
&\left. + \frac{r}{h_\chi^2}\left(-k_1^2\mathrm{sn}(\chi_1)\mathrm{cn}(\chi_1)\mathrm{sn}(\chi_2)\frac{\partial}{\partial \chi_1} + \mathrm{dn}(\chi_1)\mathrm{cn}(\chi_2)\mathrm{dn}(\chi_2)\frac{\partial}{\partial \chi_2}\right)\right],\\
\hat{p}_y &= -i\hbar\left[\mathrm{cn}(\chi_1)\mathrm{cn}(\chi_2)\frac{\partial}{\partial r} + \right.\\
&\left. + \frac{r}{h_\chi^2}\left[-\mathrm{sn}(\chi_1)\mathrm{dn}(\chi_1)\mathrm{cn}(\chi_2)\frac{\partial}{\partial \chi_1} - \mathrm{cn}(\chi_1)\mathrm{sn}(\chi_2)\mathrm{dn}(\chi_2)\frac{\partial}{\partial \chi_2}\right)\right],\\
\hat{p}_z &= -i\hbar\left[\mathrm{sn}(\chi_1)\mathrm{dn}(\chi_2)\frac{\partial}{\partial r} + \right.\\
&\left. + \frac{r}{h_\chi^2}\left(\mathrm{cn}(\chi_1)\mathrm{dn}(\chi_1)\mathrm{dn}(\chi_2)\frac{\partial}{\partial \chi_1} - k_2^2\mathrm{sn}(\chi_1)\mathrm{sn}(\chi_2)\mathrm{cn}(\chi_2)\frac{\partial}{\partial \chi_2}\right)\right].
\end{aligned} \tag{A.8}$$

Notice that under the exchange $1 \rightleftarrows 2$ of the elliptical cone arguments and parameters: $p_x \rightleftarrows p_z$ and $p_y \rightarrow p_y$ similar to the exchange of their canonically conjugate cartesian coordinates.

The Laplacian operator can also be written directly in terms of the partial derivatives with respect to r, χ_1 and χ_2 and the scale factors:

$$\nabla^2 = \frac{1}{r^2}\frac{\partial}{\partial r}\left(r^2\frac{\partial}{\partial r}\right) + \frac{1}{h_\chi^2}\left(\frac{\partial^2}{\partial \chi_1^2} + \frac{\partial^2}{\partial \chi_2^2}\right). \tag{A.9}$$

From here we can identify the square of the angular momentum operator as the angular part including its natural units and signs:

$$\hat{L}^2 = -\hbar^2\frac{r^2}{h_\chi^2}\left(\frac{\partial^2}{\partial \chi_1^2} + \frac{\partial^2}{\partial \chi_2^2}\right). \tag{A.10}$$

Notice its invariance under the exchange of $1 \rightleftarrows 2$, including in the scale factor Eq. (A.5).

A.3 Angular Momentum Operators

The angular momentum vector operator can be written as

$$\hat{\vec{L}} = \vec{r} \times \hat{\vec{p}} = -i\hbar \vec{r} \times \nabla = -\frac{i\hbar r^2}{h_\chi^2}\left(\hat{\chi}_2 \frac{\partial}{\partial \chi_1} - \hat{\chi}_1 \frac{\partial}{\partial \chi_2}\right) \tag{A.11}$$

by using Eq. (A.7) for the gradient, and the properties of the unit vectors. Obviously, there is no radial part, and the directions of the derivatives have been exchanged, including a change in sign. The cartesian components can be written at once by projecting the unit vectors $\hat{\chi}_i$ in Eq. (A.6):

$$\begin{aligned}
\hat{L}_x &= \frac{-i\hbar^2 r^2}{h_\chi^2}\left[\mathrm{dn}(\chi_1)\mathrm{cn}(\chi_2)\mathrm{dn}(\chi_2)\frac{\partial}{\partial \chi_1} + k_1^2 \mathrm{sn}(\chi_1)\mathrm{cn}(\chi_1)\mathrm{sn}(\chi_2)\frac{\partial}{\partial \chi_2}\right],\\
\hat{L}_y &= \frac{-i\hbar^2 r^2}{h_\chi^2}\left[-\mathrm{cn}(\chi_1)\mathrm{sn}(\chi_2)\mathrm{dn}(\chi_2)\frac{\partial}{\partial \chi_1} + \mathrm{sn}(\chi_1)\mathrm{dn}(\chi_1)\mathrm{cn}(\chi_2)\frac{\partial}{\partial \chi_2}\right],\\
\hat{L}_z &= \frac{-i\hbar^2 r^2}{h_\chi^2}\left[-k_2^2 \mathrm{sn}(\chi_1)\mathrm{sn}(\chi_2)\mathrm{cn}(\chi_2)\frac{\partial}{\partial \chi_1} - \mathrm{cn}(\chi_1)\mathrm{dn}(\chi_1)\mathrm{dn}(\chi_2)\frac{\partial}{\partial \chi_2}\right].
\end{aligned} \tag{A.12}$$

Not surprisingly, the elements in Eqs. (A.8) and (A.12) are the same in exchanged positions and with some changes in sign, as determined by the comparison of Eqs. (A.7) and (A.11). Notice now that the exchange $1 \rightleftarrows 2$ in Eq. (A.12) leads to $\hat{L}_x \rightleftarrows -\hat{L}_z$ and $\hat{L}_y \to -\hat{L}_y$.

It is also possible to use Eq. (A.11) to go back to the linear momentum, by taking the cross product, $\hat{r}\times$, in both sides of the equation:

$$\frac{\vec{r}}{r} \times \hat{\vec{L}} = \frac{\vec{r}}{r} \times (\vec{r} \times \hat{\vec{p}}) = \vec{r}\left(\frac{\vec{r}}{r} \cdot \hat{\vec{p}}\right) - \left(\frac{\vec{r}}{r} \cdot \vec{r}\right)\hat{\vec{p}}, \tag{A.13}$$

leading to the decomposition of the vector of our interest into its radial and transverse components

$$\hat{\vec{p}} = \hat{r}\hat{p}_r - \frac{\vec{r} \times \hat{\vec{L}}}{r} \tag{A.14}$$

which coincide with Eqs. (A.7) and (A.8).

The asymmetry distribution Hamiltonian of Eq. (7) can be written in spheroconal coordinates by using Eq. (A.12), with the result:

$$\hat{H}^* = -\frac{\hbar^2 r^2}{2h_\chi^2}\left[\left(e_1 - (e_1 - e_2)\,\mathrm{sn}^2(\chi_2)\right)\frac{\partial^2}{\partial \chi_1^2} + \left(e_3 + (e_2 - e_3)\,\mathrm{sn}^2(\chi_1)\right)\frac{\partial^2}{\partial \chi_2^2}\right] \tag{A.15}$$

provided that the geometric parameters k_i^2 of the coordinates and the dynamic parameters of asymmetry distribution e_i of the molecule are connected by the relationships:

$$k_1^2 = \frac{e_2 - e_3}{e_1 - e_3}, \; k_2^2 = \frac{e_1 - e_2}{e_1 - e_3}. \tag{A.16}$$

Otherwise, there is a term with the second partial derivative with respect to χ_1 and χ_2. Additionally, these relationships lead to the simultaneous separation of the eigenvalue problems in Eqs. (A.10) and (A.15). The values of the geometric parameters for the symmetric prolate and oblate molecules are respectively $k_1^2 = 1$, $k_2^2 = 0$ and $k_1^2 = 0$, $k_2^2 = 1$; and for the most asymmetric molecules $k_1^2 = k_2^2 = 1/2$. Notice also the exchange $k_1^2 \rightleftarrows k_2^2$ associated with the changes $e_1 \rightleftarrows -e_3$ and $e_2 \rightarrow -e_2$, discussed in Section 2.1 for molecules with complementary angular asymmetry-distribution parameters σ and $60^\circ - \sigma$. The corresponding changes $1 \rightleftarrows 2$ in Eq. (A.15) lead to the relationship between the Hamiltonians with complementary asymmetry distribution:

$$\hat{H}^*_{1\rightleftarrows 2}(e_1 \rightarrow -e_3, e_2 \rightarrow -e_2, e_3 \rightarrow -e_1) = -\hat{H}^*(e_1, e_2, e_3). \tag{A.17}$$

A.4 Simultaneous Separation and Eigenfunctions of $\hat{L}^2$ and $\hat{H}^*$

The explicit form of the eigenvalue equations for the square of the angular momentum and asymmetry-distribution Hamiltonian follow from the use of Eqs. (A.10) and (A.15) for the respective operators:

$$-\hbar^2 \frac{r^2}{h_\chi^2}\left(\frac{\partial^2}{\partial \chi_1^2} + \frac{\partial^2}{\partial \chi_2^2}\right)\Psi(\chi_1, \chi_2) = \hbar^2 \ell(\ell+1)\Psi(\chi_1, \chi_2). \tag{A.18}$$

$$\begin{aligned} -\frac{\hbar^2 r^2}{2h_\chi^2}\Bigg[&\left(e_1 - (e_1 - e_2)\,\mathrm{sn}^2(\chi_2)\right)\frac{\partial^2}{\partial \chi_1^2} + \\ &+ \left(e_3 + (e_2 - e_3)\,\mathrm{sn}^2(\chi_2)\right)\frac{\partial^2}{\partial \chi_2^2}\Bigg]\Psi(\chi_1, \chi_2) = E^*\Psi(\chi_1, \chi_2), \end{aligned} \tag{A.19}$$

where the scale factor squared, Eq. (A.5), appears as a common factor in both operators. The eigenfunctions have the factorizable form

$$\Psi(\chi_1, \chi_2) = \Lambda_1(\chi_1)\Lambda_2(\chi_2), \tag{A.20}$$

which allows the simultaneous separation of both equations, leading to ordinary differential equations of the Lamé type in each variable:

$$\left[-\frac{\mathrm{d}^2}{\mathrm{d}\chi_i^2}+\ell(\ell+1)k_i^2\mathrm{sn}^2(\chi_i)\right]\Lambda_i(\chi_i)=h_i\Lambda_i(\chi_i) \tag{A.21}$$

for $i = 1{,}2$. The separation constants, playing the role of the eigenvalues of the Lamé operator, are combinations of the angular momentum and asymmetry-distribution energy eigenvalues of Eqs. (A.18) and (A.19):

$$h_1=-\frac{\ell(\ell+1)e_3}{e_1-e_3}+\frac{2E^*}{\hbar^2(e_1-e_3)},\quad h_2=\frac{\ell(\ell+1)e_1}{e_1-e_3}-\frac{2E^*}{\hbar^2(e_1-e_3)}. \tag{A.22}$$

Their combinations

$$h_1+h_2=\ell(\ell+1)\ \text{and}\ e_1h_1+e_3h_2=2E^*, \tag{A.23}$$

provide the physical eigenvalues for the rotational eigenstates of our interest. Notice also the exchange $h_1 \rightleftarrows h_2$ accompanying that of $e_1 \rightleftarrows -e_3$; as well as the invariance of the angular momentum and the change of sign of the energy $E^* \rightleftarrows -E^*$.

Concerning the solutions of the Lamé equation Eq. (A.21), we must recognize their singularities associated with the derivatives of the Jacobi elliptical integrals, specifically of the $\mathrm{sn}^2(\chi_i)$ term in the same equation: $2\,\mathrm{sn}(\chi_i)\,\mathrm{cn}(\chi_i)\,\mathrm{dn}(\chi_i)$. The respective singularity removing factors for i =,1, 2 and their matching products are illustrated in Table A.1, using only the initial letter of the functions involved.

Their products are identified as the successive powers and products of the cartesian coordinates. Notice that all of them are harmonic functions, and can be classified in two kinds, with overall positive and negative parity

Table A.1 Spheroconal [*A*], [*B*], and Cartesian Species and Their Parities

A	1	d	c	s	dc	ds	cs	dcs
B	1	s	c	d	sc	sd	cd	scd
AB	*1*	x	y	z	xy	xz	yz	xyz
$x \to -x$	+	−	+	+	−	−	+	−
$y \to -y$	+	+	−	+	−	+	−	−
$z \to -z$	+	+	+	−	+	−	−	−

according to their numbers of cartesian coordinate factors even and odd, respectively; in turn, each kind appears in four species 1, xy, xz, yz and x, y, z, xyz belonging to eigenfunctions with even and odd values of the angular momentum ℓ, respectively.

The eigensolutions of the Lamé equation are constructed correspondingly with their singularity removing factors multiplied by a polynomial:

$$\Lambda_i(\chi_i) = A(\chi_i)\sum_{s=0}^{N_{\max}} a_s^A \mathrm{sn}^{2s}(\chi_i). \tag{A.24}$$

The polynomials involve even powers of the elliptical sine function ensuring that their parity is positive. The substitution of the form of Λ of Eq. (A.24) in Eq. (A.21) leads to three-term recurrence relations for the expansion coefficients a_s^A. For even and odd values of ℓ, the upper limits in the sum of Eq. (A.24) are respectively: $N_{\max} = \frac{\ell}{2} + 1$ for species 1, $N_{\max} = \frac{\ell}{2}$ for species xy, xz, yz with a total of $2\ell + 1$ different states; $N_{\max} = \frac{\ell+1}{2}$ for species x, y, z, $N_{\max} = \frac{\ell-1}{2}$ for species xyz with the same number of $2\ell + 1$ different states.

The three term-recurrence relations for the expansion coefficients can be written in a matrix form, which upon diagonalization yields the eigenvalues $h_n^A(k_i^2)$ and eigenvectors $a_s^A(h_n^A)$, following Hunter and Pritchard.[16]

The matching species in Table A.1 is applied to the Lamé polynomials of Eq. (A.24) in the respective coordinates to construct the rotational eigenfunctions of Eq. (A.20), and to check that the first of Eq. (A.23) is satisfied and to evaluate the energy in the second:

$$h_{n_1}^A(k_1^2) + h_{n_2}^B(k_2^2) = \ell(\ell+1), \quad e_1 h_{n_1}^A(k_1^2) + e_3 h_{n_2}^B(k_2^2) = 2E_{n_1 n_2}^{*\ell[AB]}. \tag{A.25}$$

The subindices n_1 and n_2 count the number of excitations and elliptical-cone nodes in the respective eigenfunctions, apart from the cartesian plane nodes associated with the singularity removing factors: $n^{AB} = 0, 1, 2, 3$ according to the number of cartesian coordinate factors in $[AB]$. For a given value of ℓ, all the harmonic eigenfunctions have the same number of nodes, $n^{AB} + n_1 + n_2 = \ell$. Since the Lamé polynomials in Eq. (A.24) appear with even powers of sn $^2(\chi_i)$, their nodes are pairs of elliptical cone sheets, mirrored in the $z = 0$ plane for $i = 1$, and $x = 0$ for $i = 2$, respectively. Correspondingly n_1 and n_2 take even values adding up to $\ell - n^{AB}$.

REFERENCES

1. Kramers, H. A.; Ittmann, G. P. Zur Quantelung des asymmetrischen Kreisels. *Zeits. Phys.* **1929**, *53*, 553.
2. Patera, J.; Winternitz, P. A New Basis for the Representations of the Rotation Group. Lamé and Heun Polynomials. *J. Math. Phys.* **1973**, *14*(8), 1130.
3. Kroto, H. W. *Molecular Rotation Spectra*. John Wiley & Sons: Great Britain, 1975.
4. Piña, E. Some Properties of the Spectra of Asymmetric Molecules. *J. Mol. Struct. (THEOCHEM)* **1999**, *493*, 159.
5. Valdéz, M.; Piña, E. The Rotational Spectra of the Most Asymmetric Molecules. *Rev. Mex. Fís.* **2006**, *52*(3), 220.
6. Piña, E. Algunas Propiedades de los operadores de escalera. *Rev. Mex. Fís.* **1994**, *41*(6), 913.
7. Ley-Koo, E.; Méndez-Fragoso, R. Properties of the Spectra of Asymmetric Molecules: Matrix Evaluation in Bases of Spherical Harmonics and Common Generating Function. *Rev. Mex. Fís.* **2008**, *54*(1), 69.
8. Ley-Koo, E.; Méndez-Fragoso, R. Rotational States of Asymmetric Molecules Revisited: Matrix Evaluation and Generating Function of Lamé Functions. *Rev. Mex. Fís.* **2008**, *54*(2), 162.
9. Ley-Koo, E.; Góngora, A. Common Generating Function for Three-Dimensional Hydrogen Atom Complete Wave Functions. *Int. J. Quantum Chem.* **2009**, *109*, 790.
10. Méndez-Fragoso, R.; Koo, E. L. The Hydrogen Atom in a Semi-Infinite Space with an Elliptical Cone Boundary. *Int. J. Quantum Chem.* **2011**, *111*, 2882.
11. Méndez-Fragoso, R.; Ley-Koo, E. Lamé Spheroconal Harmonics in Atoms and Molecules. *Int J. Quantum Chem.* **2010**, *110*, 2765.
12. Méndez-Fragoso, R.; Ley-Koo, E. Rotations of Asymmetric Molecules and the Hydrogen Atom in Free and Confined Configurations. *Adv. Quantum Chem.* **2011**, Vol. 62, 137.
13. Méndez-Fragoso, R.; Ley-Koo, E. Ladder Operators for Lamé Spheroconal Harmonic Polynomials. *SIGMA.* **2012**, *8*, 74.
14. Ley-Koo, E. Rotations of the most asymmetric molecules via 4-step and 1- step ladder operators. In: *Proceedings 30th International Symposium on Methods of Group Theory in Physics, IOP Conference Series*, **2015**, 597 012055. http://dx.doi.org/10.1088/1742-6596/597/1/012055.
15. Abramowitz, M.; Stegun, I. A. *Handbook of Mathematical Functions*; Dover: New York, 1965.
16. Hunter, G.; Pritchard, H. O. Born-Oppenheimer Separation for Three-Particle Systems. II. TwoCenter Wavefunctions. *J. Chem. Phys.* **1967**, *46*, 2146.

CHAPTER SEVEN

The Fourier Space Restricted Hartree–Fock Method for the Electronic Structure Calculation of One-Dimensionally Periodic Systems

Joseph G. Fripiat*[,1], Benoît Champagne*, Frank E. Harris[†,‡]
*Laboratoire de Chimie Théorique (UCPTS), Département de Chimie, University of Namur, Namur, Belgium
[†]Department of Physics, University of Utah, Salt Lake City, Utah, USA
[‡]Quantum Theory Project, University of Florida, Gainesville, Florida, USA
[1]Corresponding author: e-mail address: joseph.fripiat@unamur.be

Contents

Abstract

Computational studies of the electronic structures of periodic systems at the level of Hartree–Fock or density functional theory require the evaluation to appropriate accuracy of the lattice sums that appear in these formalisms. This chapter describes and

Advances in Quantum Chemistry, Volume 71
ISSN 0065-3276
http://dx.doi.org/10.1016/bs.aiq.2015.03.003

illustrates a method for treating systems that are periodic in one of the three dimensions (i.e., stereoregular polymers), showing how a combination of Fourier-transform techniques and an Ewald-type partitioning enables these sums to be divided between physical (direct) space and reciprocal (Fourier) space in a way that enhances their convergence rate. It is also shown how the spatial (line-group) symmetry (including rotational and screw axes, reflection, and glide planes) can be exploited to improve the efficiency of computation, extending to this domain the technique of Dupuis and King for building a complete Fock matrix from a minimal set of its matrix elements. Other issues of computational efficiency are also reviewed. The methods, implemented in our computer program FT-1D, are illustrated for a problem of significant size: A carbon single-wall (7, 0) nanotube with 56 spatial symmetry operations and bases of up to 420 atomic orbitals per unit cell.

1. INTRODUCTION

Calculations of the electronic structure of materials with one-dimensional (1D) periodicity (polyethylene, polyyne, polypyrrole, polythiophene, graphene ribbons, nanotubes, etc.) play an important role in the understanding of their electronic and optical properties.

There are two categories of methods to compute the electronic structure of polymers: One is to model these systems as oligomers containing a large though finite number of unit cells and to use molecular quantum chemistry programs: Gaussian, GAMESS, Dalton, etc. The application of molecular *ab initio* methods to the study of the electronic structure of medium- and large-size molecules has, however, two drawbacks: (1) This approach requires substantial computational resources, and (2) end-of-chain effects are introduced.

The second category of methods models polymeric materials as infinite periodic systems. That approach permits exploitation of translational symmetry, thereby reducing to manageable dimensions the otherwise formidable task of computing the electronic states of an infinite but stereoregular system. The periodic-system approach can, in principle, be carried out either in an atomic orbital (AO) basis or as an expansion in plane waves. The use of plane waves is far simpler than using an AO basis (see, for example, work by Payne[1]) but plane-wave expansions do not converge rapidly enough to yield results that are useful in current contexts.

A few years ago, works[2,3] have appeared in which comparisons are made with somewhat diverging opinions as to the merits of the two approaches. Hutchinson et al.[2] point to the limitations of oligomer extrapolation approximations, while Pomogaeva et al.[3] stress various computational drawbacks of the crystalline-orbital method, citing issues that include quasilinear dependencies and the slow convergence of exchange contributions. A common point to these arguments is that they refer to calculations that are carried out in a so-called direct-space (DS) formulation.

Electronic structure calculations[4–8] on polymers are often performed in the framework of the (restricted) Hartree–Fock method or of density functional theory (DFT), expressed in DS and using Gaussian-type atomic orbitals (GTOs). We limit discussion here to the Hartree–Fock method. The ideas which we present may however be applied to DFT, especially when the exchange-correlation functional is built from a mixing of the Hartree–Fock exchange with other terms.

In the next section, the basic Hartree–Fock equations are discussed in the context of one-dimensionally periodic chains. In these equations, the periodicity introduces lattice sums that can in some cases converge very slowly. We then show that the use of the Fourier transform coupled with the Poisson summation formula and the Ewald procedure overcomes the convergence problems that appear in the DS representation.

Section 3 shows how the use of the spatial symmetry of periodic linear chains can reduce the computer time by taking into account symmetry elements that include rotational and screw axes, and mirror and glide planes. In Section 4, we discuss a way to make optimal the calculation of the Coulomb and exchange part of the Fock matrix.

Section 5 illustrates its methods by presenting some numerical results obtained on a carbon single-wall nanotube (SWNT), and our conclusions are summarized in Section 6.

2. THEORETICAL BACKGROUND

In this section, we review the basic Hartree–Fock formalism for one-dimensional periodic chains, identifying the numerical difficulties due to the lattice sums appearing in these equations and showing how they can be overcome using the Poisson summation formula and transforming the summations by the Ewald procedure.

Consider a macroscopic one-dimensional array of three-dimensional unit cells with n_0 electrons per cell, with periodicity in the z-direction. We express energy in Hartree atomic units, but designate all lengths in units of the lattice constant (a_0 bohr), so the Coulomb repulsion of two unit charges takes the form $1/a_0 r$, the kinetic energy operator is $-(1/2a_0^2)\nabla^2$, and the lattice constant is simply the unit vector $\hat{\mathbf{z}}$. A reference cell is assumed to contain nuclei labeled A, B,... with respective charges Q_A,Q_B,..., located at points $\mathbf{A}$, $\mathbf{B}$,..., where $\mathbf{A} = (A_x,A_y,A_z)$, etc., and will contain ω AOs, usually centered on one of the nuclei. An AO χ_p centered on nucleus A is a fixed linear combination (a *contraction*) of GTO basis functions G_a with coefficients d_{ap}; the G_a and d_{ap} have values that correspond to the Slater-type orbital they are designed to mimic. Thus,

$$\chi_p(\mathbf{r}) = \sum_a G_a(\mathbf{r}) d_{ap}. \tag{1}$$

Expressions corresponding to Eq. (1) can be written for AOs at other centers; a contraction at center B will ordinarily be written using b as the summation index and GTO's G_b.

Assigning the index $\mu = 0$ to the reference cell, the other cells of the array will correspond to nonzero integer values of μ in the range $-\infty < \mu < \infty$, with $\mathbf{A}$,... replaced by $\mathbf{A} + \mu\hat{\mathbf{z}} = (A_x, A_y, A_z + \mu), \ldots$, and with AOs and GTO basis functions designated, respectively, by $\chi_p^\mu, \chi_q^\mu, \ldots$, and G_a^μ, $G_b^\mu, \ldots$.

The Gaussian function G_a^μ is assumed to have the form

$$\begin{aligned} G_a^\mu(\mathbf{r}) &= (x - A_x)^{n_a}(y - A_y)^{l_a}(z - A_z - \mu)^{m_a} e^{-\alpha_a(\mathbf{r}-\mathbf{A}-\mu\hat{\mathbf{z}})^2} \\ &= x_a^{n_a} y_a^{l_a} z_a^{m_a}(\mu) e^{-\alpha_a(\mathbf{r}_a(\mu))^2}. \end{aligned} \tag{2}$$

Here, x_a, y_a, $z_a(\mu)$, and $\mathbf{r}_a(\mu)$ denote $x - A_x$, $y - A_y$, $z - A_z - \mu$, and $\mathbf{r} - \mathbf{A} - \mu\hat{\mathbf{z}}$, respectively. Note that because lengths are in units of a_0, the parameter α_a characterizing each Gaussian in a contraction is a_0^2 times the value (in 1/bohr^2) listed for that contraction in most tabulations.

An important property of Gaussians is that the product of two Gaussians centered at A and B and with parameters α_a and α_b can be written as a linear combination of Gaussians located at a point $\mathbf{P}$ given by the relation:

$$\mathbf{P} = \frac{\alpha_a \mathbf{A} + \alpha_b \mathbf{B}}{\alpha_a + \alpha_b}. \tag{3}$$

2.1 LCAO–Hartree–Fock Equations

In the LCAO approximation, the "polymer" orbitals $\phi_n(k, \mathbf{r})$ are represented by a linear combination of Bloch sums, $\eta_p(k, \mathbf{r})$, of AOs, $\chi_p(\mathbf{r})$, centered in each unit cell:

$$\phi_n(k,\mathbf{r}) = \sum_{p=1}^{\omega} \eta_p(k,\mathbf{r}) c_{pn}(k); \quad \eta_p(k,\mathbf{r}) = \lim_{N\to\infty} (2N+1)^{-1/2} \sum_{\mu=-N}^{N} e^{2\pi i \mu k} \chi_p^{\mu}(\mathbf{r}). \tag{4}$$

In these Bloch sums, each AO is multiplied by a phase factor depending on a point k in the (1D) reciprocal space.

In the LCAO approximation, the unknowns are the LCAO coefficients $c_{pn}(k)$; they are obtained by solving the eigenvalue equation

$$\mathsf{F}(k)\mathsf{c}(k) = \mathsf{S}(k)\mathsf{c}(k)\mathsf{E}(k) \tag{5}$$

for each k point of the Brillouin zone.

In Eq. (5), $\mathsf{F}(k)$ is the Fock matrix, $\mathsf{c}(k)$ is the matrix collecting the LCAO coefficients as columns, $\mathsf{E}(k)$ is a diagonal matrix containing the eigenvalues, and $\mathsf{S}(k)$ is the overlap matrix:

$$S_{pq}(k) = \sum_{\mu=-\infty}^{\infty} \exp(2\pi i k\mu) \left\langle \chi_p(\mathbf{r}) | \chi_q^{\mu}(\mathbf{r}) \right\rangle . \tag{6}$$

The Fock matrix elements are:

$$F_{pq}(k) = T_{pq}(k) + V_{pq}(k) + C_{pq}(k) + X_{pq}(k). \tag{7}$$

In the units we use here, the Brillouin zone is over a range of unit length, $k = \left[-\frac{1}{2}, \frac{1}{2}\right]$.

The quantities T_{pq} and V_{pq} are kinetic energy and electron-nuclear interaction integrals, while $C_{pq}(k)$ and $X_{pq}(k)$ are the Coulomb and exchange parts of the electron–electron interaction.

The integrals needed for Eq. (7) are:

$$T_{pq}(k) = \sum_{\mu=-\infty}^{\infty} \exp(2\pi i k\mu) \left\langle \chi_p(\mathbf{r}) \middle| \frac{-\nabla^2}{2a_0^2} \middle| \chi_q^{\mu}(\mathbf{r}) \right\rangle, \tag{8}$$

$$V_{pq}(k) = \sum_{C}^{n_{\text{atoms}}} \sum_{\mu=-\infty}^{\infty} \exp(2\pi i k\mu) \sum_{\mu''=-\infty}^{\infty} \left\langle \chi_p(\mathbf{r}) \middle| \frac{-Q_C}{a_0 |\mathbf{r} - \mathbf{C} - \mu''\hat{\mathbf{z}}|} \middle| \chi_q^{\mu}(\mathbf{r}) \right\rangle , \tag{9}$$

$$C_{pq}(k) = \sum_{r,s=1}^{\omega} \int_{BZ} P_{sr}(k') Z_{pqrs}(0;k,k')\, dk', \tag{10}$$

and

$$X_{pq}(k) = -\frac{1}{2}\sum_{r,s=1}^{\omega}\int_{BZ} P_{sr}(k')Z_{psrq}(k-k';\, k',k)\,\mathrm{d}k' \tag{11}$$

with

$$Z_{pqrs}(k'';k,k') = \sum_{\mu=-\infty}^{\infty}\sum_{\mu'=-\infty}^{\infty} \exp(2\pi i(k\mu + k'\mu'))\Xi_{pqrs}(k'';\, \mu,\mu'). \tag{12}$$

The Z_{pqrs} are electron repulsion integrals; note that $C_{pq}(k)$ involves only $Z_{pqrs}(0;\, k,k')$ and that $X_{pq}(k)$ contains $Z_{psrq}(k - k';\, k',\, k)$. Integrals $Z_{pqrs}(k'';\, k',k)$ with more general values of k'' occur in the treatment of electron correlation. The quantities P_{sr} are elements of the first-order density matrix. Since P_{sr} depends on the solution to Eq. (5), that eigenvalue equation must be solved iteratively until convergence is achieved.

The basic two-electron integrals Ξ are

$$\Xi_{pqrs}(k'';\, \mu,\mu') = \sum_{\mu''=-\infty}^{\infty} \exp(2\pi i k''\mu'')\left\langle \chi_p^0(\mathbf{r}_1)\chi_r^{\mu''}(\mathbf{r}_2)\left|\frac{1}{a_0 r_{12}}\right|\chi_q^{\mu}(\mathbf{r}_1)\chi_s^{\mu'+\mu''}(\mathbf{r}_2)\right\rangle. \tag{13}$$

These integrals can be written in terms of the basis GTOs:

$$\begin{aligned} S_{pq}(k) &= \sum_{ab} d_{ap}d_{bq}S_{ab}(k), \quad T_{pq}(k) = \sum_{ab} d_{ap}d_{bq}T_{ab}(k), \\ V_{pq}(k) &= \sum_{ab} d_{ap}d_{bq}V_{ab}(k), \\ \Xi_{pqrs}(k'';\mu,\mu') &= \sum_{abcd} d_{ap}d_{bq}d_{cr}d_{ds}\Xi_{abcd}(k'';\mu,\mu'). \end{aligned} \tag{14}$$

The integrals with indices a, b, c, and d are obtained by replacing $\chi_p, \chi_q, \ldots$ in Eqs. (6), (8), (9), or (13) by $G_a, G_b, \ldots$, so, for example,

$$S_{ab}(k) = \sum_{\mu=-\infty}^{\infty} \exp(2\pi i k\mu)\langle G_a(\mathbf{r})|G_b^{\mu}(\mathbf{r})\rangle.$$

Note that integrals involving χ are always written with the index set $pqrs$, while those involving G have indices from the set $abcd$.

After the "polymer" orbitals are determined from the solution to Eq. (5), the occupied part of the Brillouin zone is determined (i.e., the orbitals are assigned occupation numbers that add, when integrated over k, to n_0). The density matrix elements $P_{sr}(k)$ can be then computed from

$$P_{sr}(k) = \sum_{n=1}^{\omega} c_{sn}(k)\nu_n(k)c_{rn}^*(k), \tag{15}$$

where $\nu_n(k)$ is the occupation number (= 0, 1, or 2) of the "polymer" orbital $\phi_n(k, \mathbf{r})$.

Taken by themselves, the electron-nuclear integrals diverge (i.e., predict a negative energy contribution that increases in magnitude more rapidly than linearly in N), but this divergence is offset (for neutral polymers) by a divergent positive contribution from the Coulomb part of the electron–electron repulsion. It is therefore necessary to combine the divergent parts of both these contributions in a mathematically consistent fashion before completing a numerical energy evaluation. We discuss this and other potential divergences later in this paper.

2.2 Convergence of Lattice Sums

In the periodic-system Hartree–Fock equations, the extended nature of the system appears in the form of infinite lattice summations that run in principle from $-\infty$ to ∞. However, in practice we must limit the summation to a finite number of cells, keeping in mind that the computing time is proportional to a certain power of this number.

For the overlap and kinetic-energy matrix elements, $S_{pq}(k)$ (Eq. 6) and $T_{pq}(k)$ (Eq. 8), there is only one lattice summation, and it converges exponentially, with the rate of convergence determined by the Gaussian exponents.

In the electron-nuclear attraction matrix elements $V_{pq}(k)$ (Eq. 9), two lattice summations occur; one (the μ sum) converges exponentially at a rate determined by the exponents of a pair of GTOs. However, the other sum (over μ'') diverges. In fact, for large values of $|\mu''|$, the terms of this sum behave like $1/|\mu''|$.

For the electron–electron integrals (Eqs. 12 and 13), three lattice summations occur. Two of them, those over μ and μ' in Eq. (12), converge exponentially due to the GTOs in the sums. However, the third lattice summation, over μ'', Eq. (13), behaves like

$$\sum_{\mu''=-\infty}^{\infty} \frac{\exp(2\pi i k'' \mu'')}{\mu''}$$

and diverges if $k'' = 0$.

It is necessary to group together the electron-nuclear attraction terms and the direct (Coulomb) electron–electron repulsion terms. With that grouping, the μ'' series from V_{pq} and Z_{pqrs} combine to yield an alternating harmonic series which converges, but the convergence can be painfully slow if the separation between the positive and negative parts of the charge distribution is significant. It then becomes important to accelerate the convergence of these series. One possible approach is to compute integrals exactly over a specified number of cells (a "short-range region") and outside this region, the "long-range region," to use a multipole expansion.[9] That approximation is only valid in an asymptotic regime of small overlap of the interacting charge distributions. It is also semiconvergent. A more versatile approach is to use the Ewald procedure, as discussed later in this chapter.

The exchange terms (containing Z_{psrq}) also present three lattice sums, two of which converge exponentially. The third summation also converges except when $k - k'$ is equal to zero. But in that case, the quantity k' is integrated over the Brillouin zone, and the k' integration makes the exchange contribution to the Fock matrix convergent. For polymers that are insulators, it is apparently not a real problem to reach convergence.[10,11] But the convergence rate worsens as the gap between occupied and unoccupied states becomes smaller. Moreover, for metallic chains, the function $P_{sr}(k)$ has a discontinuity in the Brillouin zone, and, according a theorem on the decay of Fourier coefficients,[12] this discontinuity causes the asymptotic decay of the exchange matrix elements $X_{pq}(k)$ to be no better than $(1/m'')^2$. Papers by Delhalle and Calais[13–15] have shown that very many terms may be needed to stabilize the exchange summations and to reproduce correctly the behavior of an orbital at the Fermi level. A technique based on a multipolar expansion has been presented by Stolarczyk et al.,[16,17] but to our knowledge, it has not been implemented in an *ab initio* program.

In summary, the DS RHF method for polymers presents several drawbacks: That of most importance is the difficulty of obtaining a satisfactory rate of convergence for the lattice summations. A second issue is the necessity to use different treatments for the long-range effects in the Coulomb and in the exchange terms.

2.3 Poisson Summation Formula

A partial solution to the problems in the DS solution of the RHF equations for polymers is provided by reformulating the problem in Fourier space (FS).

A key theorem facilitating this approach is provided by the Poisson transformation formula that converts a summation in the DS into one in FS.

The Poisson formula[18] states that the sum of a function evaluated on the points μ of a DS lattice is equal to the sum of its Fourier transform evaluated at the points of the corresponding reciprocal lattice. In the notation used in this paper, the formula is

$$\sum_{\mu=-\infty}^{\infty} f(\mu) = \sum_{\nu=-\infty}^{\infty} f^T(\nu) = \sum_{\nu=-\infty}^{\infty} \int f(\mathbf{r})\exp(2\pi i\nu z)\mathrm{d}\mathbf{r}. \tag{16}$$

A very interesting property of the Gaussian function $e^{-\alpha r^2}$ is that its Fourier transform is also a Gaussian, but with the parameter α now appearing in the denominator of the exponent. This property causes the application of the Poisson formula to a lattice sum of Gaussians to be converted as shown here:

$$S = \sum_{\mu=-M}^{+M} \exp(-\alpha\mu^2) = \sqrt{\frac{\pi}{a}} \sum_{\nu=-N}^{+N} \exp\left(-\frac{\pi^2\nu^2}{\alpha}\right), \quad M \text{ and } N \to \infty. \tag{17}$$

It can be seen from Eq. (17) that for large values of α, the DS lattice sum representing S converges rapidly but is only slowly convergent when α is small, while the opposite of this behavior is observed for the FS representation of the same sum. These convergence patterns are illustrated in Table 1.

Table 1 Number of Terms Required to Achieve an Accuracy of 10^{-6} in Summations S, Eq. (17), in Direct Space and in Fourier Space for Various Values of the Exponent α

α	S	**Direct Space** $2M+1$ ($-M$ to $+M$)	**Fourier Space** $2N+1$($-N$ to $+N$)
100	1.000000	1 (0)	41 (−20 to +20)
10	1.000091	5 (−2 to +2)	15 (−7 to +7)
1	1.772637	15 (−7 to +7)	5 (−2 to +2)
0.1	5.604991	41 (−20 to +20)	1 (0)
0.01	17.724539	123 (−61 to +61)	1 (0)

2.4 Fourier Transform of Coulomb and Exchange terms

As shown in detail elsewhere,[19–22] the Fourier transform formulas given in Table 2 lead to the following formulas for $V_{pq}(k)$ and for the electron–electron Coulomb and exchange integrals:

$$V_{pq}(k) = \frac{1}{\pi a_0}\sum_{\nu}\int \frac{\mathrm{d}\mathbf{q}_0}{q_0^2+\nu^2} S_{pq}(k,\mathbf{q}_\nu)\sum_{C} Q_C \exp(-2\pi i\mathbf{q}_\nu\cdot\mathbf{C}), \quad (18)$$

$$C_{pq}(k) = \frac{1}{\pi a_0}\sum_{\nu}\int \frac{\mathrm{d}\mathbf{q}_0}{q_0^2+\nu^2} S_{pq}(k,\mathbf{q}_\nu)\int_{BZ}\mathrm{d}k'\sum_{rs} P_{sr}(k')S_{rs}(k',-\mathbf{q}_\nu), \quad (19)$$

$$X_{pq}(k) = -\frac{1}{2}\frac{1}{\pi a_0}\int_{BZ}\mathrm{d}k'\sum_{rs} P_{sr}(k') \times\sum_{\nu}\int\frac{\mathrm{d}\mathbf{q_0}}{q_0^2+(\nu+k-k')^2} S_{ps}(k',\mathbf{q}_{\nu+k-k'})S_{rq}(k,-\mathbf{q}_{\nu+k-k'}), \quad (20)$$

with $\mathbf{q}_\nu = \mathbf{q}_0 + \nu\hat{\mathbf{z}}$ and $\mathbf{q}_{\nu+k-k'} = \mathbf{q}_0 + (\nu+k-k')\hat{\mathbf{z}}$. In these equations, the integrations in $\mathbf{q}_0 (= q_x\hat{\mathbf{x}} + q_y\hat{\mathbf{y}})$ are over the xy-plane and $S_{pq}(k,\mathbf{q})$ is a generalized overlap integral defined as

$$S_{pq}(k,\mathbf{q}) = \sum_{\mu=-\infty}^{\infty}\exp(2\pi ik\mu)\int\chi_p^0(\mathbf{r})\exp(2\pi i\mathbf{q}\cdot\mathbf{r})\chi_q^\mu(\mathbf{r})\mathrm{d}\mathbf{r}. \quad (21)$$

Table 2 Fourier Transform Relations (in the Notation of This Chapter)

Fourier transform	$f^T(\mathbf{q}) = [f(\mathbf{r})]^T(\mathbf{q}) = \int_{-\infty}^{\infty} f(\mathbf{r})\mathrm{e}^{2\pi i\mathbf{q}\cdot\mathbf{r}}\mathrm{d}\mathbf{r}$
Poisson summation formula	$\sum_{\mu=-\infty}^{\infty} f(\mu+x) = \sum_{\nu=-\infty}^{\infty} f^T(\nu)\mathrm{e}^{2\pi i\nu x}$
Fourier convolution theorem	$[f(\mathbf{r})g(\mathbf{r})]^T(\mathbf{q}) = \int f^T(\mathbf{p})g^T(\mathbf{q}-\mathbf{p})\mathrm{d}\mathbf{p}$
Lattice orthogonality	$\sum_{\mu=-\infty}^{\infty}\mathrm{e}^{2\pi iq\mu} = \sum_{\nu=-\infty}^{\infty}\delta(q-\nu)$
Fourier transform of $1/\lvert\mathbf{r}\rvert$	$\int\frac{\mathrm{e}^{2\pi i\mathbf{q}\cdot\mathbf{r}}}{\lvert\mathbf{r}\rvert}\mathrm{d}\mathbf{r} = \frac{1}{\pi q^2}$

Note that if $\mathbf{q} = 0$, the above expression reduces to the overlap integral $S_{pq}(k)$.

As we have indicated previously, the electron-nuclear attraction and the electron–electron repulsion contributions to $C_{pq}(k)$ in DS are individually described by divergent series. However, in FS, the divergences become localized in the terms $\nu = 0$ of V_{pq} and C_{pq} and the divergences cancel if the unit cell is electrically neutral.

For $X_{pq}(k)$, the integrand causes the occurrence of a logarithmic singularity when $\nu = k' - k$. However, this singularity is removed if the integration in k' over the Brillouin zone is performed before the integration over $\mathbf{q}_0$.

To complete the Fourier-space evaluation of V_{pq}, C_{pq}, and X_{pq}, we need a more explicit form for the generalized overlap integral $S_{pq}(k, \mathbf{q})$. We start by writing this integral in terms of the GTO basis:

$$S_{pq}(k, \mathbf{q}) = \sum_{ab} d_{ap} d_{bq} S_{ab}(k, \mathbf{q}), \tag{22}$$

where $S_{ab}(k, \mathbf{q})$ is the GTO equivalent of the quantity defined in Eq. (21).

The exponentials in S_{ab} combine to yield a single exponential at a point $\mathbf{P}(\mu)$ calculated according to Eq. (3), and the preexponential factors of $G_a^0 G_b^\mu$ are expanded as polynomials in x_p, y_p, and $z_p(\mu)$, as illustrated here for x_p:

$$x_a^{n_a} x_b^{n_b} = \sum_{n=0}^{n_a + n_b} h_n^{x, n_a, n_b} x_p^n. \tag{23}$$

The coefficients for y_p and z_p are denoted h_l^{y, l_a, l_b} and $h_m^{z(\mu), m_a, m_b}$.

The generalized overlap integral $S_{ab}(k, \mathbf{q})$ can then be written

$$S_{ab}(k, \mathbf{q}) = \sum_{\mu=-\infty}^{\infty} e^{2\pi i k \mu} \sum_{n=0}^{n_a + n_b} \sum_{l=0}^{l_a + l_b} \sum_{m=0}^{m_a + m_b} \mathbf{\Phi}_{nlm}^{ab}(\mu) \hat{\mathcal{D}}_P^{nlm} e^{2\pi i \mathbf{q} \cdot \mathbf{P}(\mu) - \pi^2 q^2 / \alpha_{ab}}, \tag{24}$$

where

$$\mathbf{\Phi}_{nlm}^{ab}(\mu) = h_n^{x, n_a, n_b} h_l^{y, l_a, l_b} h_m^{z(\mu), m_a, m_b} \left(\frac{\pi}{\alpha_{ab}} \right)^{3/2} e^{-\alpha_a \alpha_b (\mathbf{R}_{ab} - \mu \hat{\mathbf{z}})^2 / \alpha_{ab}}, \tag{25}$$

$$\hat{\mathcal{D}}_P^{nlm} = \left(\frac{\partial}{\partial P_x} \right)^n \left(\frac{\partial}{\partial P_y} \right)^l \left(\frac{\partial}{\partial P_z} \right)^m. \tag{26}$$

Here, $\alpha_{ab} = \alpha_a + \alpha_b$, $\mathbf{R}_{ab} = \mathbf{A} - \mathbf{B}$, and $\mathbf{P}(\mu)$ is the center of the Gaussian that is the product of $G_a^0(\mathbf{r})$ and $G_b^\mu(\mathbf{r})$. For later use, we define $\mathbf{Q}(\mu')$ to be the center resulting from the product of $G_c^0(\mathbf{r})$ and $G_d^{\mu'}(\mathbf{r})$.

With the above formula, the matrix elements $V_{ab}(k)$ and $Z_{abcd}(k''; k, k')$ can be reduced to the following forms (see Refs. 23–26 for the details):

$$V_{ab}(k) = -\sum_C \frac{Q_C}{a_0 \pi} \sum_{\mu=-\infty}^{\infty} e^{2\pi i k \mu} \sum_{n=0}^{n_a+n_b} \sum_{l=0}^{l_a+l_b} \sum_{m=0}^{m_a+m_b} \Phi_{nlm}^{ab}(\mu) W_{nlm}^{ab}(\mu, \mathbf{C}) \tag{27}$$

$$Z_{abcd}(k''; k, k') = \sum_{\mu=-\infty}^{\infty} \sum_{\mu'=-\infty}^{\infty} e^{2\pi i (k\mu + k'\mu')} \sum_{n=0}^{n_a+n_b} \sum_{l=0}^{l_a+l_b} \sum_{m=0}^{m_a+m_b} \Phi_{nlm}^{ab}(\mu) \times \sum_{n'=0}^{n_c+n_d} \sum_{l'=0}^{l_c+l_d} \sum_{m'=0}^{m_c+m_d} (-1)^{n'+l'+m'} \Phi_{n'l'm'}^{cd}(\mu') U_{n+n', l+l', m+m'}^{abcd}(k'', \mu, \mu'). \tag{28}$$

The quantities $W_{nlm}^{ab}(\mu, \mathbf{C})$ and $U_{nlm}^{abcd}(k'', \mu, \mu')$ which appear in Eqs. (27) and (28) are obtained when the $\mathbf{q}_0$ integrals in those equations are evaluated. The evaluation yields a special function that is a generalization of an incomplete Bessel function. We define

$$K_{nlm}(x, \mathbf{R}_0^2/\lambda) \equiv \left(\frac{\partial}{\partial R_x}\right)^n \left(\frac{\partial}{\partial R_y}\right)^l K_m(x, \mathbf{R}_0^2/\lambda), \tag{29}$$

where $\mathbf{R}_0$ is a vector in the xy-plane and $K_m(x,y)$ is an incomplete Bessel function whose definition for general index values m is

$$K_m(x, y) = \int_1^\infty \frac{dt}{t^{m+1}} e^{-xt - y/t}. \tag{30}$$

Using this definition,

$$W_{nlm}^{ab}(\mu, \mathbf{C}) = \sum_{\nu=-\infty}^{\infty} (2\pi i \nu)^m e^{2\pi i \nu [P_z(\mu) - C_z]} K_{nl0}\left(\pi^2 \lambda_{ab} \nu^2, \frac{|\mathbf{PC}_0^2|}{\lambda_{ab}}\right), \tag{31}$$

and

$$U_{nlm}^{abcd}(k'', \mu, \mu') = \sum_{\nu=-\infty}^{\infty} [2\pi i(\nu + k'')]^m e^{2\pi i(\nu + k'')[P_z(\mu) - Q_z(\mu')]} \times K_{nl0}\left(\pi^2 \lambda_{abcd} (\nu + k'')^2, \frac{|\mathbf{PQ}_0^2|}{\lambda_{abcd}}\right) \tag{32}$$

Here, $\mathbf{PC}_0$ and $\mathbf{PQ}_0$ are the projections of $\mathbf{P}(\mu) - \mathbf{C}$ and $\mathbf{P}(\mu) - \mathbf{Q}(\mu')$ onto the xy-plane, and λ_{ab} and λ_{abcd} are defined as

$$\lambda_{ab} = \frac{1}{\alpha_{ab}}, \quad \lambda_{abcd} = \frac{1}{\alpha_{ab}} + \frac{1}{\alpha_{cd}}. \tag{33}$$

2.5 Incomplete Bessel Functions

The function $K_m(x,y)$, with $x \geq 0$ and $y \geq 0$, occurs in treatment of the flow of water through fractured media and is called by geologists the leaky aquifer function. This function has received considerable attention in the literature.[27–34] For numerical evaluation of the K_m, see Ref. 35.

The function $K_0(x,y)$ presents a logarithmic behavior when $x \to 0$:

$$K_0(x,y) \approx -\ln(x) - \ln(y) - 2\gamma_E - E_1(y). \tag{34}$$

Here, E_1 is an exponential integral; it and a generalization thereof have definition[36]

$$E_n(z) = \int_1^{\infty} \frac{\mathrm{d}t}{t^n} \mathrm{e}^{-zt}, \quad n = 1, \ 2, \ldots, \tag{35}$$

and γ_E is the Euler–Mascheroni constant ($\gamma_E \approx 0.5772156649$). Methods for the efficient computation of $E_1(x)$ have been given by Amos.[37,38]

On the other hand, the $K_m(x, y)$ with $m \neq 0$ do not have this divergent behavior. Indeed, when x is equal to zero, $K_m(0, y)$ reduces to the incomplete gamma function $\gamma(m, y)$[34]:

$$K_m(0,y) = \frac{1}{y^m} \gamma(m,y). \tag{36}$$

When $m > 0$, Eq. (36) further reduces in the limit $y \to 0$ to

$$K_m(0,0) = \frac{1}{m}. \tag{37}$$

When the second argument y is zero and $m \geq 0$, the incomplete Bessel function $K_m(x, 0)$ reduces to a generalized exponential integral:

$$K_m(x,0) = E_{m+1}(x), \tag{38}$$

while if $x \to \infty$, this function behaves like

$$K_m(x,y) \sim \mathrm{e}^{-x-y}. \tag{39}$$

2.6 Singularities in Coulomb and Exchange Terms

The Fourier transforms of the Coulomb and exchange terms (see Eqs. 19 and 20) exhibit a singularity whose source is a zero denominator in the $\mathbf{q}_0$ integrand when $\mathbf{q}_0$ and ν are both zero. After integration over $\mathbf{q}_0$ this singularity manifests itself through the logarithmic nature of the incomplete Bessel function $K_0(x, y)$ when $x \to 0$.

For the Coulomb term, the apparently divergent character at $\mathbf{q} = 0$ is concentrated in the terms corresponding to $n = l = n' = l' = 0$ in Eqs. 27 and 28).[39] However, because of the charge neutrality of the unit cell, the grouping together of the electron-nuclear attraction and the electron–electron repulsion contributions removes the singularity, leaving only the following net contribution to the Coulomb matrix element when $\nu = 0$ and $n = l = n' = l' = 0$:

$$C^0_{abcd}(k,k') = \frac{1}{a_0}\sum_{\mu\mu'} e^{2\pi i(k\mu + k'\mu')}\left[\sum_{m=0}^{m_a+m_b}\sum_{m'=0}^{m_a+m_b}\Phi^{ab}_{00m}(\mu)\Phi^{cd}_{00m'}(\mu')\right]$$
$$\times\left[\sum_C \frac{Q_C}{n_0}\left[\ln \mathrm{PC}_0^2 + E_1\left(\frac{\mathrm{PC}_0^2}{\eta}\right)\right] - \ln \mathrm{PQ}_0^2 - E_1\left(\frac{\mathrm{PQ}_0^2}{\lambda}\right)\right]. \quad (40)$$

The situation is quite different for the exchange term $Z_{abcd}(k''; k, k')$,[39] in which for certain values of k and k' there is a logarithmic divergence that does not cancel against other contributions. This situation only occurs for the "s-part" of Z_{abcd} and then only in the ν summation in Eq. (32) for the terms $\nu = 0$ and $\nu = \pm 1$. Indeed, since k and k' belong to the interval $[-1/2, 1/2]$, the denominator can vanish in the following three situations:

- $\nu = 0,\ k = k'$
- $\nu = 1,\ k = -k' = -1/2$
- $\nu = -1,\ k = -k' = 1/2$

However, because $P_{sr}(k')$ is, at worst, a bounded piecewise continuous function, the logarithmic divergence at the points where the denominator vanishes is removed by integrating over k'.[22] However, in practice, the integration over $\mathbf{q}_0$ is performed before the integration over k', and a suitable numerical technique must be used to achieve a satisfactory accuracy in the exchange terms. A way to proceed is to subtract the divergent contribution from the incomplete Bessel function $K_0(x, y)$ and treat it in a suitable way without modifying the overall integration scheme. This approach amounts to dividing the exchange matrix element into two parts:

$$X_{pq}(k) = X_{pq}^{(R)}(k) + X_{pq}^{(S)}(k), \tag{41}$$

where the "regular" part is defined as

$$\begin{aligned} X_{pq}^{(R)}(k) &= -\frac{1}{2}\sum_{rq}\int_{BZ} P_{sr}(k')\Bigg\{\sum_{\substack{\nu=-\infty \\ \nu\neq -1,0,1}}^{\infty} Z_{psrq}(k-k'+\nu;k',k) \\ &+ \sum_{\nu=-1}^{1} Z_{psrq}^{(R)}(k-k'+\nu;k',k)\Bigg\}\mathrm{d}k'. \end{aligned} \tag{42}$$

The terms $Z_{psrq}^{(R)}(k-k'+\nu;k',k)$ correspond to the terms in the summations in Eq. (32) that are at the origin of the singularity. The superscript (R) indicates that when Z_{psrq} is evaluated we remove the divergent contribution $\ln(k-k'+\nu)$ from the function $K_0(0, y)$. Since the integrand in $X_{pq}^{(R)}$ is thereby made nonsingular, the integral over k' can be computed using a standard quadrature formula; we use Newton–Cotes formulae.[36]

The "singular" term in Eq. (41) restores the change made by modifying the $K_0(x, y)$ function:

$$X_{pq}^{(S)}(k) = -\frac{1}{2}\sum_{rs}\sum_{\nu=-1}^{1}\int_{-1/2}^{1/2} P_{sr}(k')\ln\left[(k-k'+\nu)^2\right] Z_{psrq}^{(S)}(k-k'+\nu;k,k')\,\mathrm{d}k' \tag{43}$$

with

$$\begin{aligned} Z_{psrq}^{(S)}(k'';k,k') &= \sum_{\mu=-\infty}^{\infty}\sum_{\mu'=-\infty}^{\infty} \mathrm{e}^{2\pi i(k\mu + k'\mu')}\mathrm{e}^{2\pi i k''[P_z(\mu)-Q_z(\mu')]} \\ &\times \sum_{m=0}^{m_p+m_s}\sum_{m'=0}^{m_r+m_q} [2\pi i(k'')]^{m+m'}\Phi_{0,0,m}^{\mathrm{ps}}(\mu)\Phi_{0,0,m'}^{\mathrm{rq}}(\mu'). \end{aligned} \tag{44}$$

It is necessary for practical reasons to employ the same grid points (n in number) that are used for the Coulomb and the regular part of the exchange contributions. Our approach is to develop a quadrature formula that uses those grid points, assigning each point i (for each k and ν) the weight $w_i(k, \nu)$ that will give exact results when any polynomial of degree less than n is multiplied by $[\ln(k-k'+\nu)]^2$ and integrated over k'. The recipe for the weights is

$$w_i(k,\nu) = \int_{k-1/2}^{k+1/2} \ln\left[(t+\nu)^2\right]p_i(t)\,\mathrm{d}t, \tag{45}$$

where $p_i(t)$ is related to the grid points x_i by the Lagrange formula

$$p_i(t) = \prod_{\substack{j=1 \\ j \neq i}}^{n} (t - x_j) \Big/ \prod_{\substack{j=1 \\ j \neq i}}^{n} (x_j - x_i). \tag{46}$$

Integrals of quantities $t^n \ln(t + a)$ have known tabular entries.[36] The weights were generated using Maple 9 software.[40]

Given these weights, the evaluation of $X_{pq}^{(S)}(k)$ takes the form

$$X_{pq}^{(S)}(k) = -\frac{1}{2} \sum_{rs} \sum_{\nu=-1}^{1} \sum_{i} w_i(k, \nu) P_{sr}(k - x_i) Z_{psrq}^{(S)}(x_i + \nu;\ k;\ k - x_i). \tag{47}$$

2.7 Ewald Procedure

The lattice sums over ν in Eqs. (31) and (32) can exhibit slow convergence when the Gaussian exponents are large, thereby generating small values of λ_{ab} or λ_{abcd}. Table 3 shows the number of terms needed to reach convergence to 10^{-12} for some values of λ and for some values of the orbital quantum numbers. Data are shown both for orbitals centered on and off the z-axis. In this table, k'' is set to zero and the logarithmic singularity of the incomplete Bessel function is excluded since it is canceled by the electron-nuclear contribution to the Coulomb potential.

Table 3 Convergence of the ν-Lattice Summation in Eq. (32) as a Function of the Magnitude of λ, as Characterized by the Number (N_{FS}) of Terms Larger than 10^{-12} in Absolute Value (Here k'' Was Set to Zero)

P −Q	$n + l + m$	λ	N_{FS}
{0,0,0}	0	10^{-6}	3145
	0	10^{-3}	99
	4	2.0	7
	8	0.15	21
{0.2,0,0.25}	0	10^{-6}	43
	0	10^{-3}	43
	4	2.0	7
	8	0.15	17

The values of λ equal to 10^{-6}, 2, and 0.15 respectively correspond roughly to the most localized component of the 1s AO, to the outer 3sp AOs, and to 3d AOs in the 6-311G** basis for Si in polysilane ($a_0 = 3.95$ bohr). The value $\lambda = 10^{-3}$ corresponds to the most localized Gaussian of the 1s AO in the 3-21G basis set. The table illustrates the fact that the FS approach transfers the region of slow convergence to a different part of the parameter space, and that the problem of convergence in practical computations is not eliminated.

It is therefore important to use the Ewald procedure,[41] which partitions the integrals into two parts, with one part computed in FS and the other part in DS. In the current context, the starting point of the Ewald procedure is the partitioning of the one-dimensional integration that led to the incomplete Bessel function K_m. The essence of the procedure can be illustrated by writing a prototype FS sum in the following way:

$$S = \sum_{\nu} \int_{1}^{\infty} \frac{\mathrm{d}t}{t^{n+1}} \mathrm{e}^{-\nu^2 \alpha t - \beta/t} = S_1 + S_2, \tag{48}$$

with

$$S_1 = \sum_{\nu} \int_{\tau}^{\infty} \frac{\mathrm{d}t}{t^{n+1}} \mathrm{e}^{-\nu^2 \alpha t - \beta/t}, \quad S_2 = \sum_{\nu} \int_{1}^{\tau} \frac{\mathrm{d}t}{t^{n+1}} \mathrm{e}^{-\nu^2 \alpha t - \beta/t}. \tag{49}$$

Note that the separation is at a parameter $\tau > 1$. By a change of variable to $t' = \tau t$, integral S_1 can be brought to the same form as S, but with different arguments, including in particular the replacement of α by $\tau\alpha$, thereby yielding an FS lattice sum with an improved convergence rate.

Summation S_2 is subjected to a Poisson transformation, converting it into a DS sum in which the summed exponential is $\exp(-\pi^2\mu^2/\alpha\tau)$. After manipulation, we retrieve a summation which converges rapidly for small α.

This procedure is described in detail in Refs. 25 and 26. The formulas obtained in this way for W^{ab}_{nlm}, Eq. (31), and U^{abcd}_{nlm}, Eq. (32), consist of sums of two contributions, one with a lattice summation in the DS and the other in FS:

$$W^{ab}_{nlm}(\mu, \mathbf{C}) = W^{ab}_{\mathrm{DS},nlm}(\mu, \mathbf{C}) + W^{ab}_{\mathrm{FS},nlm}(\mu, \mathbf{C}), \tag{50}$$

$$U^{abcd}_{nlm}(k'', \mu, \mu') = U^{abcd}_{\mathrm{DS},nlm}(k'', \mu, \mu') + U^{abcd}_{\mathrm{FS},nlm}(k'', \mu, \mu'), \tag{51}$$

where

$$W_{\mathrm{DS},nlm}^{ab}(\mu,\mathbf{C}) = \frac{2}{\sqrt{\pi}}\sum_{\nu}\left[\frac{1}{\sqrt{\lambda_{ab}}}F_{nlm}\left(\frac{[\mathbf{P}(\mu)-\mathbf{C}+\nu\hat{\mathbf{z}}]^2}{\lambda_{ab}}\right) - \frac{1}{\sqrt{\lambda_{ab}+\tau'}}F_{nlm}\left(\frac{[\mathbf{P}(\mu)-\mathbf{C}+\nu\hat{\mathbf{z}}]^2}{\lambda_{ab}+\tau'}\right)\right], \tag{52}$$

$$W_{\mathrm{FS},nlm}^{ab}(\mu,\mathbf{C}) = \sum_{\nu=-\infty}^{\infty}(2\pi i\nu)^m e^{2\pi i\nu(P_z(\mu)-C_z)} \times K_{nl0}\left(\pi^2(\lambda_{ab}+\tau')\nu^2, \frac{|\mathbf{PC}_0^2|}{(\lambda_{ab}+\tau')}\right), \tag{53}$$

$$U_{\mathrm{DS},nlm}^{abcd}(k'',\mu,\mu') = \frac{2}{\sqrt{\pi}}\sum_{\nu}e^{2\pi i\nu k''} \times\left[\frac{1}{\sqrt{\lambda_{abcd}}}F_{nlm}\left(\frac{[\mathbf{P}(\mu)-\mathbf{Q}(\mu')+\nu\hat{\mathbf{z}}]^2}{\lambda_{abcd}}\right) - \frac{1}{\sqrt{\lambda_{abcd}+\tau''}}F_{nlm}\left(\frac{[\mathbf{P}(\mu)-\mathbf{Q}(\mu')+\nu\hat{\mathbf{z}}]^2}{\lambda_{abcd}+\tau''}\right)\right], \tag{54}$$

$$U_{\mathrm{FS},nlm}^{abcd}(k'',\mu,\mu') = \sum_{\nu=-\infty}^{\infty}[2\pi i(\nu+k'')]^m e^{2\pi i(\nu+k'')[P_z(\mu)-Q_z(\mu')]} \times K_{nl0}\left(\pi^2(\lambda_{abcd}+\tau'')(\nu+k'')^2, \frac{|\mathbf{PQ}_0^2|}{(\lambda_{abcd}+\tau'')}\right) \tag{55}$$

The quantities τ' and τ'' are partitioning constants (not necessarily equal) which control the division of the lattice summations into DS and FS components. The function $F_{nlm}(x)$ is the same function that arises in the formulation of Gaussian integrals for molecular problems[42]:

$$F_0(x) = \int_0^1 dt\, e^{-xt^2} = \frac{1}{2}\sqrt{\frac{\pi}{x}}\,\mathrm{erf}(\sqrt{x}), \tag{56}$$

$$F_{nlm}(|\mathbf{R}|^2/\lambda) = \hat{\mathcal{D}}_R^{nlm}F_0(|\mathbf{R}|^2/\lambda). \tag{57}$$

The values of U or W do not depend upon τ' or τ'', so the values of these parameters can be chosen to optimize either the convergence rate or the overall computational efficiency. In the limit of large τ (a partitioning which would be entirely DS), the use of Eq. (52) or (54) would lead to the known result of summing Gaussians over the entire DS lattice; finite values of τ diminish the contribution of the DS sum; in the limit $\tau = 0$, the DS sum completely vanishes.

One way to select the partitioning parameter τ' or τ'' is to require that the summations over ν appearing in the series U_{DS} (Eq. 54) and U_{FS} (Eq. 55) [or in series W_{DS} (Eq. 52) and W_{FS} (Eq. 53)] have the same rate of convergence. On the basis of the properties of the incomplete Bessel and error functions, which have similar asymptotic rates of decay for sufficiently large values of their arguments,[34–36] this choice of τ corresponds to

$$\pi^2(\lambda+\tau)=\frac{1}{\lambda+\tau}, \tag{58}$$

which leads to an optimum $\tau^* = \pi^{-1} - \lambda$ if $0 < \lambda \leq \pi^{-1}$, where λ is defined in Eq. (33). If $\lambda > \pi^{-1}$, τ^* is set to zero and only Eq. (53) or (55) is used.

Table 4 lists the number of terms needed in Eqs. (54) (N_{DS}) and (55) (N_{FS}) that contribute to the lattice sums with an absolute value greater than 10^{-12}, for three different values of λ (10^{-6}, 2.0, and 0.15), four values of τ (0.0, $\pi^{-1} - \lambda = \tau^*$, 1.0, and 1000.0), and two values of $\mathbf{P} - \mathbf{Q}$ ({0.0,0.0,0.0} and ({0.2,0.0,0.25}). For $\lambda = 2.0$, the value of τ^* is negative and that value must be excluded; the best τ is then 0.0. In this table, k'' is set to zero and the logarithmic singularity of the incomplete Bessel function is excluded as in Table 3.

From Table 4, it can be seen that the theoretical optimum value of τ, τ^*, based on asymptotic considerations, does fairly well at equalizing the two expansion lengths. The entry for $\tau = 0$, which corresponds to a computation entirely in FS, shows that for small λ, the lattice summation in Eq. (54) exhibits a rather slow convergence as already noted in Table 3, while for $\tau = 1000.0$ (computation entirely in DS), the series requires a large number of terms to reach convergence. For $\lambda = 2$, the best convergence is obtained by the lattice sums in the FS ($\tau = 0$), while for $\lambda = 0{,}15$, the best value of τ is formally $\tau^* = \pi^{-1} - \lambda$, but since the calculation of the functions $F_n(x)$ is faster than the computation of the incomplete Bessel functions K_m, the value $\tau = 1.0$ would be a reasonable choice.

3. SPATIAL SYMMETRY IN LINEAR CHAINS

Stereoregular chain molecules often possess, in addition to translational periodicity, symmetry elements such as rotational or screw axes or glide planes of arbitrary order relative to the translational axis, and/or axes of order 2 perpendicular to that axis. The full symmetry group of such a system is known as a *line group*,[43] defined as the symmetry group of a three-dimensional object that is translationally periodic along a line.

Table 4 Number of Terms in Absolute Value Greater than 10^{-12} in the Lattice Summations in Eqs. (54) (N_{DS}) and (55) (N_{FS}) for Different Values of λ and τ ($\tau^* = \pi^{-1} - \lambda$, $k'' = 0$)

P − Q	$n+l+m$	λ	τ	N_{DS} in Eq. (54)	N_{FS} in Eq. (55)
{0.0,0.0,0.0}	0	10^{-6}	0.0	–	3145
	0	10^{-6}	τ^*	5	5
	0	10^{-6}	1.0	9	3
	0	10^{-6}	1000.0	285	0
{0.2,0.0,0.25}	0	10^{-6}	0.0	–	43
	0	10^{-6}	τ^*	6	5
	0	10^{-6}	1.0	10	3
	0	10^{-6}	1000.0	286	0
{0.0,0.0,0.0}	0	2.0	0.0	–	7
	0	2.0	τ^*	–	–
	0	2.0	1.0	37	3
	0	2.0	1000.0	479	3
{0.2,0.0,0.25}	0	2.0	0.0	–	7
	0	2.0	τ^*	–	–
	0	2.0	1.0	35	3
	0	2.0	1000.0	444	3
{0.0,0.0,0.0}	0	0.15	0.0	–	21
	0	0.15	τ^*	13	13
	0	0.15	1.0	27	7
	0	0.15	1000.0	377	3
{0.2,0.0,0.25}	0	0.15	0.0	–	17
	0	0.15	τ^*	15	11
	0	0.15	1.0	25	7
	0	0.15	1000.0	353	3

The symmetry elements of a line group leave invariant the straight line connecting the centers of mass of the unit cells; in this paper that line is designated the z-axis. Line groups have been studied extensively by Božović and coworkers.[44–49]

Exploitation of the line-group symmetry of a polymer makes possible a significant reduction in the computational effort needed to construct its Fock matrix. The procedure we use for that purpose is an extension to line groups of a technique originally developed for molecular systems with point-group symmetry by Dupuis and King.[50] The main advantage of that type of procedure is that it reduces most of the effort in constructing and manipulating the two-electron integrals to processes that scale as the cube, rather than as the fifth power of the orbital basis. To proceed in this direction, we need to develop some appropriate line-group formulas.

The symmetry operations of a line group can be written $\{\hat{R}|n+\tau\}$, where $\hat{R}$ is a proper or improper rotation about a line through the origin of the translational axis, n is an integer number of unit-cell translations, and τ $(0 \le \tau < 1)$ is a translation by a proper fraction of a unit cell. Application of a symmetry operation affects a point $\mathbf{A}$ as follows:

$$\{\hat{R}|n+\tau\}\mathbf{A} = \hat{R}\mathbf{A} + (\tau + n)\hat{\mathbf{z}}. \tag{59}$$

If a line-group operation is applied to an AO with specified angular-momentum quantum numbers l (azimuthal) and m (magnetic), here designated $\chi_{p;lm}$, the result will be an AO of the same l centered at a point that is symmetry-equivalent to $\mathbf{A}_p$, but with a linear combination of m values determined by $\hat{R}$ and the quantum numbers. If we require that all $\mathbf{A}_p$ be within the reference unit cell (Cell 0), the point equivalent to $\mathbf{A}_p$ can be identified as $\mathbf{A}_{p'} + \delta_p$, where δ_p is the (signed) number of unit-cell translations (if any) needed to make $\mathbf{A}_{p'}$ lie within the reference unit cell. Some line groups include rotations $\hat{R}$ that convert z into $-z$; to deal with this possibility we define ε_R as ± 1, with the minus sign chosen if the sign of z is reversed by $\hat{R}$.

Based on the above considerations, we can write

$$\{\hat{R}|\tau\}\chi_{p;lm}(\mathbf{r} - \mathbf{A}_p - \mu\hat{\mathbf{z}}) = \sum_{m'} \chi_{p';lm'}(\mathbf{r} - \mathbf{A}_{p'} - \mu'\hat{\mathbf{z}})\mathsf{R}^{p}_{m'm}, \tag{60}$$

where

$$\mu' = \varepsilon_R\,\mu + \delta_p, \tag{61}$$

and R^p is a matrix whose elements describe the effect of performing a rotation $\hat{R}$ on the shell of symmetry-equivalent AOs that includes AO p (of angular momentum l). The nonzero elements of these rotation matrices can be found in the literature (see, for example, Ref. 51).

Following Dovesi,[52] we can now write the result of applying a line-group operation to a Bloch function of specified quantum numbers:

$$\{\hat{R}|\tau\}\eta_{p;lm}(k,\mathbf{r}) = \lim_{N\to\infty}(2N+1)^{-1/2}\sum_{\mu=-\infty}^{\infty} e^{2\pi i\mu k}\{\hat{R}|\tau\}\chi_{p;lm}(\mathbf{r}-\mathbf{A}_p-\mu\hat{\mathbf{z}})$$
$$= e^{-2\pi i\varepsilon_R\,\delta_p k}\sum_{m'}\eta_{p';lm}(\varepsilon_R\,k,\mathbf{r})\,\mathsf{R}^p_{m'm}. \tag{62}$$

To reach the last line of the above equation, we replaced μ by $\varepsilon_R\mu' - \varepsilon_R\delta_p$, changed the summation variable to μ', and identified most of the resulting expression as $\eta_{p';lm'}$.

There will be a matrix of the type R^p for each set of symmetry-equivalent Bloch orbitals; it is convenient to collect them for the complete set of Bloch orbitals (for each $\{\hat{R},\tau\}$, abbreviated $\hat{R}\tau$, and k) into a single $\omega \times \omega$ matrix $\mathsf{W}(\hat{R}\tau,k)$ which is entirely zero except for blocks containing (for each symmetry-equivalent set, here labeled p) the elements

$$\mathsf{W}_p(\hat{R}\tau,k) = e^{-2\pi i\varepsilon_R\delta_p k}\mathsf{R}^p_{m'm}. \tag{63}$$

For each k, the set of matrices $\mathsf{W}(\hat{R}\tau,k)$ constitute a unitary representation of the line group with a basis consisting of the Bloch sums.

Following along the lines of the procedure of Dupuis and King, one can use the fact that the Fock matrix F commutes with all the line-group operations to simplify its computation. To do so, we need to make the following observations, considering initially the kinetic-energy part of F:

(1) For any $\hat{R}\tau$ and k, $\mathsf{W}^\dagger(\hat{R}\tau,k)\mathsf{T}(k)\mathsf{W}(\hat{R}\tau,k) = \mathsf{T}(k)$, and for a group with g members,

$$\mathsf{T}(k) = \frac{1}{g}\sum_{\hat{R}\tau}\mathsf{W}^\dagger(\hat{R}\tau,k)\mathsf{T}(k)\mathsf{W}(\hat{R}\tau,k). \tag{64}$$

(2) If $P'Q'$ denotes a pair of shells that (in either order) are symmetry-equivalent to the pair PQ, then the $P'Q'$ contribution to $\mathsf{T}(k)$ consists of a block of integrals $T_{P'Q'}(k)$ that is equivalent to the block of integrals from PQ (or to its adjoint). The adjoint is produced by the permutation of PQ to QP and by rotations that change the sign of z, and it will therefore occur in half the set of symmetry-equivalent contributions.

(3) The number of line-group operations leaving invariant (except for order) a shell pair PQ can be counted; we assign it the symbol $n(PQ)$. This means that the number of times a specific block (or its adjoint) occurs in the summation of Eq. (64) is $2g/n(PQ)$.

(4) Based on (1)–(3) above, the summation in Eq. (64) will yield the same result as a summation in which the right-hand-side $\mathsf{T}(k)$ is replaced by a "skeleton" matrix $\overline{\mathsf{T}}(k)$ that contains only the symmetry-inequivalent contributions, each multiplied by $2g/n(PQ)$, as follows:

$$\mathsf{T}(k) = \frac{1}{2g}\sum_{\hat{R}\tau} \mathsf{W}^{\dagger}(\hat{R}\tau, k)\left(\overline{\mathsf{T}}(k) + \overline{\mathsf{T}}^{\dagger}(k)\right)\mathsf{W}(\hat{R}\tau, k). \tag{65}$$

By averaging $\overline{\mathsf{T}}(k)$ and $\overline{\mathsf{T}}^{\dagger}(k)$, we have caused the adjoint contributions to be entered properly. For more detail on the use of the skeleton matrix, see Ref. 50.

The above discussion corresponds to the definition

$$\bar{T}_{PQ}(k) = \begin{cases} \dfrac{2g}{n(PQ)} T_{pq}(k), & PQ \ \epsilon \ P_2 \\ 0, & \text{otherwise.} \end{cases} \tag{66}$$

The notation in Eq. (66) indicates that P and Q are to be assigned all values in the designated shells; P_2 has as members all the symmetry-inequivalent shell pairs.

Similar considerations apply to the two-electron contributions to $\mathsf{F}(k)$, namely $\mathsf{C}(k)$ and $\mathsf{X}(k)$. For integrals involving shells P, Q, R, and S, we again note that half the integrals of each symmetry-equivalent set will occur as complex conjugates of the others; we also introduce the notation $n(PQ,RS)$ to denote the number of times the shell pairs PQ and RS remain unchanged under the full set of line-group operations, and thereby observe that the number of times a particular two-electron integral (or its conjugate) with these shell pairs occurs is $8g/n(PQ,RS)$. Based on the above, the overall two-electron contributions to F take the form

$$\mathsf{C}(k) + \mathsf{X}(k) = \frac{1}{2g}\sum_{\hat{R}\tau} \mathsf{W}^{\dagger}(\hat{R}\tau, k)\left(\overline{\mathsf{C}}(k) + \overline{\mathsf{C}}^{\dagger}(k) + \overline{\mathsf{X}}(k) + \overline{\mathsf{X}}^{\dagger}(k)\right)\mathsf{W}(\hat{R}\tau, k), \tag{67}$$

where

$$\bar{C}_{PQ}(k) = \begin{cases} \displaystyle\sum_{rs} \frac{8g}{n(PQ,RS)} \int_{BZ} Z_{PQRS}(0;\, k, k') P_{SR}(k')\, \mathrm{d}k', & (PQ,RS) \, \epsilon \, P_4 \\ 0, & \text{otherwise;} \end{cases} \tag{68}$$

$$\bar{X}_{PQ}(k) = \begin{cases} -\frac{1}{2}\sum_{rs} \frac{8g}{n(PS,RQ)} \int_{BZ} Z_{PSRQ}(k-k';\, k',k) P_{SR}(k')\, \mathrm{d}k', & (PS,RQ) \in P_4 \\ 0, & \text{otherwise.} \end{cases} \tag{69}$$

Here, P_4 has as members the symmetry-inequivalent four-shell sets (note that they are different for **C** than for **X**).

We do not develop the formulas here in detail, but a similar approach can be applied to the electron-nuclear interactions. The contributions to V_{pq} from nuclei C can be identified as terms with multiplicities given by $2g/n(PQ,CC)$ for only those P, Q, and C that are in the set $(PQ,CC) \in P_4$.

The advantages of the exploitation of symmetry are now particularly apparent from the two-electron part of **F**. The process described here involves only two indices, not the four that would be required for a full transformation of all the two-electron integrals. The overall burden of integral computation has been reduced by a factor proportional to g^2, while that of forming the Fock matrix (in each iteration) has been reduced by a factor proportional to g.

4. COMPUTATIONAL DETAILS

4.1 Symmetry Properties of the Integrals

There are symmetries additional to the spatial symmetry that can be used to reduce the number of integrals to be computed and stored. For example, in the overlap matrix $\mathbf{S}(k)$, Eq. (6), the $\left\langle \chi_p(\mathbf{r})\chi_q^{\mu}(\mathbf{r}) \right\rangle$ integrals are invariant with respect to any integer shift of the origin for $\mathbf{r}$, so that

$$S_{pq}(k) = \sum_{\mu=-\infty}^{\infty} \exp(2\pi i\mu k)\left\langle \chi_p(\mathbf{r})\chi_q(\mathbf{r}-\mu\hat{\mathbf{z}}) \right\rangle$$
$$= \sum_{\mu=-\infty}^{\infty} \exp(2\pi i\mu k)\left\langle \chi_p(\mathbf{r}+\mu\hat{\mathbf{z}})\chi_q(\mathbf{r}) \right\rangle.$$

By setting $\mu \to -\mu$ and taking into account that the summation over μ runs from $-\infty$ to $+\infty$, we get

$$\begin{aligned} S_{pq}(k) &= \sum_{\mu=-\infty}^{\infty} \exp(-2\pi i \mu k)\left\langle \chi_q(\mathbf{r})\chi_p(\mathbf{r}-\mu\hat{\mathbf{z}})\right\rangle \\ &= S_{qp}(-k) = S_{qp}^{*}(k). \end{aligned} \tag{70}$$

In the same way, it is possible to show that $\mathsf{T}(k)$ (Eq. 8) and $\mathbf{V}(k)$ (Eq. 9) have the similar properties:

$$T_{pq}(k) = \ T_{qp}(-k) = T_{qp}^{*}(k), \tag{71}$$

$$V_{pq}(k) = \ V_{qp}(-k) = V_{qp}^{*}(k). \tag{72}$$

The two-electron integrals between real AOs are also invariant when applying an integer shift to $\mathbf{r}_1$ or $\mathbf{r}_2$. For example, a shift of μ units to both $\mathbf{r}_1$ and $\mathbf{r}_2$ yields

$$\begin{aligned} &\left\langle \chi_p^0(\mathbf{r}_1)\chi_r^{\mu''}(\mathbf{r}_2)\left|\frac{1}{r_{12}}\right|\chi_q^{\mu}(\mathbf{r}_1)\chi_s^{\mu'+\mu''}(\mathbf{r}_2)\right\rangle \\ &= \left\langle \chi_p^{-\mu}(\mathbf{r}_1)\chi_r^{\mu''-\mu}(\mathbf{r}_2)\left|\frac{1}{r_{12}}\right|\chi_q^{0}(\mathbf{r}_1)\chi_s^{\mu'+\mu''-\mu}(\mathbf{r}_2)\right\rangle \\ &= \left\langle \chi_q^0(\mathbf{r}_1)\chi_s^{\mu''+\mu'-\mu}(\mathbf{r}_2)\left|\frac{1}{r_{12}}\right|\chi_p^{-\mu}(\mathbf{r}_1)\chi_r^{\mu''-\mu}(\mathbf{r}_2)\right\rangle. \end{aligned} \tag{73}$$

Similar transformations include[53]:

$$\begin{aligned} &\left\langle \chi_p^0(\mathbf{r}_1)\chi_r^{\mu''}(\mathbf{r}_2)\left|\frac{1}{r_{12}}\right|\chi_q^{\mu}(\mathbf{r}_1)\chi_s^{\mu'+\mu''}(\mathbf{r}_2)\right\rangle \\ &= \left\langle \chi_p^0(\mathbf{r}_1)\chi_s^{\mu'+\mu''}(\mathbf{r}_2)\left|\frac{1}{r_{12}}\right|\chi_q^{\mu}(\mathbf{r}_1)\chi_r^{\mu''}(\mathbf{r}_2)\right\rangle \\ &= \left\langle \chi_q^0(\mathbf{r}_1)\chi_r^{\mu''-\mu}(\mathbf{r}_2)\left|\frac{1}{r_{12}}\right|\chi_p^{-\mu}(\mathbf{r}_1)\chi_s^{\mu''+\mu'-\mu}(\mathbf{r}_2)\right\rangle \\ &= \left\langle \chi_q^0(\mathbf{r}_1)\chi_s^{\mu''+\mu'-\mu}(\mathbf{r}_2)\left|\frac{1}{r_{12}}\right|\chi_p^{-\mu}(\mathbf{r}_1)\chi_r^{\mu''-\mu}(\mathbf{r}_2)\right\rangle \\ &= \left\langle \chi_r^0(\mathbf{r}_1)\chi_p^{-\mu''}(\mathbf{r}_2)\left|\frac{1}{r_{12}}\right|\chi_s^{\mu'}(\mathbf{r}_1)\chi_q^{\mu-\mu''}(\mathbf{r}_2)\right\rangle \\ &= \left\langle \chi_r^0(\mathbf{r}_1)\chi_q^{\mu-\mu''}(\mathbf{r}_2)\left|\frac{1}{r_{12}}\right|\chi_s^{\mu'}(\mathbf{r}_1)\chi_p^{-\mu''}(\mathbf{r}_2)\right\rangle \\ &= \left\langle \chi_s^0(\mathbf{r}_1)\chi_p^{-\mu'-\mu''}(\mathbf{r}_2)\left|\frac{1}{r_{12}}\right|\chi_r^{-\mu'}(\mathbf{r}_1)\chi_q^{\mu-\mu'-\mu''}(\mathbf{r}_2)\right\rangle \\ &= \left\langle \chi_s^0(\mathbf{r}_1)\chi_q^{\mu-\mu'-\mu''}(\mathbf{r}_2)\left|\frac{1}{r_{12}}\right|\chi_r^{-\mu'}(\mathbf{r}_1)\chi_p^{-\mu'-\mu''}(\mathbf{r}_2)\right\rangle. \end{aligned} \tag{74}$$

Applying these identities, the integrals $\Xi_{pqrs}(k''; \mu, \mu')$ defined in Eq. (13) satisfy the following relations:

$$\begin{aligned}\Xi_{pqrs}(k''; \mu, \mu') &= \Xi^*_{pqrs}(-k''; \mu, \mu') \\ &= \exp(-2\pi i k'' \mu')\Xi_{pqsr}(k''; \mu, -\mu') \\ &= \exp(2\pi i k'' \mu)\Xi_{qprs}(k''; -\mu, \mu') \\ &= \exp(2\pi i k''(\mu - \mu')\Xi_{qpsr}(k''; -\mu, -\mu') \\ &= \Xi_{rspq}(-k''; \mu', \mu) = \Xi^*_{rspq}(k''; \mu', \mu) \\ &= \exp(2\pi i k'' \mu)\Xi_{rsqp}(-k''; \mu', -\mu) \\ &= \exp(-2\pi i k'' \mu')\Xi_{srpq}(-k''; -\mu', \mu) \\ &= \exp(2\pi i k''(\mu - \mu')\Xi_{srqp}(-k''; -\mu', -\mu).\end{aligned} \tag{75}$$

These integrals are also invariant under translation in k'' by any integer n:

$$\Xi_{pqrs}(k''; \mu, \mu') = \Xi_{pqrs}(k'' + n; \mu, \mu'). \tag{76}$$

These equalities make it necessary to compute only integrals $\Xi_{pqrs}(k''; \mu, \mu')$ with $k'' \geq 0$, $p \geq q$, $r \geq s$, and $(pq) \geq (rs)$ (meaning that (pq) does not precede (rs) in lexical order).

4.2 Symmetry Reduction of the Matrix Elements

The symmetry properties of Ξ_{pqrs} lead to corresponding properties for the matrix elements $Z_{pqrs}(k''; k, k')$ (Eq. 12). In the special case $k'' = 0$ (relevant for the Coulomb terms), we note the following identities:

$$\begin{aligned}Z_{pqrs}(0; k, k') &= Z_{pqsr}(0; k, -k') = Z_{qprs}(0; -k, k') = Z_{qpsr}(0; -k, -k') \\ &= Z_{rspq}(0; k', k) = Z_{rsqp}(0; k', -k) = Z_{srpq}(0; -k', k) \\ &= Z_{srqp}(0; -k', -k).\end{aligned} \tag{77}$$

A complete set of the independently valued $Z_{pqrs}(0; k, k')$ thus consists of those for $k \geq 0$ and $-\frac{1}{2} < k' \leq \frac{1}{2}$, with $p \geq q$, $r \geq s$, and $(pq) \geq (rs)$. Consequently, the rs double sum in $C_{pq}(k)$ may be restricted to values $r \geq s$:

$$
\begin{aligned}
C_{pq}(k) &= \sum_{r\geq s} \int_{-1/2}^{1/2} [P_{sr}(k')Z_{pqrs}(0;k,k') + (1-\delta_{rs})P_{sr}(-k')Z_{pqrs}(0;k,-k')]\,\mathrm{d}k' \\
&= \sum_{r\geq s} \int_{0}^{1/2} [P_{sr}(k')Z_{pqrs}(0;k,k') + (1-\delta_{rs})P_{sr}(-k')Z_{pqrs}(0;k,-k') \\
&\quad + P_{sr}(-k')Z_{pqrs}(0;k,-k') + (1-\delta_{rs})P_{sr}(k')Z_{pqrs}(0;k,k')]\,\mathrm{d}k' \\
&= \sum_{r\geq s} (2-\delta_{rs}) \int_{0}^{1/2} [P_{sr}(k')Z_{pqrs}(0;k,k') + P_{sr}(-k')Z_{pqrs}(0;k,-k')]\,\mathrm{d}k'
\end{aligned}
\tag{78}
$$

The exchange part of the Fock matrix contains integrals $Z_{psrq}(k'';k',k)$ with k'' generally nonzero:

$$
X_{pq}(k) = -\frac{1}{2}\sum_{rs} \int_{-1/2}^{1/2} P_{sr}(k')Z_{psrq}(k-k';k',k)\mathrm{d}k'. \tag{79}
$$

These terms present the following permutation symmetries:

$$
\begin{aligned}
Z_{psrq}(k-k';k',k) &= Z_{spqr}(k-k';-k,-k') = Z_{rqps}(-(k-k');k,k') \\
&= Z_{qrsp}(-(k-k');-k',-k),
\end{aligned}
\tag{80}
$$

and four additional equivalencies based on the identity

$$
Z_{psrq}(-(k-k'),-k,-k') = Z^{*}_{psrq}(k-k';k,k').
$$

Note that $Z_{psrq}(k-k';k,k')$ has only half as many symmetry equivalences as $Z_{psrq}(0;k,k')$.

One way of specifying an independently valued set of $Z_{psrq}(k-k';k,k')$ would be to require $k \geq 0$, $-\frac{1}{2} < k' \leq \frac{1}{2}$, with $p \geq q$, $r \geq s$, $(rs) \leq (pq)$, but store values of both Z_{psrq} and Z_{prsq}. Based on the above, we only need $X_{pq}(k)$ with $p \geq q$ and $k \geq 0$.

We can write

$$\begin{aligned} X_{pq}(k) &= \sum_{rs} \int_0^{1/2} \left[P_{sr}(k') Z_{psrq}(k-k';k',k) + P_{sr}(-k') Z_{psrq}(k+k';-k',k) \right] \mathrm{d}k' \\ &= \sum_{r \geq s} \int_0^{1/2} \Big[P_{sr}(k') Z_{psrq}(k-k';k',k) + P_{sr}(-k') Z_{psrq}(k+k';-k',k) \\ &\quad + (1-\delta_{rs}) (P^*_{sr}(k') Z_{prsq}(k-k';k',k) + P^*_{sr}(-k') Z_{prsq}(k+k';-k',k) \Big] \mathrm{d}k'. \end{aligned} \tag{81}$$

Note that Z_{psrq} and Z_{prsq} will contribute to both $X_{pq}(k)$ and $X_{rs}(k')$.

4.3 Construction of C_{pq} and X_{pq} matrix elements

To make optimal use of symmetry-inequivalent sets of $Z_{psrq}(k'';k,k')$, it is useful to consider how they might be used efficiently to form the $C_{pq}(k)$ and $X_{pq}(k)$ matrix elements. Assuming that these matrix elements are to be computed on a mesh of points k_i, with the corresponding values indicated $C_{pq}[i]$ and $X_{pq}[i]$, and defining $w_{sr}[j]$ as the combination of $P_{sr}(k_j)$ and the weight factor associated with the point k_j when the k' integration is approximated by a summation over j, a given pair of members of the symmetry-inequivalent set of $Z_{pqrs}(0;k_i, k'_j)$ and $Z_{pqrs}(0;k_i, -k'_j)$, henceforth denoted $Z_{pqrs}[0;i,j]$ and $Z_{pqrs}[0;i,-j]$, will contribute to $C_{pq}[i]$ and $C_{rs}[j]$ by respective amounts (shown by the notation $\Leftarrow$)

$$C_{pq}[i] \Leftarrow (2-\delta_{rs})((w_{sr}[j]\, Z_{pqrs}[0;i,j] + w^*_{sr}[j]\, Z^*_{pqrs}[0;i,-j]), \tag{82}$$

$$C_{rs}[j] \Leftarrow (1-\delta_{(pq)(rs)})((2-\delta_{pq})(w_{qp}[i]\, Z_{pqrs}[0;i,j] + w^*_{qp}[i]\, Z^*_{pqrs}[0;i,-j]). \tag{83}$$

The above formulas are for $i,j \geq 0$, $p \geq q$, $r \geq s$, and $(pq) \geq (rs)$. Note that the two terms on the right-hand side of Eq. (82) are identical when $j = 0$ but both must be kept; a similar remark applies to the case $i = 0$ in Eq. (83).

For $X_{pq}(k)$, we have the following contributions, again for $i,j \geq 0$, $p \geq q$, $r \geq s$, and $(pq) \geq (rs)$, from the group of four symmetry-inequivalent quantities $Z_{psrq}[i \mp j);\pm j,i]$ and $Z_{prsq}[i \mp j;\pm j,i]$:

$$\begin{aligned} X_{pq}[i] \Leftarrow\, & w_{sr}[j](Z_{psrq}[i-j;j,i] + (1-\delta_{rs}) Z_{prsq}[i+j;-j,i]) \\ & + w^*_{sr}[j](Z^*_{psrq}[i+j;-j,i] + (1-\delta_{rs}) Z^*_{prsq}[i-j;j,i]), \end{aligned} \tag{84}$$

$$\begin{aligned} X_{rs}[j] \Leftarrow\, & (1-\delta_{(pq)(rs)})[w_{qp}[i](Z_{psrq}[i-j;j,i] + (1-\delta_{qp}) Z_{prsq}[i+j;-j,i]) \\ & + w^*_{qp}[i](Z^*_{psrq}[i+j;-j,i] + (1-\delta_{pq}) Z^*_{prsq}[i-j;j,i])]. \end{aligned} \tag{85}$$

5. A NUMERICAL EXAMPLE

The Fourier transform procedure presented in this chapter was implemented in a computer program called FT-1D[54] that carries out Hartree–Fock electronic structure computations of stereoregular polymers using Ewald-type convergence acceleration methods. That code also takes full advantage of all line-group symmetries to calculate only the minimal set of two-electron integrals and to optimize the computation of the Fock matrix.

This section reports the results obtained for a carbon zigzag single-walled carbon nanotube (7, 0) called in this chapter SWNT(7, 0). The notation (n,m) was introduced by Hamada et al.[55] These two numbers represent a lattice point on the graphene sheet that defines the chiral vector with respect to the origin. To build the SWNT, a graphene sheet is rolled in a way such that this (n,m) point is superposed on the origin. In our case, the translational unit cell contains 28 carbon atoms.

This structure, represented in Fig. 1, belongs to the $L14_7/mcm$ line group. This line group contains the isogonal point group D_{14h}[44,57] and has a total of 56 symmetry operations: $\{\hat{C}_{14}^{2r}|0\}$, $\{\hat{C}_{14}^{2r+1}|\frac{1}{2}\}$, $\{\sigma_v\hat{C}_{14}^{2r}|0\}$,

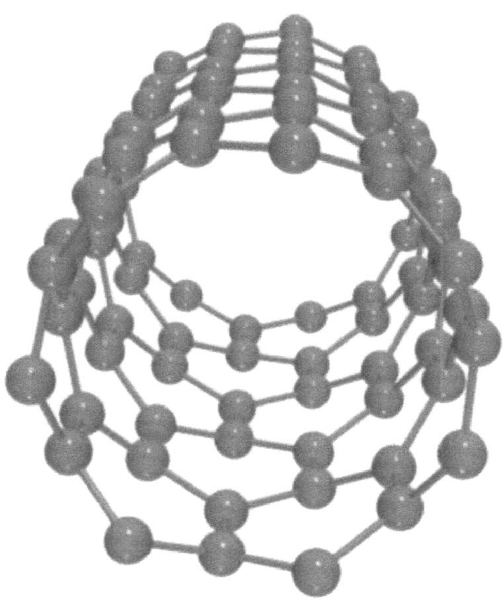

Figure 1 Molecular geometry of SWNT(7, 0) (with three repeating units along the z-axis) produced with the DRAWMOL program.[56]

$\{\sigma_v \hat{C}_{14}^{2r+1}|\frac{1}{2}\}$, $\{\sigma_h \hat{C}_{14}^{2r}|0\}$, $\{\sigma_h \hat{C}_{14}^{2r+1}|\frac{1}{2}\}$, $\{\sigma_h \sigma_v \hat{C}_{14}^{2r}|0\}$, $\{\sigma_h \sigma_v \hat{C}_{14}^{2r+1}|\frac{1}{2}\}$ with $r = 0,1,\ldots,6$. In our calculations the C—C bond length is set to 1.43 Å, causing the cell parameter a_0 to be 4.32 Å. The atomic coordinates were generated using the TUBEASP program.[58]

Three different atomic basis sets were used in the calculations: STO-3G,[59] 3-21G,[60] and 6-31G**.[61] The number of AOs per unit cell for these bases are, respectively, 140, 252, and 420. The number of k points used in the integration over the Brillouin zone ($-1/2 \leq k \leq 1/2$) was set to 33.

In the FT-1D program, the extents of the lattice sums were set to include all overlap terms exceeding 10^{-12} and all two-electron integrals larger than 10^{-10} hartree. In the 3-21G calculation, these thresholds cause the lattice summations to run between −2 and 2 for indices μ and μ' that depend on the AO overlap (see Eqs. 27 and 28). For the summations over ν appearing in Eqs. (52)–(55), the thresholds cause the number of 3-21G terms explicitly computed not to exceed 7 in DS and 6 in FS. For the 6-31G** calculation, these limits are, respectively, −3 to +3 for μ and μ', and 8 for ν. These narrow summation ranges illustrate the efficiency of the Ewald acceleration when the optimum partitioning parameter τ^* is used. A discussion of τ^* appears just after Eq. (58).

Table 5 lists the total energy per unit cell (in hartree), the energy of the top of the valence bands (band 84), and the width of the forbidden gap (these two quantities expressed in eV). They are compared with the values obtained with the program Gaussian-09[62] using periodic boundary conditions (G09-PBC). There is good agreement between the values

Table 5 Total Energy, Energy of the Top of the Valence Bands, and Bandgap of the SWNT(7.0) for Different Basis Sets Obtained with the FT-1D and G09-PBC Programs

Atomic Basis Set	Program	Total Energy (in hartree)	Top of the Valence Band ($\epsilon_{84}(k=0)$) (in eV)	Bandgap $\epsilon_{85}-\epsilon_{84}$ at k=0 (in eV)
STO-3G	FT-1D	−1047.436815	−3.21435	5.28278
	G09-PBC	−1047.436880	−3.21459	5.28322
3-21G	FT-1D	−1054.323375	−6.19104	4.73779
	G09-PBC	−1054.323655	−6.19209	4.73759
6-31G**	FT-1D	−1060.297185	−5.64859	4.74745
	G09-PBC	−1060.297217	−5.64729	4.74470

obtained with the two programs. The observed small differences can be explained in part by the differences in the last figures of the atomic coordinates. Indeed, in the FT-1D program, the coordinates are corrected at machine precision to take into account the action of the symmetry operators on the atomic positions. The default thresholds used in the two programs are also different.

Table 5 shows that this SWNT has a nonzero bandgap. That is consistent with the rule that a (*n*,0) nanotube is not metallic provided *n* is not a multiple of 3.[63–66] Still, the calculated bandgap, ranging between 5.28 and 4.74 eV depending on the basis set, is much larger than the values reported in Refs. 67–69. For example, employing a tight-binding model where the parameters are adjusted based on experimental data, the value found by Song et al.[68] is 1.0463 eV. Then, adopting DFT and the PBE exchange-correlation functional to describe finite (7, 0) nanotubes, Buonocore et al.[69] reported a value of −4.128 eV for the HOMO and −4.127 eV for the LUMO, yielding a bandgap of 0.001 eV. The overestimated HF bandgap results from missing electron correlation effects, whereas the DFT/PBE values are underestimated due to missing self-interaction effects.

Comparisons with bandgaps of π-conjugated polymers provide a basis for assessing the importance of delocalization in SWNT(7, 0) because the smaller the bandgap, the stronger the π-electron delocalization. Note however that the bandgap of a conjugated polymer depends also on the regularity of its backbone structure, which can be described by the so-called bond length alternation (BLA), the average bond length difference for the consecutive C—C bonds of the backbone (in the case of polyacetylene, the BLA reduces to the difference between the single and double CC bond lengths). The general rule attributes smaller bandgaps to structures with smaller BLA. When considering polyacetylene, the HF/3-21G (HF/6-311G**) bandgaps are 7.06 and 6.20 eV (7.22 and 6.35 eV) for BLAs of 0.123 and 0.082 Å, respectively.[70] These results confirm that increasing the basis set from 3-21G to 6-311G** has a small effect on the bandgap and that smaller BLA leads to smaller bandgap. Moreover, polyacetylene, which is known to display strong electron delocalization, has a bandgap larger than SWNT(7, 0). Then, for polythiophene, the calculated bandgap is 7.26 eV for the 3-21G basis and 7.35 eV for the 3-31G* basis.[71] For planar polyparaphenylene, the HF/3-21G bandgap is 8.16 eV, while for that basis it is 7.84 eV for polydiacetylene and 5.43 eV for polybutatriene.[70] All these results show that π-electron delocalization is very large in SWNT(7, 0).

Figures 2–4 display the band structures of SWNT(7, 0) obtained in this work with the STO-3G, 3-21G, and 6-31G** basis sets. The bands are labeled in Table 6 according to their symmetry, using the notation adopted by Božović and Vujičić,[45,46] who analyzed the irreducible representations and reported character tables for line groups for general points k in the Brillouin zone. The character table for the line group of SWNT(7, 0), $L14_7/mcm$, is listed in Table 7.

The figures are limited to bands 57–98. For these bands, there are no significant differences between the 3-21G and the 6-31G** computations except some inversions near $k = 0$. Such inversions were noted for the degenerate occupied bands (61,62) and (63,64), the unoccupied bands 94 and 95, and the unoccupied band 96 and the two degenerate unoccupied bands (97,98). A comparison between the STO-3G and the 3-21G (or the 6-31G**) bands shows more significant differences in the low-energy valence bands (bands 57–64) and in the uppermost conduction bands (94–98). The three calculations are in agreement that the highest occupied bands are the degenerate bands (83, 84) between $k = 0$ and $k = 1/8$ and (81, 82) between $k = 1/8$ and $k = 1/2$. In the interval [11/32,1/2], the (76, 77) degenerate bands present a higher energy than the (83,84) bands. All these bands have π-symmetry character with respect to the carbon sheet. Figure 5 displays the convoluted density of states as determined using the 6-31G** basis set, characterized by a first peak at −6.86 eV and a maximum intensity at −17.14 eV.

6. CONCLUSION

In 1972, one of the present authors (F.E.H.) demonstrated for the first time that Fourier transform methods could be used at the restricted Hartree–Fock level of approximation to make fully *ab initio* computations of the electronic structure of linear polymers.[20] On the other hand, in 1971, J.G.F., J. Delhalle. and J.M. André[72] had attempted to use Fourier series in the description of the electronic structure of periodic systems. This review shows that the work that started 40 years ago has now arrived at a point such that the Ewald–Fourier transform method can make useful *ab initio* computations on chainlike systems of chemical interest (e.g., with unit cells containing up to several hundred AOs and with basis members of s, p, and d symmetries).

In this review, we have shown that the Fourier representation combined with the Ewald procedure provides expressions for Coulomb and exchange

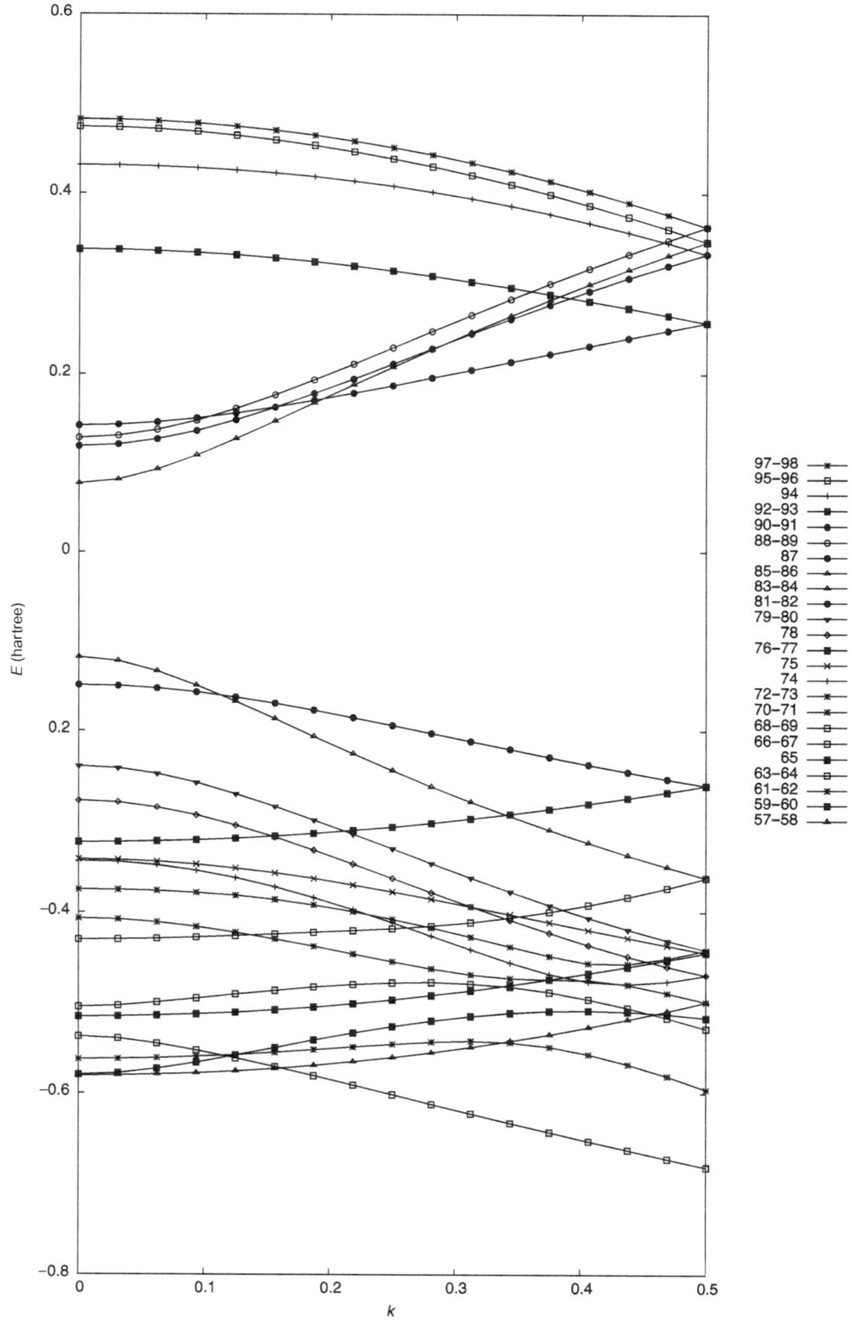

Figure 2 STO-3G band structure of SWNT(7, 0). Only bands 57–98 are represented.

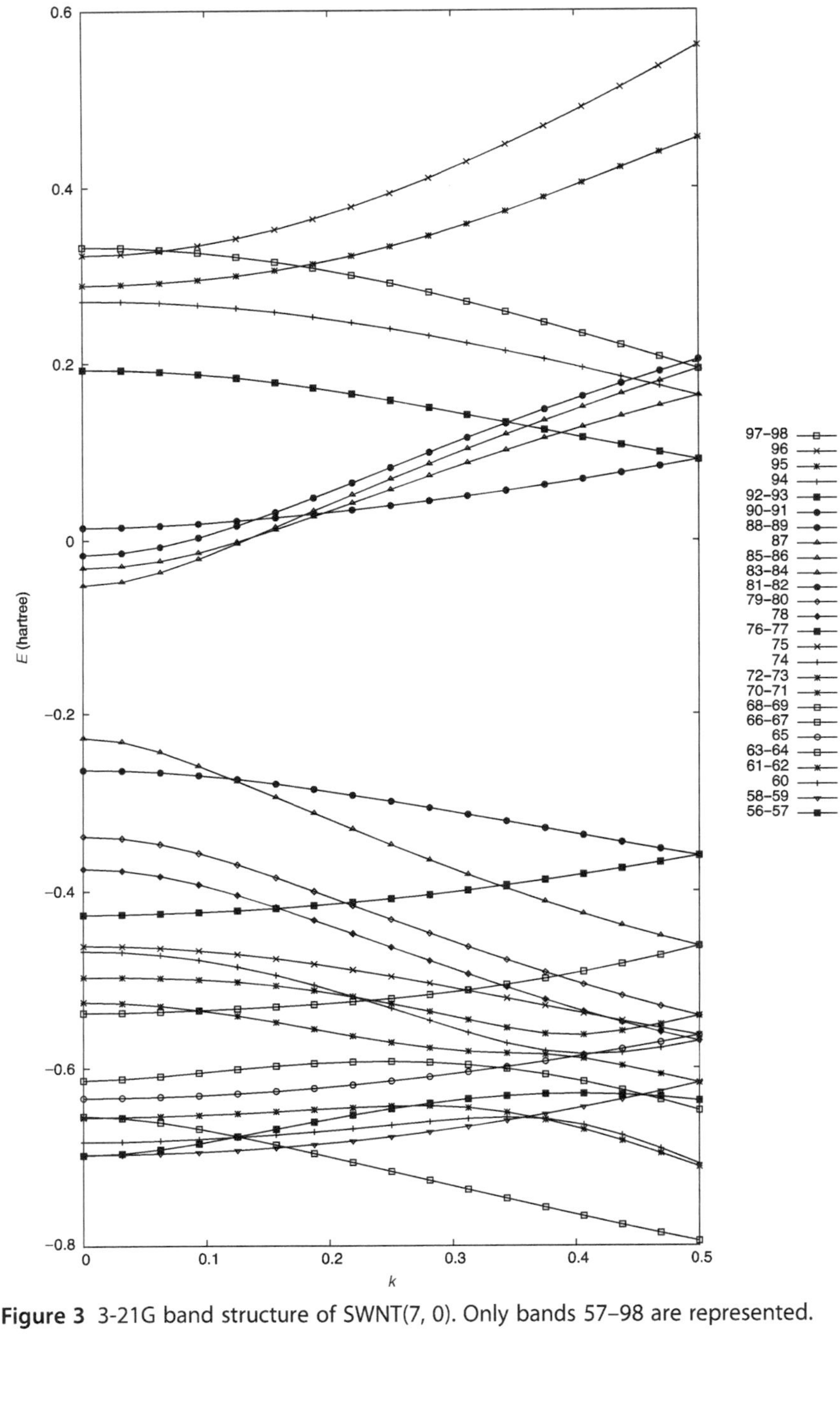

Figure 3 3-21G band structure of SWNT(7, 0). Only bands 57–98 are represented.

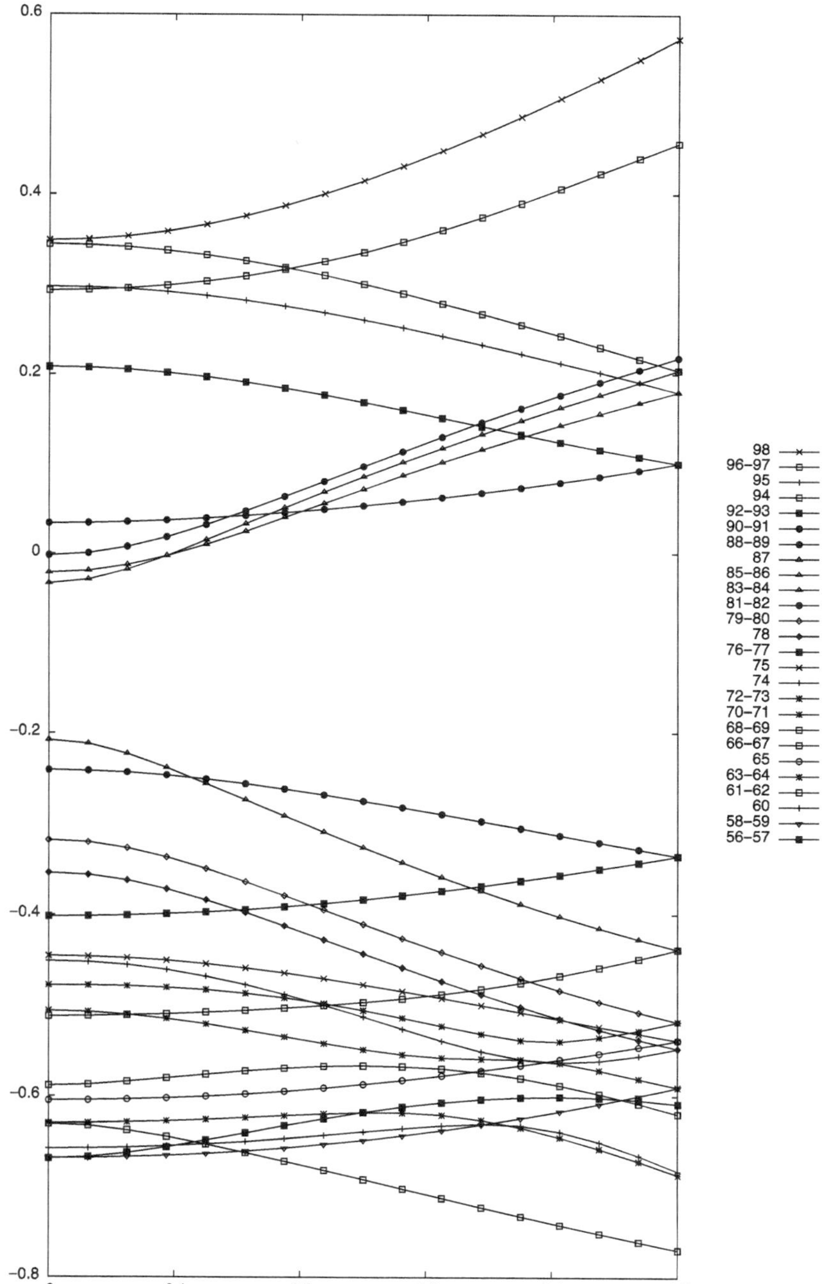

Figure 4 6-31G** band structure of SWNT(7, 0). Only bands 57–98 are represented.

Table 6 The Symmetry of the Valence and the First Unoccupied Bands Obtained with the STO-3G, 3-21G, and 6-31G** Basis Sets Using the Notation of Božović and Vujičić[45,46]

Symmetry	STO-3G Band Number	3-21G Band Number	6-31G** Band Number
$\left\{{}_0A_0^+, {}^{-k}_{k}E_{A_0}, {}_{1/2}E_{A_0}^{A_7}\right\}$	29,56,74	29,60,74	29,60,74
$\left\{{}_0A_0^-, {}^{-k}_{k}E_{A_0}, {}_{1/2}E_{A_0}^{A_7}\right\}$	94	94	95
$\left\{{}_0B_0^+, {}^{-k}_{k}E_{B_0}, {}_{1/2}E_{B_0}^{A_7}\right\}$	–	96	98
$\left\{{}_0B_0^-, {}^{-k}_{k}E_{B_0}, {}_{1/2}E_{B_0}^{A_7}\right\}$	75	75	75
$\left\{{}_0A_7^+, {}^{-k}_{k}E_{A_7}, {}_{1/2}E_{A_0}^{A_7}\right\}$	43, 87	43, 87, 95	43, 87, 94
$\left\{{}_0A_7^-, {}^{-k}_{k}E_{A_7}, {}_{1/2}E_{A_0}^{A_7}\right\}$	36, 78	36, 78	36, 78
$\left\{{}_0B_7^+, {}^{-k}_{k}E_{B_7}, {}_{1/2}E_{B_0}^{B_7}\right\}$	65	65	65
$\left\{{}_0B_7^-, {}^{-k}_{k}E_{B_7}, {}_{1/2}E_{B_0}^{B_7}\right\}$	–	–	–
$\left\{{}_0E_{1,-1}^+, {}^{-k}_{k}G_{1,-1}, {}_{1/2}G_{1,-1}^{6,-6}\right\}$	(30,31), (61,62), (72,73)	(30,31), (61,62), (72,73)	(30,31), (63,64), (72,73)
$\left\{{}_0E_{1,-1}^-, {}^{-k}_{k}G_{1,-1}, {}_{1/2}G_{1,-1}^{6,-6}\right\}$	(70,71), (97,98)	(70,71)	(70,71)
$\left\{{}_0E_{2,-2}^+, {}^{-k}_{k}G_{2,-2}, {}_{1/2}G_{2,-2}^{5,-5}\right\}$	(32,33), (66,67), (68,69)	(32,33), (66,67), (68,69)	(32,33), (66,67), (68,69)
$\left\{{}_0E_{2,-2}^-, {}^{-k}_{k}G_{2,-2}, {}_{1/2}G_{2,-2}^{5,-5}\right\}$	(63,64), (95,96)	(63,64), (97,98)	(61,62), (96,97)
$\left\{{}_0E_{3,-3}^+, {}^{-k}_{k}G_{3,-3}, {}_{1/2}G_{3,-3}^{4,-4}\right\}$	(34,35), (59,60), (76,77)	(34,35), (56,57), (76,77)	(34,35), (56,57), (76,77)
$\left\{{}_0E_{3,-3}^-, {}^{-k}_{k}G_{3,-3}, {}_{1/2}G_{3,-3}^{4,-4}\right\}$	(54,55), (92,93)	(54,55), (92,93)	(52,53), (92,93)
$\left\{{}_0E_{4,-4}^+, {}^{-k}_{k}G_{4,-4}, {}_{1/2}G_{3,-3}^{4,-4}\right\}$	(39,40), (52,53), (81,82)	(39,40), (52,53), (81,82)	(39,40), (54,55), (81,82)
$\left\{{}_0E_{4,-4}^-, {}^{-k}_{k}G_{4,-4}, {}_{1/2}G_{3,-3}^{4,-4}\right\}$	(48,49), (90,91)	(48,49), (90,91)	(48,49), (90,91)
$\left\{{}_0E_{5,-5}^+, {}^{-k}_{k}G_{5,-5}, {}_{1/2}G_{2,-2}^{5,-5}\right\}$	(46,47), (50,51), (85,86)	(46,47), (50,51), (85,86)	(46,47), (50,51), (85,86)

Table 6 The Symmetry of the Valence and the First Unoccupied Bands Obtained with the STO-3G, 3-21G, and 6-31G** Basis Sets Using the Notation of Božović and Vujičić[45,46]—(cont'd)

Symmetry	STO-3G Band Number	3-21G Band Number	6-31G** Band Number
$\left\{{}_0E^-_{5,-5},\ {}^{-k}_{k}G_{5,-5},\ {}_{1/2}G^{5,-5}_{2,-2}\right\}$	(41,42), (83,84)	(41,42), (83,84)	(41,42), (83,84)
$\left\{{}_0E^+_{6,-6},\ {}^{-k}_{k}G_{6,-6},\ {}_{1/2}G^{6,-6}_{1,-1}\right\}$	(44,45), (57,58), (88,89)	(44,45), (58,59), (88,89)	(44,45), (58,59), (88,89)
$\left\{{}_0E^-_{6,-6},\ {}^{-k}_{k}G_{6,-6},\ {}_{1/2}G^{6,-6}_{1,-1}\right\}$	(37,38), (79,80)	(37,38), (79,80)	(37,38), (79,80)

The three labels between curly brackets give the band symmetries in the notation for $k = 0$, $k \neq 0$, and $k = 1/2$.

Table 7 Character Table of Line Group $L14_7/mcm$, in the Notation of Božović and Vujičić[46]; Possible Values of r: 0, 1,…, 6; of m: 1, 2,…, 6; of m': 1, 2, 3; of n: 0,± 1, ± 2,… (See the Table Footnote for Operations R_i and Symbols in the Characters

Symmetry	R_1	R_2	$\sigma_v R_1$	$\sigma_v R_2$	R_5	R_6	$\sigma_v R_5$	$\sigma_v R_6$
${}_0A_0^{\pm}$	1	1	1	1	± 1	± 1	± 1	± 1
${}_0B_0^{\pm}$	1	1	-1	-1	± 1	± 1	∓ 1	∓ 1
${}_0A_7^{\pm}$	1	-1	1	-1	± 1	∓ 1	± 1	∓ 1
${}_0B_7^{\pm}$	1	-1	-1	1	± 1	∓ 1	∓ 1	± 1
${}_0E^{\pm}_{m,-m}$	a_{mr}	d'_{mr}	0	0	$\pm\, a_{mr}$	$\pm\, d'_{mr}$	0	0
${}^{-k}_{k}E_{A_0}$	a_{kn}	d'_{kn}	a_{kn}	d'_{kn}	0	0	0	0
${}^{-k}_{k}E_{B_0}$	a_{kn}	d'_{kn}	$-a_{kn}$	$-d'_{kn}$	0	0	0	0
${}^{-k}_{k}E_{A_7}$	a_{kn}	$-d'_{kn}$	a_{kn}	$-d'_{kn}$	0	0	0	0
${}^{-k}_{k}E_{B_7}$	a_{kn}	$-d'_{kn}$	$-a_{kn}$	d'_{kn}	0	0	0	0
${}^{-k}_{k}G_{m,-m}$	$a_{kn}a_{mr}$	$d'_{kn}d'_{mr}$	0	0	0	0	0	0
${}_{1/2}E^{A_7}_{A_0}$	$2(-1)^n$	0	$2(-1)^n$	0	0	0	0	0
${}_{1/2}E^{A_7}_{B_0}$	$2(-1)^n$	0	$-2(-1)^n$	0	0	0	0	0
${}_{1/2}G^{7-m',-7+m'}_{m',-m'}$	$2(-1)^n a_{m'r}$	0	0	0	0	0	0	0

$R_1 = \{\hat{C}_{14}^{2r}|n\}$, $R_2 = \{\hat{C}_{14}^{2r+1}|n+\frac{1}{2}\}$, $R_5 = \sigma_h\{\hat{C}_{14}^{2r}|n\}$, $R_6 = \sigma_h\{\hat{C}_{14}^{2r+1}|n+\frac{1}{2}\}$.
$a_{kn} = 2\cos(2\pi nk)$, $d'_{kn} = 2\cos(2\pi[n+\frac{1}{2}]k)$, $a_{mr} = 2\cos(2\pi mr/7)$,
$a_{m'r} = 2\cos(2\pi m'r/7)$, $d'_{mr} = 2\cos(2\pi m[r+\frac{1}{2}]/7)$.

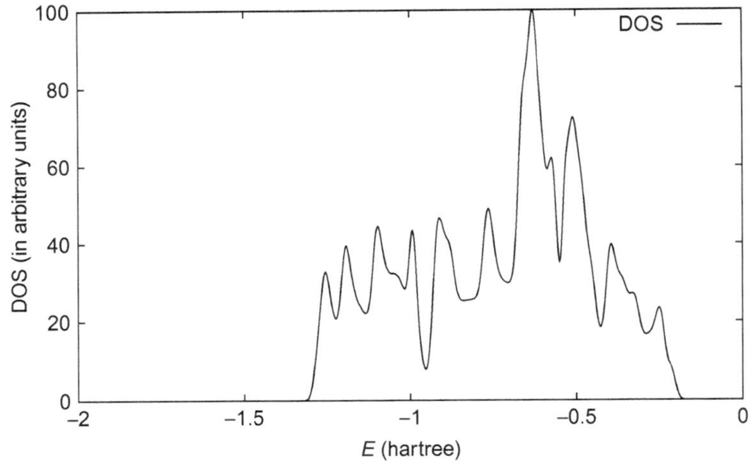

Figure 5 6-31G** convoluted density of states of SWNT(7, 0).

terms in which the lattice sums are brought to forms that converge rapidly irrespective of the compactness of the AOs involved. Moreover, in this method, the Coulomb and exchange terms are treated on an equal footing. That is not the case in "DS" formulations of the Hartree–Fock equations. In the DS methods, a multipolar expansion is often used to treat the long range part of the Coulomb potential, while the lattice sum convergence of exchange terms is controlled only by the magnitude of the Fourier transforms of the density matrix elements.

A computer program (FT-1D) based on this methodology is in the development phase, and tests show that the number of terms in the series is reduced significantly with respect to those needed in existing programs.[73]

The so-called linear-dependency problem which appears in DS programs when treating systems with delocalized basis sets (e.g., 6-31G**) disappears when using the Fourier-space algorithm.

The full inclusion of line-group symmetry in our FT-1D program was found to yield a meaningful increase in computational efficiency for systems such as nanotubes. Effective treatment of the symmetry makes it possible to treat this class of macromolecules with a reasonable level of computational effort.

ACKNOWLEDGMENTS

This achievement would not have been possible without the important contributions of Professor Joseph Delhalle and Dr. Isabelle Flamant to the development of the method. The authors wish to thank Professor J.M. André for his support. F.E.H. was supported by U.S. National Science Foundation Grant PHY-0601758.

Part of this research has been funded by BELSPO (IAP P7/05 network "Functional Supramolecular Systems"). The calculations were performed on the computing facilities of the Consortium des Équipements de Calcul Intensif (CÉCI), in particular those of the Plateforme Technologique de Calcul Intensif (PTCI) installed in the University of Namur, for which we gratefully acknowledge financial support from the FNRS-FRFC (Conventions No. 2.4.617.07.F and 2.5020.11) and from the University of Namur.

REFERENCES

1. Segall, M. D.; Lindan, P. L. D.; Probert, M. J.; Pickard, C. J.; Hasnip, P. J.; Clark, S. J.; Payne, M. C. First-Principles Simulation: Ideas, Illustrations and the CASTEP Code. *J. Phys. Condens. Matter* **2002**, *14*, 2717.
2. Hutchison, G. R.; Zhao, Y. J.; Delley, B.; Freeman, A. J.; Ratner, M. A.; Marks, T. J. Electronic Structure of Conducting Polymers: Limitations of Oligomer Extrapolation Approximations and Effects of Heteroatoms. *Phys. Rev. B* **2003**, *68*, 035204.
3. Pomogaeva, A.; Kirtman, B.; Gu, F. L.; Aoki, Y. Band Structure Built from Oligomer Calculations. *J. Chem. Phys.* **2008**, *128*, 074109.
4. André, J. M. Quantum Mechanical Methods for Regular Polymers. In: Löwdin, P. Ed.; Advances in Quantum Chemistry, 12; Academic Press: New York, 1980; p 65.
5. Ladik, J. *Quantum Theory of Polymers as Solids*; Plenum: New York, 1988.
6. André, J. M.; Delhalle, J.; Brédas, J. L. *Quantum Chemistry-Aided Design of Organic Polymers for Molecular Electronics*; World Scientific Publishing Company: London, 1991.
7. Champagne, B. Ab Initio Polymer Quantum Theory. In: *Molecular Simulation Methods for Predicting Polymer Properties*; Galiatsatos, V., Ed.; Wiley & Sons: New York, 2005; pp 1–46.
8. Hirata, S. Quantum Chemistry of Macromolecules and Solids. *Phys. Chem. Chem. Phys.* **2009**, *11*, 8397.
9. Delhalle, J.; Piela, L.; Brédas, J. L.; André, J. M. Multipole Expansion in Tight-Binding Hartree-Fock Calculations for Infinite Model Polymers. *Phys. Rev. B* **1980**, *22*, 6254.
10. Fripiat, J. G.; Delhalle, J.; André, J. M.; Calais, J. L. Dependence of RHF Properties of Hydrogen and Helium Chains on the Exchange Lattice Summations. *Int. J. Quantum Chem.* **1990**, *S24*, 593.
11. Jacquemin, D.; Fripiat, J. G.; Champagne, B. Convergence of Exchange Lattice Summations in Direct-Space Polymer Calculations. *Int. J. Quantum Chem.* **2002**, *89*, 452.
12. Kufner, A.; Kadlec, J. *Fourier Series*. Iliffe Books: London, 1971, p 41.
13. Delhalle, J.; Calais, J. L. Convergence of Direct-Space Exchange Lattice Sums in Polymer Band Structure Calculations. *J. Chem. Phys.* **1986**, *85*, 5286.
14. Delhalle, J.; Calais, J. L. Direct-Space Analysis of the Hartree-Fock Energy Bands and Density of States for Metallic Extended Systems. *Phys. Rev. B* **1987**, *35*, 9460.
15. Delhalle, J.; Calais, J. L. Pathological Aspects of RHF Band Calculations for Metallic Chains. *Int. J. Quantum Chem.* **1987**, *S21*, 115.
16. Stolarczyk, L. Z.; Jeziorska, M.; Monkhorst, H. J. Exact Hartree-Fock Exchange in One-Dimensional Metals. *Phys. Rev. B* **1988**, *37*, 10646.
17. Jeziorska, M.; Stolarczyk, L. Z.; Paldus, J.; Monkhorst, H. J. Exact Hartree-Fock Exchange in One-Dimensional Metals. II. *Phys. Rev. B* **1990**, *41*, 12473.
18. Olver, F. W. J.; Lozier, D. W.; Bisvert, R. F.; Clark, C. W. *NIST Handbook of Mathematical Functions*. National Institute of Standards and Technology (USA) and Cambridge University Press: Cambridge, UK, 2010.
19. Harris, F. E.; Monkhorst, H. J. Electronic Structure Studies of Solids. I. Fourier Representation Method for Madelung Sums. *Phys. Rev. B* **1970**, *2*, 4400.

20. Harris, F. E. Fourier Representation Methods for Electronic Structures of Linear Polymers. *J. Chem. Phys.* **1972**, *56*, 4422.
21. Harris, F. E. Hartree-Fock Studies of Electronic Structures of Crystalline Solids. In: Henderson, D., Eyring, H., Eds.; Theoretical Chemistry, Advances and Perspectives, 1; Academic Press: New York, 1975; pp 147–218.
22. Delhalle, J.; Harris, F. E. Fourier-Representation Method for Electronic Structure of Chainlike Systems: Restricted Hartree-Fock Equations and Applications to the $(H)_x$ Chain in a Basis of Gaussian Functions. *Phys. Rev. B* **1985**, *31*, 6755.
23. Flamant, I. Fourier Space Restricted Hartree-Fock Method for the Electronic Structure Calculation of Stereoregular Polymers. Ph.D. thesis, University of Namur, 1998.
24. Flamant, I.; Fripiat, J. G.; Delhalle, J.; Harris, F. E. Efficient Electronic Structure Calculations for Systems of One-Dimensional Periodicity with the Restricted Hartree-Fock-Linear Combination of Atomic Orbitals Method Implemented in Fourier Space. *Theor. Chem. Acc.* **2000**, *104*, 350.
25. Fripiat, J. G.; Delhalle, J.; Flamant, I.; Harris, F. E. Ewald-Type Formulas for Gaussian-Basis Bloch States in One-Dimensionally Periodic Systems. *J. Chem. Phys.* **2010**, *132*, 044108.
26. Fripiat, J. G.; Harris, F. E. Ewald-Type Formulas for Gaussian-Basis Studies of One-Dimensionally Periodic Systems. *Theor. Chem. Acc.* **2012**, *131*, 1257.
27. Terras, R. A Miller Algorithm for an Incomplete Bessel Function. *J. Comput. Phys.* **1981**, *39*, 233.
28. Agrest, M. M.; Maksimov, M. S. *Theory of Incomplete Cylindrical Functions and Their Applications*; Springer-Verlag: Berlin, 1971.
29. Hantush, M. S.; Jacob, C. E. Non-Steady Radial Flow in an Infinite Leaky Aquifer. *Trans. Am. Geophys. Union* **1955**, *36*, 95.
30. Thomas, G. W.; Keys, R. G.; Reynolds, A. C., Jr. The Computation of Leaky Aquifer Functions. *J. Hydrol.* **1978**, *36*, 173.
31. Chaudhry, M. A.; Zubair, S. M. Generalized Incomplete Gamma Functions with Applications. *J. Comput. Appl. Math.* **1994**, *55*, 99.
32. Chaudhry, M. A.; Temme, N. M.; Veling, E. J. M. Asymptotics and Closed Form of a Generalized Incomplete Gamma Function. *J. Comput. Appl. Math.* **1996**, *67*, 371.
33. Alford, J. A. Calculation of the Generalized Leaky Aquifer Integral. *Comput. Phys. Commun.* **2005**, *173*, 1.
34. Harris, F. E. Incomplete Bessel, Generalized Incomplete Gamma, or Leaky Aquifer Functions. *J. Comput. Appl. Math.* **2008**, *215*, 260.
35. Harris, F. E.; Fripiat, J. G. Methods for Incomplete Bessel Function Evaluation. *Int. J. Quantum Chem.* **2009**, *109*, 1728.
36. Abramowitz, M.; Stegun, I. *Handbook of Mathematical Functions*; Dover: New York, 1964.
37. Amos, D. E. Computation of Exponential Integrals. *ACM Trans. Math. Softw.* **1980**, *6*, 365.
38. Amos, D. E. Algorithm 556: Exponential Integrals. *ACM Trans. Math. Softw.* **1980**, *6*, 420.
39. Fripiat, J. G.; Delhalle, J. Efficient Calculation of the Exchange in the Fourier Representation of HF-LCAO-SCF Equations for 1D Periodic Systems. *Int. J. Quantum Chem.* **2009**, *109*, 2960.
40. Maple Is a Product of Maplesoft, Headquartered in Waterloo, Ontario, Canada.
41. Ewald, P. P. Die Berechnung optischer und elektrostatischer Gitterpotentiale. *Ann. Phys.* **1921**, *64*, 253.
42. Shavitt, I. The Gaussian Function in Calculations of Statistical Mechanics and Quantum Mechanics. In Alder, B., Fernbach, S., Rotenberg, M., Eds.; Methods in Computational Physics, 2; Academic Press: New York, 1963; p 1.

43. McCubbin, W. L. Effect of Symmetry on the Band Structure of Polymers. In: *NATO Advanced Study Institute Series: Series B, Physics*; André, J. M., Ladik, J., Eds.; Electronic Structure of Polymers and Molecular Crystals, vol. 9; 1974; pp 171–198.
44. Vujičić, M.; Božović, I. B.; Herbut, F. Construction of the Symmetry Groups of Polymer Molecules. *J. Phys. A* **1977**, *10*, 1271.
45. Božović, I. B.; Vujičić, M.; Herbut, F. Irreducible Representations of the Symmetry Groups of Polymer Molecules. I. *J. Phys. A* **1978**, *11*, 2133.
46. Božović, I. B.; Vujičić, M. Irreducible Representations of the Symmetry Groups of Polymer Molecules. II. *J. Phys. A* **1981**, *14*, 777.
47. Božović, I. B.; Božović, N. Irreducible Representations of the Symmetry Groups of Polymer Molecules. III. Consequences of Time Reversal Symmetry. *J. Phys. A* **1981**, *14*, 1825.
48. Božović, I. B.; Delhalle, J.; Damnjanović, M. A Posteriori Symmetry Analysis of Polymer Electronic Band Structures. *Int. J. Quantum Chem.* **1981**, *20*, 1143.
49. Božović, I. B.; Delhalle, J. Symmetry-Adapted Linear Combination of Atomic Orbitals Bases and Band-Structure Computation for Quasi-One Dimensional Solids. *Phys. Rev. B* **1984**, *29*, 4733.
50. Dupuis, M.; King, H. F. Molecular Symmetry and Closed-Shell SCF Calculations. *Int. J. Quantum Chem.* **1977**, *11*, 613.
51. Edmonds, A. R. *Angular Momentum in Quantum Mechanics*; Princeton University Press: Princeton, NJ, 1957.
52. Dovesi, R. On the Role of Symmetry in the Ab Initio Hartree-Fock Linear Combination of Atomic Orbitals Treatment of Periodic Systems. *Int. J. Quantum Chem.* **1986**, *29*, 1755.
53. André, J. M.; Mosley, D. H.; Champagne, B.; Delhalle, J.; Fripiat, J. G.; Brédas, J. L.; Vanderveken, D. J.; Vercauteren, D. P. LCAO Ab Initio Band Structure Calculations for Polymers. In Clementi, E. Ed.; Methods and Techniques in Computational Chemistry: METECC-94; STEF: Cagliari, Italy, 1993; pp 423–480.
54. Fripiat, J. G.; Harris, F. E. *ft_1d Program, Version 5.2*; 2014.
55. Hamada, N.; Sawada, S.; Oshiyama, A. New One-Dimensional Conductors: Graphitic Microtubules. *Phys. Rev. Lett.* **1992**, *68*, 1579.
56. Liègeois, V. *drawmol Program*; 2014.
57. Damnjanović, M.; Milošević, I.; Vuković, T.; Sredanović, R. Symmetry and Lattices of Single-Wall Nanotubes. *J. Phys. A* **1999**, *32*, 4097.
58. Veiga, R. G. A. TubeASP. Carbon Nanotube Generation Applet. Corrected by David Tomanek, Michigan State University; ported to PHP by Nicholas Frederick, Michigan State University, URL: http://www.nanotube.msu.edu/tubeASP/.
59. Hehre, W. J.; Stewart, R. F.; Pople, J. A. Self Consistent Molecular Orbital Methods. I. Use of Gaussian Expansions of Slater Type Atomic Orbitals. *J. Chem. Phys.* **1969**, *51*, 2657.
60. Binkley, J. S.; Pople, J. A.; Hehre, W. J. Self-Consistent Molecular Orbital Methods. 21. Small Split-Valence Basis Sets for First-Row Elements. *J. Am. Chem. Soc.* **1980**, *102*, 939.
61. Hehre, W. J.; Ditchfield, R.; Pople, J. A. Self Consistent Molecular Orbital Methods. XII. Further Extensions of Gaussian Type Basis Sets for Use in Molecular Orbital Studies of Organic Molecules. *J. Chem. Phys.* **1972**, *56*, 2257.
62. Frisch, M. J.; Trucks, G. W.; Schlegel, H. B.; Scuseria, G. E.; Robb, M. A.; Cheeseman, J. R.; Scalmani, G.; Barone, V.; Mennucci, B.; Petersson, G. A.; Nakatsuji, H.; Caricato, M.; Li, X.; Hratchian, H. P.; Izmaylov, A. F.; Bloino, J.; Zheng, G.; Sonnenberg, J. L.; Hada, M.; Ehara, M.; Toyota, K.; Fukuda, R.; Hasegawa, J.; Ishida, M.; Nakajima, T.; Honda, Y.; Kitao, O.; Nakai, H.; Vreven, T.; Montgomery, J. A., Jr.; Peralta, J. E.; Ogliaro, F.; Bearpark, M.; Heyd, J. J.; Brothers, E.; Kudin, K. N.; Staroverov, V. N.;

Kobayashi, R.; Normand, J.; Raghavachari, K.; Rendell, A.; Burant, J. C.; Iyengar, S. S.; Tomasi, J.; Cossi, M.; Rega, N.; Millam, J. M.; Klene, M.; Knox, J. E.; Cross, J. B.; Bakken, V.; Adamo, C.; Jaramillo, J.; Gomperts, R.; Stratmann, R. E.; Yazyev, O.; Austin, A. J.; Cammi, R.; Pomelli, C.; Ochterski, J. W.; Martin, R. L.; Morokuma, K.; Zakrzewski, V. G.; Voth, G. A.; Salvador, P.; Dannenberg, J. J.; Dapprich, S.; Daniels, A. D.; Farkas, O.; Foresman, J. B.; Ortiz, J. V.; Cioslowski, J.; Fox, D. J.Gaussian 09, Revision A.2; 2009.
63. Mintmire, J. W.; Dunlap, B. I.; White, C. T. Are Fullerene Tubules Metallic? *Phys. Rev. Lett.* **1992**, *68*, 631.
64. White, C. T.; Robertson, D. H.; Mintmire, J. Helical and Rotational Symmetries of Nanoscale Graphitic Tubules. *Phys. Rev. B* **1993**, *47*, 5485.
65. Dresselhaus, M. S.; Dresselhaus, G.; Saito, R. Physics of Carbon Nanotubes. *Carbon* **1995**, *33*, 883.
66. Mintmire, J. W.; White, C. T. Electronic and Structural Properties of Carbon Nanotubes. *Carbon* **1995**, *33*, 893.
67. Charlier, J. C. Defects in Carbon Nanotubes. *Acc. Chem. Res.* **2002**, *35*, 1063.
68. Song, W.; Ni, M.; Lu, J.; Gao, Z.; Nagase, S.; Yu, D.; Ye, H.; Zhang, X. Electronic Structure of Semiconducting Double-Walled Carbon Nanotubes: Important Effect of Interlay Interaction. *Chem. Phys. Lett.* **2005**, *414*, 429.
69. Buonocore, F.; Trani, F.; Ninno, D.; Matteo, A. D.; Cantele, G.; Iadonisi, G. Ab Initio Calculations of Electronic Affinity and Ionization of Carbon Nanotubes. Nanotechnology **2008**, *19*, 025711.
70. Champagne, B. unpublished.
71. Mosley, D. H.; Fripiat, J. G.; Champagne, B.; André, J. M. Ab Initio Investigation of the Static Polarizability of Planar and Twisted Infinite Polythiophene Chains. *Int. J. Quantum Chem.* **1994**, *28*, 451.
72. André, J. M.; Delhalle, J.; Fripiat, J. G.; Leroy, G. Computation of LCAO Wave Functions for Ground States of Polymers and Solids. *Int. J. Quantum Chem.* **1971**, *5*, 67.
73. Fripiat, J. G.; Harris, F. E. The Fourier Space Restricted Hartree-Fock Method for the Electronic Structure Calculation of Linear Poly(Tetrafluoroethylene). *Sci. China Chem.* **2014**, *57*, 1355.

CHAPTER EIGHT

Generalized Response Theory for a Photoexcited Many-Atom System

David A. Micha[1]
Quantum Theory Project, Department of Chemistry, University of Florida, Gainesville, Florida, USA
Quantum Theory Project, Department of Physics, University of Florida, Gainesville, Florida, USA
[1]Corresponding author: e-mail address: micha@qtp.ufl.edu

Contents

Abstract

A new treatment is presented for the generalized response of a physical system in a nonequilibrium state resulting from photoexcitation, using reduced density operators to account for dissipative dynamics, and developed so that it can be applied to physical systems starting from their atomic composition and their electronic structure obtained by *ab initio* calculations. This requires generation of a steady-state density operator and a perturbation treatment for driving force effects. Expressions are presented for the density and flux in terms of a generalized response function. A generalized fluctuation–dissipation relation is also given. This presentation also describes some general results on the dielectric function and on electronic relaxation of excited silicon slabs, from our recent work.

Advances in Quantum Chemistry, Volume 71
ISSN 0065-3276
http://dx.doi.org/10.1016/bs.aiq.2015.03.004

1. INTRODUCTION

The response of a physical system initially at equilibrium and driven by external forces is well known from treatments based on the perturbation theory of wave functions of isolated systems, or of density matrices for thermodynamically open systems in contact with a medium. The linear response for an open system is described by a dynamical susceptibility calculated for a thermal equilibrium distribution of states, and it is found to relate to dissipation of energy through a dissipation–fluctuation theorem.[1–3] This relates a nonequilibrium property, the dissipative rate, to an equilibrium property, the average of a time-correlation function (TCF) in the susceptibility, and is the basis for the theoretical interpretation of many experimental measurements in spectroscopy,[4,5] transport kinetics,[6–8] and hydrodynamics.[9]

Optical absorbance and photoconductivity in materials are examples of situations where a physical system is excited by light absorption and is then subject to external forces such as applied voltages. The response of the photoexcited system is different from the usual response near thermal equilibrium and requires reconsideration of response theory to include cases where the system is away from thermal equilibrium to begin with. This subject has recently received attention in treatments considering different aspects of response from nonequilibrium states and related extensions of the fluctuation–dissipation relations familiar near thermal equilibrium. These treatments have dealt with mathematical aspects[10–13] and with physical descriptions based on density matrix theory,[14] on nonequilibrium Green functions,[15–17] and on time-dependent density functional theory (TD-DFT),[18–21] with applications to spectroscopy and to electron transport through molecular junctions, among others.

The purpose here is to present a treatment of the generalized response of a physical system in a nonequilibrium state resulting from photoexcitation, using reduced density operators (RDOps) to account for dissipative dynamics, and developed so that it can be applied to physical systems starting from their atomic composition and their electronic structure obtained by *ab initio* calculations. This allows prediction of response properties as atomic constituents are varied. Light absorption can lead to formation of a steady state if the photoexcitation is stable, which responds to external driving forces, or to formation of a transient excited state created by a pulse of light that decays in response to a cutoff of the exciting light. Specifically, the treatment can be applied to the important practical matter of photoconductivity in semiconductor materials resulting from visible and near-infrared light absorption. Photoconductivity in semiconductor slabs occurs under two perturbations: a steady light pulse that excites electrons, and an applied external voltage that extracts excited electrons. This requires a treatment using a density operator to account for the dissipative dynamics of the excited electrons as they interact with a medium, and a double perturbation treatment for light and voltage effects. This presentation derives some new relations giving density and current responses to applied fields in terms of time-correlation functions (TCFs) of an excited system and also illustrates some of the general results on optical properties and conductivity of excited silicon slabs, from our recent work.[22–31]

Light absorption moves the physical system away from thermal equilibrium (eq) and into a new state which for steady light excitation can reach a steady state (ss) involving dissipation in a medium. This results from continuous light excitation in the primary or p-region of interest concurrent with its de-excitation as the medium, or s-region, absorbs energy and is electronically excited in a time scale much faster than the alternative de-excitation by light emission. This is illustrated in Fig. 1.

The following treatment deals with a steady states driven by an external applied voltage and is described in terms of an RDOp for a system of interest interacting with its medium, satisfying an equation of motion (EOM), which contains fluctuation and dissipative rates. The treatment of response to an applied voltage differs from standard response theory in that the

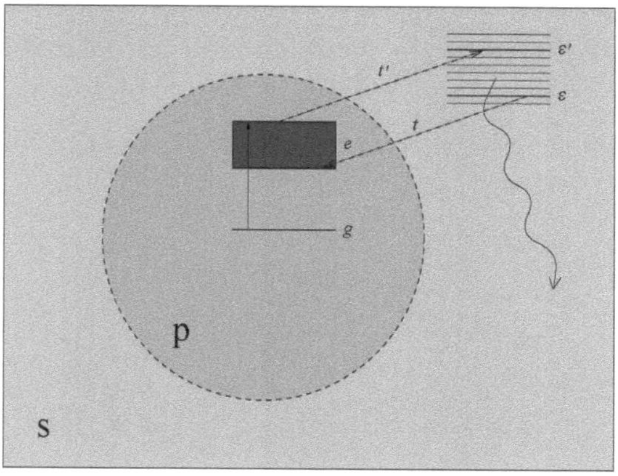

Figure 1 Pictorial representation of delayed dissipation of energy in a secondary (s-) region following excitation by light of a primary (p-) region. For steady light excitation, a combination of de-excitation in the p-region and dissipation during p- and s-region interactions leads to formation of steady states in the p-region. *Reproduced with permission from Ref. 22. Copyright 2009 American Institute of Physics.*

reference state is not thermal equilibrium, but instead a steady state, and also in that it involves a dissipative dynamics due to medium effects.

A second application is to a system prepared with a light pulse in an excited state (exc), which is also interacting with a medium, and then responds to a cutoff of the pulse by relaxing into a final equilibrium state. The treatment here also differs from the standard relaxation theory where the prepared excited system does not account for dissipation effects due to interaction with its medium.

2. EQUATIONS OF MOTION FOR A DISSIPATIVE DYNAMICS

We consider a physical system composed of electrons and atomic ion cores in an extended structure with either a periodic lattice or amorphous, forming a slab with surfaces open to a vacuum. It is factorized into a primary or p-region (the region of interest) containing active electrons and possibly adsorbed atoms as well as their nearest neighbors, and a secondary or s-region (a medium) containing inactive electrons and other atomic ions (vibrating around equilibrium). Figure 2 shows electronic energy regions below the highest occupied molecular orbital (HOMO) and above the lowest unoccupied MO (LUMO), where electrons are activated by light absorption and emission and undergo transitions, while electrons in lower MOs are part of a photoinactive medium and contribute to electronic density fluctuations (excitons).

The density operator for the whole system is time dependent $\hat{\Gamma}(t)$ on all position variables of p- and s-regions, from which a p-region RDOp $\hat{\Gamma}_{\mathrm{p}}(t) = \mathrm{tr}_{\mathrm{s}}\left[\hat{\Gamma}(t)\right]$ is constructed by taking the trace over all s-region

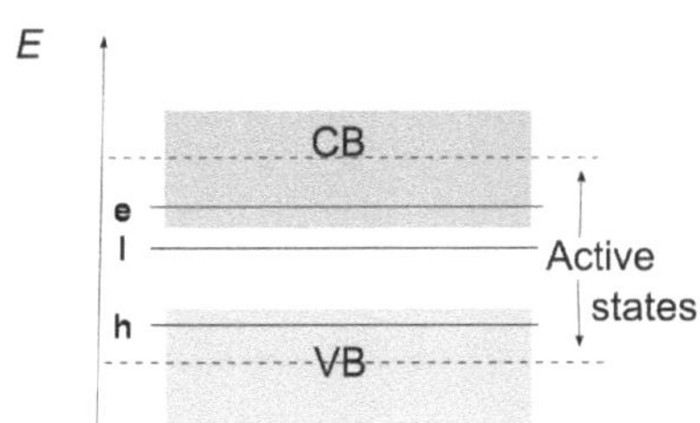

Figure 2 Schematic representation of energy levels activated by light absorption in a semiconductor, as treated in what follows. Electrons (e) are promoted to the conduction band (CB) leaving behind holes (h) in the valence band (VB). Localized (l) electron levels appear due to the presence of adsorbates or dopants. The range of active energy levels can be increased in calculations to verify convergence of calculated properties. *Reproduced with permission from Ref. 30. Copyright 2014 American Chemical Society.*

variables. Its EOM is obtained starting from the Liouville–von Neumann equation for $\hat{\Gamma}(t)$, solving it formally and reintroducing it on the right side to display a time-correlation of operators, and then taking the same s-trace on both sides of the EOM, which leads to an EOM for $\hat{\Gamma}_{\mathrm{p}}(t)$. The equation contains a fluctuation rate around the initial state and a delayed dissipative rate. This is valid for initial conditions involving a set of initial variables, and it is convenient to work with density operators averaged over a distribution of initial variables, indicated here with a top bar. The EOM for the p-region is simplified when it is possible to assume that upon averaging, the total density operator is factorized into p- and s-factors as in

$$\bar{\hat{\Gamma}} = \bar{\hat{\Gamma}}_{\mathrm{p}} \otimes \bar{\hat{\Gamma}}_{\mathrm{s}}, \quad \bar{\hat{\Gamma}}_{\mathrm{p}}(t) = \hat{\rho}(t).$$

Furthermore, the averaging eliminates the fluctuation rate term and simplifies the dissipative rate so that the EOM is given by

$$\frac{\partial \hat{\rho}}{\partial t} = -\frac{\mathrm{i}}{\hbar}\left[\hat{F}_{\mathrm{p}} + \hat{F}_{\mathrm{pL}}, \hat{\rho}(t)\right] - \frac{1}{\hbar^2}\int_0^t \mathrm{d}t' \widehat{\mathcal{M}}_{\mathrm{p}}(t, t')\hat{\rho}(t')$$

where the first commutation term contains effective Hamiltonians for the p-region and for its interaction with light, and the second term contains a delayed dissipation rate with a p-region dissipative superoperator, which can be obtained from TCFs of electronic densities and vibrational displacements in the s-region.[23]

After the initial averaging, dissipative rates from fast electronic motions and slow atomic motions are usually separated into fast and slow terms in

$$\widehat{\mathcal{M}}_{\mathrm{p}}(t, t') = \widehat{\mathcal{M}}_{\mathrm{p}}^{(\mathrm{el})}(t, t') + \widehat{\mathcal{M}}_{\mathrm{p}}^{(\mathrm{at})}(t, t').$$

Fast (instantaneous) electronic dissipation is due to electronic fluctuations (excitons) in the medium and can be described by the Lindblad rate expression constructed from transition rates between vibronic states in the p-region induced by interactions with the s-region. Slow (delayed) dissipation is due to atomic lattice vibrations. It can be obtained for bilinear displacement couplings of p- and s-regions in terms of TCFs of s-region displacements. For measurement times long compared with TCF relaxation times, it is given by the Redfield approximation for dissipative rates, which leads to differential equations containing the RDOp only at time t.[5,6]

In the case of photoconductivity, the EOM of the RDOp contains an effective Hamiltonian for the p-region from its interaction with the s-region, the Hamiltonian for light, Hamiltonians for interaction of the p-region with light (pL) and with an external electric voltage (pV), and a dissipative rate from excitons and phonons in the s-region. The rate is given by a superoperator $\hat{\mathcal{R}}$ which become time independent after long times compared with relaxation times of those excitations

$$\frac{d\hat{\rho}}{dt} = -\frac{i}{\hbar}\left[\hat{F}_p + \hat{F}_L + \hat{F}_{pL} + \hat{F}_{pV}, \hat{\rho}(t)\right] - \hat{\mathcal{R}}\hat{\rho}\,(t)$$

$$\hat{F}_{pL} = --\int d^3r \hat{\vec{D}}^{(p)}(\vec{r}) \cdot \hat{\vec{E}}_L(\vec{r})$$

$$\hat{F}_{pV} = -[c_e/(m_e\omega_V)]\hat{\vec{P}}^{(p)} \cdot \vec{E}_V(t), \omega_V \neq 0; \quad \hat{F}_{pV} = -c_e\hat{\vec{R}}^{(p)} \cdot \vec{E}_V(0), \ \omega_V = 0$$

in terms of the p-region electric dipole operator at position $\vec{r}$ and total electronic momentum operator, the electric field of light and of the applied voltage of frequency ω_V, and with c_e and m_e are the charge and mass of the electron. The dissipative rate superoperator accounts for vibronic transitions in the s-region leading to dissipation of energy from the p-region. It can be decomposed into two terms corresponding to vibronic excitations from the ground state to excited electronic states plus vibrational transitions in the ground electronic state, giving the decomposition $\hat{\mathcal{R}} = \hat{\mathcal{R}}^{(el)} + \hat{\mathcal{R}}^{(at)}$. This superoperator can be constructed for both instantaneous and delayed dissipation, within the Lindblad and Redfield formulations, respectively. This leads to a time-independent dissipative rate for phenomena observed long after electronic and vibrational excitations have settled the whole physical system into a steady state in the absence of external forces, a situation we are considering here.

It is convenient to introduce a quantized light field and treat it as a time-independent perturbation acting on matter-field states of the photoexcited system, while the voltage is treated as a time-dependent additional perturbation. Using photon creation and annihilation operators for photons of given momentum and polarization $(\vec{\kappa}, \sigma)$, the light Hamiltonian and the interaction pL energy operators are

$$\hat{F}_L = \sum_{\vec{\kappa},\sigma} \hbar\omega_{\vec{\kappa}\sigma} \hat{a}^+_{\vec{\kappa}\sigma} \hat{a}_{\vec{\kappa}\sigma}$$

$$\hat{F}_{pL} = -\int d^3r \hat{\vec{D}}^{(p)}(\vec{r}) \cdot \hat{\vec{E}}_L(\vec{r}),$$

$$\hat{\vec{E}}_L(\vec{r}) = i\sum_{\vec{\kappa},\sigma} \left[2\pi\omega_{\vec{\kappa}\sigma}/(\varepsilon_0 V_L)\right] \left[\hat{a}^+_{\vec{\kappa}\sigma}\exp(i\vec{\kappa}\cdot\vec{r}) - \hat{a}_{\vec{\kappa}\sigma}\exp(-i\vec{\kappa}\cdot\vec{r})\right]\vec{e}_\sigma$$

for light quantized in a volume V_L. When the light wavelength is long compared with the size of the slab supercell, the magnitude of κ is very small and the $\vec{r}$-dependence can be ignored so that the electric field is needed only at $\vec{r}=\vec{0}$.

The following treatment is done using a basis set of matter-field states obtained from the vibronic states $|I\rangle = |I_{el}\mathbf{v}_I\rangle$ of the p-regions, specified by the electronic and vibrational quantum numbers of states in the p-region, multiplying radiation field states $|\mathbf{r}\rangle = |\{n_{\vec{\kappa}\sigma}\}\rangle$ specified by the field occupation numbers, from

$$\hat{F}_p|I\rangle = E_I|I\rangle \quad \text{and} \quad \hat{F}_L|\mathbf{r}\rangle = E_\mathbf{r}|\mathbf{r}\rangle.$$

The following treatment involves a basis set of matter-field states $|\alpha\rangle = |I, \mathbf{r}\rangle$. Two special cases of interest here are: (1) excitation by steady light of a single frequency ω_L, leading to appearance of a steady state for the system and (2) relaxation after a steady light pulse is cut off at time $t=0$, leading to transient electronic density fluctuations and their relaxation into a final thermal equilibrium state.

3. OPTICAL AND PHOTOCONDUCTIVITY RESPONSES

3.1 Excitation by Steady Light

The electronic states of the p-region are many-electron states constructed from one-electron Bloch states for occupied and unoccupied energy bands. For the weak light fields on present interest, it is enough to include among the light states only those with one or no photons, while neglecting multiphoton excitations or decays. The radiation states are then $|\{n_{\vec{\kappa}\sigma}\}\rangle = |0_L\rangle$ and $|1_L\rangle$ for frequency ω_L. Furthermore, a physically meaningful basis set can be constructed including the ground electronic states times one-photon states and electronically excited states with no photons present. This neglects the possibility of light emission of photons away from energy resonance. The approximation is equivalent to the so-called rotating wave approximation (RWA)[5,6], which can be justified after equations of motion are averaged over times much larger than the periods of light-induced transitions. Operator equations become matrix equations in this basis set, with time-dependent matrix elements containing oscillating factors. The time averaging shows that nonresonance transitions have smaller coupling matrix elements and can be doped to begin with. They can be treated as a perturbation to estimate their magnitude if desired. Within these assumptions, matter-field states α follow from

$$\hat{F}^0|\alpha\rangle = E_\alpha|\alpha\rangle$$
$$E_\alpha = E_G + \hbar\omega_L,\quad |\alpha\rangle = |G\rangle\otimes|1_L\rangle,\quad \text{ground state}$$
$$E_\alpha = E_I,\quad |\alpha\rangle = |I\rangle\otimes|0_L\rangle,\quad \text{excited state}$$

involving the ground electronic state G and a photon to begin with, or an excited electronic state I and no photons after photoexcitation by light absorption. Here, $\hat{F}^0 = \hat{F}_p + \hat{F}_L$ is the Hamiltonian of a reference unperturbed state of matter in the p-region (active electrons) and photons.

The dissipative rate superoperator has been written as a sum of fast electronic and slow atomic vibration terms. A double perturbation treatment can be done in a compact notation using Liouvillian superoperators defined by $\hat{\mathcal{L}}\hat{A} = \left(-\frac{i}{\hbar}\right)\left[\hat{H}, \hat{A}\right]$ acting on any operators, for each term in the bracket above, so that

$$\frac{d\hat{\rho}}{dt} = \left(\hat{\mathcal{L}}^0 + \hat{\mathcal{L}}_{pL} + \hat{\mathcal{L}}_{pV}(t)\right)\hat{\rho}(t) - \hat{\mathcal{R}}\hat{\rho}(t)$$
$$\hat{\mathcal{L}}^0 = \hat{\mathcal{L}}_p + \hat{\mathcal{L}}_L$$

which contains a term for the unperturbed system, one for a time-independent perturbation (pL) of active electrons by light, and a second, time-dependent perturbation (pV) of active electrons by the applied voltage. It is convenient to group time-independent and electronic dissipation terms into a new superoperator accounting for the steady perturbation of the RDOp, as in

$$\frac{d\hat{\rho}}{dt} = \hat{\mathcal{K}}^{(ss)}\hat{\rho}(t) + \hat{\mathcal{L}}_{pV}(t)\hat{\rho}(t)$$
$$\hat{\mathcal{K}}^{(ss)} = \hat{\mathcal{L}}^0 + \hat{\mathcal{L}}_{pL} - \hat{\mathcal{R}}$$

where the second line is a time-independent Liouvillian superoperator containing counterbalancing rates of light excitation and dissipative de-excitation. We consider a case where a persistent light perturbation of frequency ω_L leads to formation of a steady photoexcited state in the p-region as this interacts with its medium, which is then driven by the small voltage force. A steady-state RDOp $\hat{\rho}^{(ss)}$ satisfies

$$0 = \hat{\mathcal{K}}^{(ss)}\hat{\rho}^{(ss)}$$

to be solved for the RDOp. This is done in a basis set of states containing one or no photons, equivalent to making the RWA in a time-dependent treatment of the light effects in a rotating states frame. An application to absorption of light by a semiconductor within a one-electron treatment of the p-region is given in Ref. 24. The RDOp matrix elements are $\rho_{\alpha\beta}^{(ss)} = \langle\alpha|\hat{\rho}^{(ss)}|\beta\rangle$, with diagonal elements providing steady population of states, and off-diagonal elements accounting for quantum coherence. This steady state contains the persistent effect of perturbation by the absorbed steady light.

The next step is calculation of the voltage perturbation of the RDOp, by adding a term to the steady-state reduced density matrix (RDM) and solving formally for its EOM, to obtain

$$\hat{\rho}(t) = \hat{\rho}^{(ss)} + \hat{\rho}'(t)$$
$$\hat{\rho}'(t) = \hat{\rho}'(t_{in}) + \int_{t_{in}}^{t} dt' \exp\left[(t-t')\hat{\mathcal{K}}^{(ss)}\right]\hat{\mathcal{L}}_{pV}(t')\hat{\rho}(t')$$

which separates the steady state from additional time-dependent events induced by the applied voltage. Here, the exponential superoperator applies to all factors to its right. To first order in the (pV) interaction,

$$\hat{\rho}_{pV}(t) = \hat{\rho}'(t_{in}) + \int_{t_{in}}^{t} dt' \exp\left[(t-t')\hat{\mathcal{K}}^{(ss)}\right]\hat{\mathcal{L}}_{pV}(t')\hat{\rho}^{(ss)}$$

From this, it is possible to obtain the electronic current flux. Its form is derived for an initial state as $t_{in} \to -\infty$ given by the steady-state RDOp, so that $\hat{\rho}'(t_{in}) = 0$, and for the usual case where the flux is null in the absence of a driving electric potential, when $\mathrm{tr}\left[\hat{P}_{\xi}^{(p)}\hat{\rho}^{(ss)}\right] = 0$. The electronic charge flux is given by the velocity average along a direction ξ multiplying the electron density for N_e electrons within the volume V_{SC} of a periodic solid supercell, as $\langle J_{\xi}(t)\rangle^{(ss)} = \left(N_e V_{SC}^{-1} m_e{}^{-1}\right)\mathrm{tr}\left[\hat{P}_{\xi}^{(p)}\hat{\rho}_{pV}(t)\right]$, and from the previous equation with a voltage field of components along η,

$$\boxed{\langle J_{\xi}(t)\rangle^{(ss)} = \int_{-\infty}^{t} dt' \sum_{\eta} \sigma_{\xi\eta}^{(ss)}(t-t')\cdot\mathcal{E}_{V\eta}(t')}$$

$$\boxed{\sigma_{\xi\eta}^{(ss)}(t-t') = -\frac{c_e N_e}{V_{SC} m_e \omega_V}\mathrm{tr}\left\{\hat{P}_{\xi}^{(p)}\exp\left[(t-t')\hat{\mathcal{K}}^{(ss)}\right]\left(-\frac{i}{\hbar}\right)\left[\hat{P}_{\eta}^{(p)},\hat{\rho}^{(ss)}\right]_{-}\right\}}$$

The second line provides the desired generalized response function and shows two important differences between the standard response around thermal equilibrium and the one here for response from a steady state. In both cases, the delayed response depends only on the difference between the time for present observation and the previous times, and involves two momentum operators correlated over the time delay. But now, the average in the trace must be done for the steady-state RDOp instead of the thermal equilibrium RDOp. And the second difference is that the time dependence of the delay comes from a time evolution superoperator which contains in addition to the unperturbed Hamiltonian of matter plus light, also two counterbalancing contributions due to light excitation and dissipation in the exponent. This response function depends parametrically on the light frequency.

The connection with the dynamical conductivity as a function of frequency follows by introducing the retarded response function $\sigma_{\xi\eta}^{(ss)}(\tau)^{(ret)} = \theta(\tau)\sigma_{\xi\eta}^{(ss)}(\tau)$ using the time step function, which allows extension of the upper integration limit to infinity, and gives by Fourier transform from time to frequency

$$\left\langle \widetilde{J}_{\xi}(\omega)\right\rangle^{(ss)} = \sum_{\eta} \widetilde{\sigma}_{\xi\eta}^{(ss)}(\omega)^{(ret)} \widetilde{\mathcal{E}}_{V\eta}(\omega)$$

in terms of the generalized dynamical response $\widetilde{\sigma}_{\xi\eta}^{(ss)}(\omega)^{(ret)}$ and the transform $\widetilde{\mathcal{E}}_{V\eta}(\omega)$ of the voltage field. It also follows from the above derivation that in the absence of light and with the system being driven by a voltage near thermal equilibrium, the above equations reduce to the known result for the response function and the induced flux.[2,5]

A generalized fluctuation–dissipation relation is obtained for a steady applied voltage field $\mathcal{E}(t) = \mathcal{E}_V^0 \cos(\omega_V t)$ from the time derivative of the (pV) interaction energy, after it is averaged over a time T long compared with the period $\tau_V = 2\pi/\omega_V$ of the applied voltage. The energy dissipation rate $Q_V^{(ss)}$ is obtained in terms of the generalized dynamical response starting from

$$Q_V^{(ss)} = \frac{1}{T}\int_0^T dt \ \mathrm{tr}\left(\hat{\rho}_{pV} \partial \hat{F}_{pV}/\partial t\right)$$

and using the generalized dynamical response in $\langle J_\xi(t)\rangle^{(ss)}$ above, which leads after integration over time to the desired generalized fluctuation–dissipation relation

$$\boxed{\begin{aligned}&Q_V^{(ss)}(\omega_V;\omega_L)\\&=\frac{c_e^2}{m_e^2\omega_V}\sum_{\xi\eta}\mathcal{E}_{V\xi}^0\mathcal{E}_{V\eta}^0\int_0^\infty dt\sin(\omega_V t)\mathrm{tr}\left\{\hat{P}_\xi^{(p)}\exp\left(t\hat{\mathcal{K}}^{(ss)}\right)\left(-\frac{i}{\hbar}\right)\left[\hat{P}_\eta^{(p)},\hat{\rho}^{(ss)}\right]_-\right\}\end{aligned}}$$

giving the dissipation of energy in terms of the sin transform of a TCF of momenta. Here again, the average is over a steady-state RDOp instead of the standard equilibrium RDOp. It is shown to depend on the frequency of the driving voltage field and now also explicitly on the frequency of the steady light creating the steady state. This relation reduces to the standard fluctuation–dissipation relation[1,2] in the absence of the steady light field and ignoring the dissipative rate in the EOM.

3.2 Relaxation after a Light Pulse

In this case, a long-lasting steady light pulse is applied to the system, with a field operator average $\mathcal{E}(t)=\mathcal{E}_V^0\theta(-t)\cos(\omega_V t-\pi/2)$ describing a steady pulse of frequency ω_L cutoff at time $t=0$, obtained from a field operator formed by superposition of stationary radiation states of varying frequency with time-dependent amplitude coefficients. For a pulse chosen with a central frequency ω_L resonant with a vibronic excitation frequency $\omega(G\rightarrow E)$ from the ground state to an excited state E, chosen for its large oscillator strength, the prepared state at the cutoff time is an RDOp $\hat{\rho}(0;G\rightarrow E)=\hat{\rho}^{(0)}(\omega_L)$, with a large population of the excited state E. It is obtained from the p-region dynamics at $t<0$ and is close to the steady state $\rho^{(ss)}(\omega_L)$ if the pulse has lasted long enough.

The response of the system is described by transient electronic densities and currents, relaxing to thermal equilibrium values. With no light or voltage present, the EOM is now

$$\frac{d\hat{\rho}}{dt}=\hat{\mathcal{K}}^{(rel)}\hat{\rho}(t)$$
$$\hat{\mathcal{K}}^{(rel)}=\hat{\mathcal{L}}^0-\hat{\mathcal{R}}$$

with a starting state at $t=0$ given by $\hat{\rho}^{(0)}=\hat{\rho}^{(exc)}(\omega_L)$. The formal solution for the RDOp is then

$$\hat{\rho}(t) = \hat{\rho}^{(0)} + \int_0^t dt' \exp\left[(t-t')\hat{\mathcal{K}}^{(\mathrm{rel})}\right]\hat{\rho}(t')$$

which can be solved by iteration.

Alternatively, the differential equation can be solved introducing eigenoperator solutions of $\hat{\mathcal{K}}^{(\mathrm{rel})}\hat{\rho}_A = -\lambda_A\hat{\rho}_A$, with complex eigenvalues λ_A, and their complementary adjoints satisfying orthonormality conditions $\mathrm{tr}\left(\bar{\hat{\rho}}_B^+\hat{\rho}_A\right) = \delta_{AB}$. Replacing the expansion $\hat{\rho}(t) = \sum_A \Lambda_A(t)\hat{\rho}_A$, with $\Lambda_A(t) = \mathrm{tr}\left[\bar{\hat{\rho}}_A^+\hat{\rho}(t)\right]$ in the differential equation gives $\Lambda_A(t) = \exp(-\lambda_A t)\Lambda_A^{(\mathrm{exc})}$, with $\Lambda_A^{(\mathrm{exc})}(\omega_\mathrm{L}) = \mathrm{tr}\left[\bar{\hat{\rho}}_A^+\hat{\rho}^{(\mathrm{exc})}(\omega_\mathrm{L})\right]$, and

$$\boxed{\hat{\rho}^{(\mathrm{rel})}(t) = \sum_A \exp(-\lambda_A t)\Lambda_A^{(\mathrm{exc})}\hat{\rho}_A}$$

from which the time evolution of properties can be obtained after $t=0$ as a superposition of oscillating and decaying states insofar the real part of λ_A, coming from the dissipative rate, is positive. The amplitude of each term is obtained from the known RDOp at the cutoff time. The eigenstates $\hat{\rho}_A$ of the relaxation superoperator are not very different from the steady state $\hat{\rho}^{(\mathrm{ss})}$ when the light field is weak. Since the steady state corresponds to zero eigenvalues of the (ss) superoperator, it follows that one of the relaxation eigenstates should have an amplitude nearly equal to one and a small value for λ_A, and this one would give the leading term for the relaxation of the RDOp.

This time-dependent RDOp can be used to calculate the transient electronic currents which appear during relaxation, with a flux given by $\langle J_\xi(t)\rangle^{(\mathrm{rel})} = (N_\mathrm{e} V_\mathrm{SC}{}^{-1} m_\mathrm{e}{}^{-1})\mathrm{tr}\left[\hat{P}_\xi^{(\mathrm{p})}\hat{\rho}^{(\mathrm{rel})}(t)\right]$ obtained from the relaxing RDOp equation above. It takes the form

$$\boxed{\langle J_\xi(t)\rangle^{(\mathrm{rel})} = \left(c_\mathrm{e} N_\mathrm{e} V_\mathrm{SC}^{-1} m_\mathrm{e}{}^{-1}\right)\sum_A \exp(-\lambda_A t)\mathrm{tr}\left[\hat{P}_\xi^{(\mathrm{p})}\hat{\rho}_A\right]\mathrm{tr}\left[\bar{\hat{\rho}}_A^+\hat{\rho}^{(\mathrm{exc})}\right]}$$

involving projections on the eigendensity operators of the starting RDOp and of the momentum. The difference between these equations and the standard ones for relaxation is that the starting RDOp at $t=0$ is not just for a single excited state but for the nearly steady-state RDOp $\hat{\rho}^{(\mathrm{exc})}$ prepared by a pulse of light in the presence of dissipation due to the p- and s-region interactions.

4. MODEL CALCULATIONS OF STEADY STATES AND RELAXING STATES AT A PHOTOEXCITED SEMICONDUCTOR SURFACE

4.1 Basis Set of One-Electron Orbitals from *Ab Initio* Electronic Structure

The derived equations are quite general and include some of the special cases involved in our previous model calculations for light absorption and for transient currents in silicon slabs, in the absence of an applied voltage. Two recent calculations have dealt with the dielectric function response for a steady state of silicon slabs,[29] and with calculations of relaxation times and of decaying currents following photoexcitation by light pulses, also for silicon slabs.[26] Calculations have been done using an electronic basis set of electron orbitals generated within DFT. Here, results and comments are presented to illustrate the present formalism with results of previous work, leaving details in the listed publications.

In model calculations based on *ab initio* electronic structure, the electronic states of the p-region are constructed from one-electron states for occupied and unoccupied energy bands. Many-electron states with single-electron excitations are labeled by the band index, and vibrational states of the nuclei can be introduced for each energy band index. Electronic Bloch states $\left|\vec{k}j\right\rangle$ for an electron of given wavevector in a periodic structure are constructed from a basis set of Kohn–Sham spin-orbitals (KSOs) with energies $\varepsilon_{\vec{k}j}$ for each band j, occupied (oc) or unoccupied (un), obtained from the KS Hamiltonian for the p-region, with KS orbitals expanded in plane waves as

$$\varphi_{\vec{k}j}(\vec{r}) = \exp\left(\mathrm{i}\,\vec{k}\cdot\vec{r}\right)\sum\nolimits_{\left|\vec{G}+\vec{k}\right|<G_{\mathrm{cut}}} C_{\vec{k}j,\vec{G}} \cdot \exp\left(-\mathrm{i}\,\vec{G}\cdot\vec{r}\right)$$

The RDOp is yet a many-electron object with all the position and spin variables of electrons in the p-region. To proceed, it is convenient to further reduce the treatment introducing effective one-electron operators. These can then be expanded in a basis set of vibronic-light field states composed of one-electron states times vibrational states for the atoms in the p-region. Equations for the one-electron density operator and a given atomic configuration $\{Q\}$ follow from the contraction

$$\rho^{(1)}(x;x') = \mathrm{tr}_{N-1}\left[\rho\left(x, X^{N-1}; x', X^{N-1}\right)\right].$$

RDM elements are obtained in a basis set of K–S orbitals satisfying the equation $\hat{F}^{\mathrm{KS}}\phi_{\vec{k}j} = \varepsilon_{\vec{k}j}\phi_{\vec{k}j}$ and given by $\rho_{\vec{k}j,\vec{k}'j'} = \left\langle \phi_{\vec{k}j} \middle| \hat{\rho}^{(1)} \middle| \phi_{\vec{k}'j'} \right\rangle$. This is yet a function of atomic position variables and of photon variables and must be expanded to include vibrational states of the lattice and photon field states. For photoinduced transitions involving only one excited electron and one photon absorption or emission, it is sufficient to construct matter-field states from one-photon and no-photon states as in

$$\{|a\rangle\} = \left\{ \left| \vec{k}j \right\rangle \otimes \left| n_{\vec{\kappa}\sigma} \right\rangle \right\}, \quad \text{electron} - \text{photon field basis set}$$

$$\left| n_{\vec{\kappa}\sigma} \right\rangle = \left| 1_{\vec{\kappa}\sigma} \right\rangle, \text{ one photon in occ states}; = \left| 0_{\vec{\kappa}\sigma} \right\rangle, \text{ no photons in unocc states}.$$

$$\{|\alpha\rangle\} = \left\{ \left| \vec{k}j\mathbf{v}_j \right\rangle \otimes \left| n_{\vec{\kappa}\sigma} \right\rangle \right\}, \quad \text{vibronic} - \text{field basis set}$$

Expansion of operators with this choice of photon states gives the same results as working in the RWA for the EOM of the RDM.[3] Vibrational matrix elements of an operator $\hat{A}$ can be obtained in the Frank–Condon approximation as $\langle \alpha | \hat{A} | \alpha' \rangle = \langle \alpha | \alpha' \rangle_{\mathrm{FC}} \left\langle \vec{k}j \middle| \hat{A}(Q_{\mathrm{ave}}) \middle| \vec{k}'j' \right\rangle$.

Matter-field energies for states a follow from

$$\hat{F}^{0}_{1,\mathrm{L}} |a\rangle = \varepsilon_a |a\rangle$$

$$\varepsilon_a = \varepsilon^{(\mathrm{oc})}_{\vec{k}j\mathbf{v}_j} + \hbar\omega_{\vec{\kappa}\sigma}, \quad |a\rangle = \left| \vec{k}j\mathbf{v}_j \right\rangle_{\mathrm{oc}} \otimes \left| 1_{\vec{\kappa}\sigma} \right\rangle, \quad \text{occupied}$$

$$\varepsilon_a = \varepsilon^{(\mathrm{un})}_{\vec{k}j\mathbf{v}_j}, \quad |a\rangle = \left| \vec{k}j\mathbf{v}_j \right\rangle_{\mathrm{un}} \otimes \left| 0_{\vec{\kappa}\sigma} \right\rangle, \quad \text{unoccupied}$$

involving an occupied state and a photon to begin with, or an unoccupied state and no photons after photoexcitation.

The EOM for the one-electron RDOp $\hat{\rho}^{(1)}(t)$ are obtained by taking the math trace of the original quantities over all electron variables but one.[27] It contains one-electron operators and superoperators for dissipation rates as in

$$\frac{\mathrm{d}\hat{\rho}^{(1)}}{\mathrm{d}t} = -\frac{\mathrm{i}}{\hbar}\left[\hat{F}^{(1)} + \hat{H}_{1,\mathrm{L}} + \hat{H}_{1,\mathrm{V}}, \hat{\rho}^{(1)}(t) \right] - \hat{\mathcal{R}}^{(1)}\hat{\rho}^{(1)}(t)$$

$$\hat{H}_{1,\mathrm{L}} = -\hat{\vec{d}} \cdot \hat{\vec{E}}_{\mathrm{L}}$$

$$\hat{H}_{1,\mathrm{V}} = -[c_{\mathrm{e}}/(m_{\mathrm{e}}\omega_{\mathrm{V}})]\hat{\vec{p}} \cdot \vec{E}_{\mathrm{V}}(t)$$

$$\hat{\mathcal{R}}^{(1)} = \hat{\mathcal{R}}^{(1,\mathrm{el})} + \hat{\mathcal{R}}^{(1,\mathrm{at})}$$

where $\hat{F}^{(1)} = \hat{F}_1 + \hat{F}_L$ is the effective Hamiltonian for noninteracting electrons and photons. The dissipative rate superoperator has been written as a sum of fast electronic and slow atomic vibration terms. A double perturbation treatment can be done as above in a compact notation using Liouvillian superoperators. It is convenient to group time-independent and electronic dissipation terms into a new superoperator accounting for the steady perturbation of the RDOp, as in

$$\frac{d\hat{\rho}^{(1)}}{dt} = \hat{\mathcal{K}}^{(1,\,ss)}\hat{\rho}^{(1)}(t) + \hat{\mathcal{L}}_{1V}(t)\hat{\rho}^{(1)}(t)$$
$$\hat{\mathcal{K}}^{(1,\,ss)} = \hat{\mathcal{L}}^{(1)} + \hat{\mathcal{L}}_{1L} - \hat{\mathcal{R}}^{(1)}$$

Now a steady state corresponds to $0 = \hat{\mathcal{K}}^{(1,\,ss)}\hat{\rho}^{(1,\,ss)}$, an equation to be solved for the RDOp. This operator equation is transformed following expansions in the vibronic basis sets, into a matrix of coupled equations. They give reduced density matrix (RDM) elements $\rho_{ab} = \langle a|\hat{\rho}^{(1)}|b\rangle$, with diagonal elements providing population of states and off-diagonal elements accounting for quantum coherence. This steady state contains the persistent effect of perturbation by the absorbed light.

We consider the behavior of a photoexcited silicon slab in two cases: (1) a persistent light perturbation of frequency ω_L leads to formation of a steady photoexcited state in the p-region and a dielectric function of a frequency ω in the presence of light and (2) a cutoff of a steady light field leads to electronic charge relaxation from a starting state, after transient currents have settled. Furthermore, the examples are given for one-electron excitations and in an approximation where vibrational motions in the p-region can be treated as classical and are obtained from *ab initio* molecular dynamics.

The dissipative rates can be derived to second order in couplings between p- and s-regions and expressed in terms of TCFs of the s-region electronic density fluctuations and lattice positions. For properties measured at long times compared to the relaxation times of the TCFs in the s-region, the RDOp in the dissipative rates can be replaced by its value at time *t*, and under these conditions, the form of the dissipative rates take the Redfield form,[5] with a time-independent dissipative rate $\hat{\mathcal{R}}^{(1)}$. Working in the matter-field basis set, the EOM become a matrix of coupled differential equations, and the dissipative rate is given by a four-index supermatrix as in

$$\left\langle a\right|\hat{\mathcal{R}}^{(1)}\hat{\rho}^{(1)}(t)\left|b\right\rangle = \sum_{c,d}\exp[it(\omega_{ab}-\omega_{cd})]R_{ab,cd}\rho_{cd}^{(1)}(t)$$

with oscillations in time which give zero values to the right when averaged over times long compared with $2\pi/\omega_L$, except for $a=b$ and $c=d$, or for $a=c$ and $b=d$, which give so-called secular nonzero terms. This time average is assumed in what follows.

To simplify, the equations are given below after making a quantum mechanical unitary transformation from the Schroedinger picture (SP) to the intermediate picture (IP) generated by $\hat{F}^{(1)}$, for fixed atomic positions, and in a RWA where photoinduced rapidly oscillating terms have been eliminated after taking a time average of the EOM over a measurement time $T_m >> 2\pi/\omega_L$.[5,6] Further treating the light field as a reservoir with a density operator $\hat{\rho}^L$, the RDOp is factored as $\hat{\rho}^{(1)} = \widetilde{\hat{\rho}}^{(1)}\hat{\rho}_L$, and an EOM follows for the purely electronic RDOp $\widetilde{\hat{\rho}}^{(1)}$. Its matrices are given in a basis set of purely one-electron states, for fixed electronic momentum, for a light field of frequency ω_L, and in the absence of an applied voltage.[24]

$$\frac{d\widetilde{\rho}_{jj}}{dt} = \frac{i}{2}\sum_k \Omega_{jk}\left(\widetilde{\rho}_{kj} - \widetilde{\rho}_{jk}\right) - \gamma_{jj}\left(\widetilde{\rho}_{jj} - \widetilde{\rho}_{jj}^{(eq)}\right), \quad \text{population change}$$

$$\frac{d\widetilde{\rho}_{jk}}{dt} = -i\Delta_{jk}\widetilde{\rho}_{jk} - i\Omega_{jk}\left(\widetilde{\rho}_{kk} - \widetilde{\rho}_{jj}\right) - \gamma_{jk}\left(\widetilde{\rho}_{jk} - \widetilde{\rho}_{jk}^{(eq)}\right), j \neq k, \text{ coherence change}$$

$$\gamma_{jj} = \sum_{k\neq j}\kappa_{jk}, \quad \gamma_{jk} = \sum_l \left(\kappa_{jl} + \kappa_{lk}\right)/2 + \gamma_{jk}^0, \quad \text{relaxation rates}$$

$$\kappa_{jk} = \hbar^{-2}\left|V_{jk}\right|^2 J\left(\omega_{jk}\right) f_{BE}\left(\omega_{jk}, T\right), \quad \text{state} - \text{to} - \text{state transition rate}$$

The symbols here are

$\Omega_{jk} = -D_{jk}\mathcal{E}_0$: Rabi frequency, $\Delta_{jk} = \omega_L - \omega_{jk}$: Detuning frequency
$\hbar\omega_{jk}$: Electronic excitation energy
$J(\omega)$: Vibrational frequency density
V_{jk} : Medium − induced electronic transition
$f_{BE}(\omega, T)$: Thermal distribution of vibrational frequencies

These coupled differential equations can be solved numerically for any given external light field. In particular, for steady light excitation, it is possible to concentrate on steady-state solutions obtained by setting the left-hand side of the EOM equal to zero:

$$\text{Steady} - \text{state solution for light of frequency } \omega_L$$
$$\frac{d\widetilde{\rho}_{jk}}{dt} = 0 \rightarrow \widetilde{\rho}_{jk}^{(ss)}(\omega_L) \text{ (in a rotating frame)}$$

This matrix solution provides populations of excited states from its diagonal elements, and also a way for calculating electronic current averages. Explicit solutions for the matrix elements in terms of the Rabi frequency and state-to-state transition rates are given in Ref. 24.

4.2 Steady States and the Dielectric Response in a Silicon Slab

The RWA, which is equivalent to using a basis set of electron-field states including one or no photons as we have already introduced, is sufficient for a treatment of optical properties.[29] The RDM elements for the electrons in a semiconductor slab are $\widetilde{\rho}_{jk}(t)=\sigma_{jk}(t)$, and in the SP, they lead to $\rho_{jk}(t)=\sigma_{jk}(t)\exp\left(-\mathrm{i}s_{jk}\omega_{\mathrm{L}}t\right)$, $s_{jk}=\mathrm{sgn}\left(\varepsilon_j-\varepsilon_k\right)$, with $s_{jj}=0$. The elements are obtained in a steady-state limit $\mathrm{d}\sigma_{jk}/\mathrm{d}t=0$, where they do not change with time and have been calculated to the lowest order in the strength of the electric field. Their values are given for a system initially at thermal equilibrium by

$$\sigma_{jj}^{(\mathrm{ss})}(\omega_{\mathrm{L}})=\Gamma_j^{-1}\sum_k \Omega_{jk}^2\gamma_{jk}\left[\gamma_{jk}^2+\Delta^2{}_{jk}\right]^{-1}\left(\sigma_{kk}^{\mathrm{eq}}-\sigma_{jj}^{\mathrm{eq}}\right)+\sigma_{jj}^{(\mathrm{eq})}$$

$$\sigma_{jk}^{(\mathrm{ss})}(\omega_{\mathrm{L}})=\Omega_{jk}\left[\Delta_{jk}+\mathrm{i}\gamma_{jk}\right]\left[\gamma_{jk}^2+\Delta^2{}_{jk}\right]^{-1}\left(\sigma_{kk}^{\mathrm{ss}}-\sigma_{jj}^{\mathrm{ss}}\right),\ j\neq k$$

where Γ_j is a depopulation rate and γ_{jk} is a quantum decoherence rate, both resulting from interaction of the p-region with medium phonons and excitons, and available from measurements or from molecular dynamics simulations. At the temperature of 0 K and also with good accuracy at room temperature, $\sigma_{jj}^{\mathrm{eq}}=2$ for $j\leq\mathrm{HOMO}$ and $\sigma_{jj}^{\mathrm{eq}}=0$ for $j\geq\mathrm{LUMO}$.

The steady-state RDM can be used to obtain the average polarizability $P(t)$, or induced electric dipole $D(t)$ per unit volume, of a semiconductor slab in a steady state and also subject to an electric field $\mathcal{E}(t)$ of frequency ω, generally different from ω_{L}, parallel or perpendicular to its surface pointing along its direction. It is obtained from a retarded response function $\chi^{(r)}(t')$ which accounts for delay effects from time t' to t, as in

$$P(t)=\int_{-\infty}^{t}\mathrm{d}t'\chi^{(r)}(t')\mathcal{E}(t-t')=\int_{-\infty}^{+\infty}\mathrm{d}t'\chi^{(+)}(t-t')\mathcal{E}(t')$$

where $\chi^{(+)}(t')=\chi^{(r)}(t')\theta(t')$ is the dynamical susceptibility in terms of the step function $\theta(t)$. Using a symmetric Fourier transform from time to frequency, the transform of the above functions of time, indicated with a tilde, gives $\widetilde{P}(\omega) = \sqrt{2\pi}\widetilde{\chi}^{(+)}(\omega)\widetilde{\mathcal{E}}(\omega)$, where $\chi(\omega) = \sqrt{2\pi}\widetilde{\chi}^{(+)}(\omega)$ is the susceptibility function of frequency ω with χ', χ'' the real and imaginary parts of the susceptibility function. From the literature [7], the relation between the frequency-dependent dielectric function $\varepsilon(\omega)$ and the dielectric susceptibility $\chi(\omega)$ is, in the cgs Gaussian system of units, $\varepsilon(\omega) = 1 + 4\pi\chi(\omega)$. Here, we omit the dielectric tensor indices and equations apply equally to all three-diagonal components. This is also a complex quantity $\varepsilon(\omega) = \varepsilon'(\omega) + \mathrm{i}\varepsilon''(\omega)$ such that $\sqrt{\varepsilon(\omega)} = \eta(\omega) + \mathrm{i}\kappa(\omega)$ where $\eta(\omega)$ is the refractive index of the medium and $\kappa(\omega)$ relates to the light absorption coefficient. From the above two equations, $\varepsilon'(\omega) = \eta^2(\omega) - \kappa^2(\omega)$ and $\varepsilon''(\omega) = 2\kappa(\omega)\eta(\omega)$.

The time-dependent average electric dipole of the slab is calculated from $\langle D(t)\rangle = \mathrm{tr}\left[\hat{D}\hat{\rho}(t)\right] = \sum_{j,k} D_{jk}\rho_{kj}(t)$ and, returning to the original SP of the RDM, the polarization in the steady state is $P(t) = (V_{\mathrm{SC}})^{-1}\sum_{j,k} D_{jk}\sigma_{kj}^{(\mathrm{ss})}\exp\left(-\mathrm{i}s_{jk}\omega_{\mathrm{L}}t\right)$ where V_{SC} is the volume of the supercell. This time dependence can be rewritten as a retarded response for our chosen periodic electric field, using that

$$\exp(\mathrm{i}\omega_{\mathrm{L}}t) = (2\mathcal{E}_0)^{-1}\int_{-\infty}^{+\infty} \mathrm{d}t' \left[\delta(t'-t) + \mathrm{i}\delta\left(t' - t + \frac{\pi}{2\omega_{\mathrm{L}}}\right)\right]\mathcal{E}_{\mathrm{L}}(t').$$

This gives, with $D_{jj} = 0$,

$$\chi^{(+)}(t;\omega_{\mathrm{L}}) = \frac{1}{V_{\mathrm{SC}}2E_0}\sum_{j\neq k} D_{jk}\sigma_{kj}^{(\mathrm{ss})}\delta\left(t + \frac{\pi}{2s_{jk}\omega_{\mathrm{L}}}\right)$$

and after a change of time variable and a Fourier transform to frequency,

$$\begin{aligned}\widetilde{\chi}^{(+)}(\omega;\omega_{\mathrm{L}}) &= \left(V_{\mathrm{SC}}2\mathcal{E}_0\sqrt{2\pi}\right)^{-1}\sum_{j,k}^{j\neq k} D_{jk}\sigma_{kj}^{(\mathrm{ss})}(\omega_{\mathrm{L}})\exp\left(-\mathrm{i}\omega s_{jk}2\pi/\omega_{\mathrm{L}}\right) \\ &= \frac{1}{V_{\mathrm{SC}}2\sqrt{2\pi}\hbar}\sum_{j,k}^{j\neq k}\left|D_{jk}\right|^2\frac{\Delta_{jk} + \mathrm{i}\gamma_{jk}}{\gamma_{jk}^2 + \Delta^2{}_{jk}}\left(\sigma_{kk}^{\mathrm{ss}} - \sigma_{jj}^{\mathrm{ss}}\right)\exp\left(-\mathrm{i}s_{jk}\omega 2\pi/\omega_{\mathrm{s}}\right)\end{aligned}$$

with the double summations extending over initially occupied orbitals k and unoccupied ones j. This expression displays the generalized form of the susceptibility when the system is in steady state, as shown by the dependence on both a fixed steady-state frequency and the variable frequency of a driving field. It reduces to the standard form of the susceptibility when $\Omega = \omega_L$.

Calculated values for the dielectric function $\varepsilon(\omega) = \varepsilon'(\omega) + i\varepsilon''(\omega)$ are given in Ref. 29 at the frequency $\omega = \omega_L > 0$ for silicon slabs of varying thickness, obtained using a basis set of orbitals from the HSE density functional.[29] Real and imaginary parts are shown in Fig. 3. Results are similar to the ones from the standard response treatment of the dielectric function,[29] but show noticeable differences due to the more accurate treatment of dissipative effects in the present RDM treatment. The system absorbance, α, is related to the imaginary part of the susceptibility and to components of the dielectric function as

$$\alpha(\omega_L) = 4\pi\omega_L V_{SC}\chi''(\omega_L)/[c\eta(\omega_L)] = 2V_{SC}\omega_L\kappa(\omega_L)/c.$$

4.3 Relaxation of Electronic Densities and Transient Currents in Silicon Slabs

The previous treatment has also been used to calculate relaxation times in photoexcited silicon slabs, in terms of the eigenoperators and their eigenvalues introduced above to expand the relaxing RDM. They are obtained from the Redfield relaxation supermatrix, which is constructed from

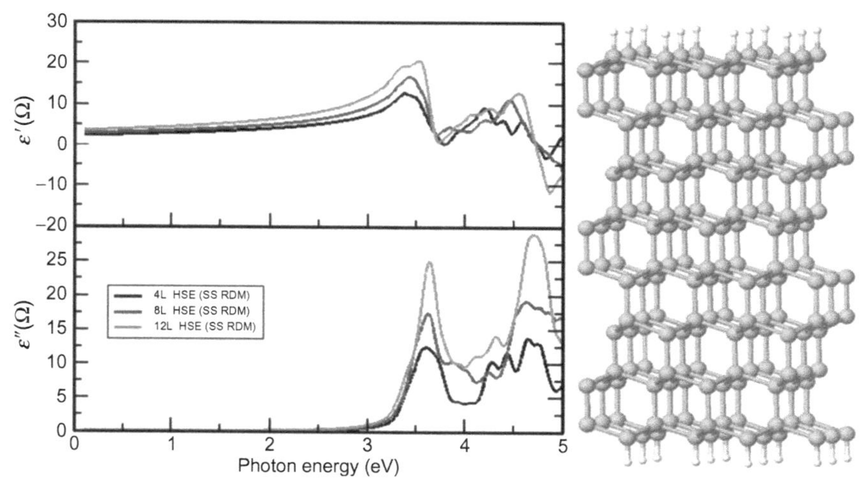

Figure 3 Real and imaginary parts of the dielectric function for Si(111) slabs of varying thickness, with four, eight, and 12 layers and surface bonds saturated with hydrogen atoms versus photon energy $\varepsilon_{ph} = \hbar\Omega$. Here, the frequency of light is $\Omega = \omega_L$ and results were obtained with the HSE density functional. *Reproduced with permission from Ref. 29. Copyright 2014 American Chemical Society.*

state-to-state electronic transition rates induced by interaction of the active electrons in the p-region with all the vibrating ions in the physical system. Relaxation times can be obtained from a calculation of electronic transition energies by means of an *ab initio* classical molecular dynamics of lattice vibrations.

The dissipative rate terms in the EOM originate from coupling of electrons with bosonic excitations in the medium, taken here to be lattice vibrations described by normal vibrational modes ξ with frequency ω_ξ, and position and momentum variables $\left\{\hat{q}_\xi, \hat{p}_\xi\right\}$. Associated vibrational decay times for their momentum–momentum TCFs are of the order of 10 fs. The electronic nonadiabatic momentum couplings appearing in the dissipative rates are given by $\hat{V}_{jk} = -\sum_\xi \left(\hat{p}_\xi / m_\xi\right) \cdot \langle j | \partial/\partial q_\xi | k\rangle$, and lead to electronic relaxation times τ_{el} of the order of 100 fs.

We consider here processes occurring well after the vibrational TCFs decay times, so that the dissipative supermatrix becomes time independent, as in the Redfield treatment of relaxation rates. Averaging over an initial thermal distribution of vibrations, with small excitation energies compared with electronic transitions energies, allows us to approximate the square of the electron–phonon coupling potential energy with a classical average obtained from an *ab initio* molecular dynamics, which treat the vibrational motions as classical. The average of the square of the nonadiabatic coupling that appears in the state-to-state transition rate κ_{jk} gets replaced by the semiclassical expression $\overline{|V_{jk}|^2} = \left(\hbar^2/N_t\right) \sum_n (|\langle j | \partial/\partial t | k\rangle|^2)_n$, where the summation is over N_t time steps n along a trajectory $\{q_\xi(t), p_\xi(t)\}$ run for the lattice at thermal equilibrium, and we have used that $\partial/\partial t = \sum_\xi (\mathrm{d}q_\xi/\mathrm{d}t) \cdot \partial/\partial q_\xi$. Although our system is photoexcited, the fraction of excited electrons is very small and the molecular dynamics calculated for the thermal lattices are effectively the same as the ones for the photoexcited systems, within our needed accuracy.[26]

In what follows, we assume that the system is optically excited at $t < 0$ by steady light of frequency $\omega_{\mathrm{L}} \cong (\varepsilon_b - \varepsilon_a)/\hbar$, that promotes an electron from occupied orbital a to unoccupied orbital b, and that at $t \geq 0$ its electric field is zero. The resulting relaxing RDM elements are $\rho_{jk}^{(ab)}(t)$ and the related changes of electronic population in the energy domain read $\Delta n^{(ab)}(\varepsilon, t) = \sum_j \rho_{jj}^{(ab)}(t) \delta(\varepsilon_j - \varepsilon) - n^{(\mathrm{eq})}(\varepsilon)$. It represents an increase of excited electrons (or particles) when $\Delta n^{(ab)} > 0$ or a decrease (increase of holes) when $\Delta n^{(ab)} < 0$. The RDM elements also provide the electronic charge density over space in terms of electronic orbitals and their populations.

The RDM matrix elements for the steady state created by photons in energy resonance with a chosen state-to-state transition are very similar to the elements of the RDM for a system in a pure excited state before light is cut off and the system relaxes. Results for relaxation times obtained for a Si slab are of the same magnitude as measured times in similar materials.[32] Calculations with Ag_3 adsorbed on the Si slab show relaxation times about one-half shorter, as expected due to the addition of energy levels in the gap region. Figure 4 shows an isocontour of the change in population versus electron energy and versus time for a cluster Ag_3 adsorbed on a Si(111) slab composed of four layers, when photons of energy 1.8 eV excite electrons from the HOMO of the slab to the LUMO+12, a transition with large oscillator strength. Red regions show increased electron population, and are seen to decay over time to lower energy levels in the conduction band, while blue regions show holes created in the valence energy band and limited inside it.

A similar treatment has been applied recently to photocurrents resulting from relaxation of electrons at a Au/Si interface, with the Si nanocrystal codoped with Al and P. The photocurrents have been calculated from the electron charge continuity equation for a nanocrystal in contact with

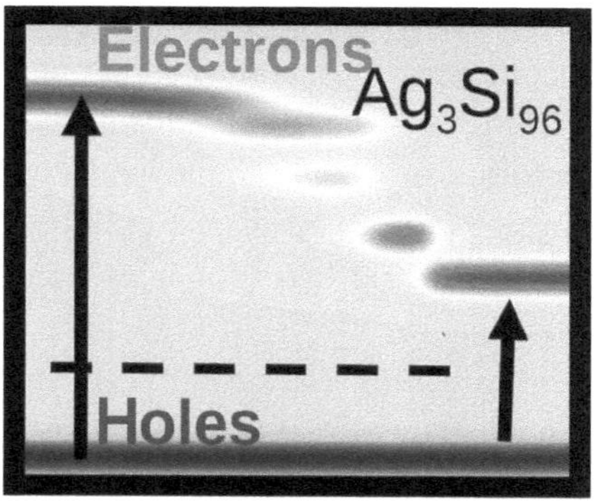

Figure 4 Isocontours of change in population $\Delta n^{(ab)}(\varepsilon, t)$ of a system with Ag_3 adsorbed on a Si(111) slab with four layers, prepared by steady light absorption and then cutoff at time $t = 0$. The vertical axis is the electron energy between −0.5 and 2.5 eV and the horizontal axis the log 10 of time from 0.001 to 10.0 ps. Red (gray in the print version) regions show increased electron populations and blue (dark gray in the print version) regions decreased populations (increased populations of holes). *Reproduced with permission from Ref. 26. Copyright 2010 American Chemical Society.*

two Au electrodes along a z-direction and have been obtained from the gradient of charge densities $\Delta c^{(ab)}(z, t) = c_e \Delta n^{(ab)}(z, t)/V$.[30]

5. CONCLUSIONS

A generalization of the standard linear response function has been obtained by constructing first a steady-state RDOp as a stationary solution of an EOM, which contains a dissipative rate. The full time-dependent solution to the equations has been written in a compact way using Liouville superoperators, and the related integral equations have provided linear responses including here dissipative effects. The treatment could formally be extended to account for nonlinear effects applicable to cases where the driving field is very large.

The expression derived for the flux induced by an applied external voltage in a system in a steady state provides a generalized response function and shows two important differences between the standard response around thermal equilibrium and the one here for response from a steady state. In both cases, the delayed response is a momentum–momentum TCF, which depends only on the difference between the time for present observation and the previous times. But now, the average in the TCF must be done for the steady-state RDOp instead of the thermal equilibrium RDOp. This response function depends parametrically on the frequency of the steady light which maintains a steady state. Furthermore, the time dependence of the delay comes from a time evolution superoperator which contains in addition to the unperturbed Hamiltonian of matter plus light, also two counterbalancing contributions due to light excitation and dissipation in the exponent. When these two are small, they can be treated using perturbation theory, preferably keeping them together since they both work to establish the steady state.

A new fluctuation–dissipation relation applicable to a system in a steady state has also been derived, giving the dissipation of energy in terms of the *sin* transform of a new TCF of momenta. Here again, the average is over a steady-state RDOp instead of the standard equilibrium RDOp. And the time dependence in the TCF accounts for the effect of the dissipative rate operator in the time evolution. The energy dissipation rate is shown to depend on the frequency of the driving voltage field and now also explicitly on the frequency of the steady light creating the steady state. This relation reduces to the standard fluctuation–dissipation relation in the absence of the steady light field and ignoring the dissipative rate in the EOM.

In the case of response to a light pulse cutoff, a relaxing RDOp $\hat{\rho}^{(\mathrm{rel})}(t)$ has been constructed as a superposition of oscillating and decaying terms with amplitudes given by eigenoperators of the relaxation superoperator from which the time evolution of properties can be obtained after $t=0$. The eigenoperators $\hat{\rho}_A$ are not very different from the steady state $\hat{\rho}^{(\mathrm{ss})}$ when the light field is weak, and since the steady state corresponds to zero eigenvalues of the (ss) superoperator, it follows that one of the relaxation eigenstates should have an amplitude nearly equal to one and a small relaxation rate λ_A, and this one gives the longer lasting term for the relaxation of the RDOp to equilibrium. Related transient electronic currents can be constructed from projections on the eigendensity operators of the electronic momentum, and of the starting RDOp $\hat{\rho}^{(\mathrm{exc})}$ after light cutoff. The difference between these equations and the standard ones for relaxation is that here the starting RDOp at $t=0$ is not just for a single excited state but for the nearly steady state RDOp $\hat{\rho}^{(\mathrm{exc})}$ prepared by a pulse of light in the presence of dissipation due to the p- and s-region interactions. In addition, the relaxing RDOp is found to contain several relaxing terms with relaxation rates related here to the dissipative rate superoperator in the EOM, which can be constructed for both slow and fast medium excitations.

The general formulation has been presented for a system of many electrons and atomic nuclei. In applications, it is necessary to introduce basis sets of electronic states to convert operator equations into matrices of coupled integrodifferential equations. This has been shown to be possible by developing, as special cases, formulations in terms of one-electron vibronic states, with electron orbitals obtained from *ab initio* electronic structure calculations. This is the same formulation we have implemented in recent calculations of optical and conductivity properties of silicon slabs, in compounds with adsorbed metal clusters and with dopants.

Two examples included here show how properties can be calculated and compared with experimental measurements for systems in steady states induced by light absorption. The steady-state RDM can be used to obtain the average polarizability $P(t)$, or induced electric dipole $D(t)$ per unit volume, of a semiconductor slab in a steady state and also subject to an electric field $\mathcal{E}(t)$ of frequency ω, generally different from ω_{L}. It has produced results for the dielectric function of silicon slabs of varying thickness accounting for dissipative phenomena, as shown above. Steady-state RDMs have also been used to calculate relaxation times in photoexcited silicon slabs, in terms of the eigenoperators of a dissipative rate supermatrix and their eigenvalues introduced above to expand the relaxing RDM. They are obtained from the Redfield relaxation supermatrix, which is constructed from state-to-state electronic transition rates induced by interaction of the photoactive electrons with all the vibrating ions in the physical system, recently obtained from *ab initio* classical molecular dynamics calculations of lattice vibrations.

The new derivations we have presented here relate to a subset of nonequilibrium systems, where steady light absorption forms steady states with a statistical distribution different from a thermal one. A similar procedure using the RDOp could be introduced to deal with more general nonequilibrium systems, replacing the steady-state RDOp with a more general time-dependent reference RDOp to be constructed for a system unperturbed by external fields. This can then be changed by addition of a term providing the response to an external field. Alternative promising approaches to nonequilibrium response have been developed using a time-dependent statistical distribution constructed from a maximal entropy with given constrains.[33–35]

ACKNOWLEDGMENTS

This chapter is based on work done in collaboration with Drs. A. Leathers, D. Kilin, T. Vazhappilly, and R. Hembree. It is a pleasure to dedicate this publication to Frank Harris to celebrate his active research at the age of 85.

Work partly supported by the Chemistry Division of the National Science Foundation of the USA, grants NSF 1011967 and NSF 1445825, and by the Dreyfus Foundation.

REFERENCES

1. Landau, L. D.; Lifshitz, E. M. *Statistical Physics*; Pergamon Press: London, England, 1958, Chapter XII.
2. Kubo, R.; Toda, M.; Hasitsume, N. *Statistical Physics II: Nonequilibrium Statistical Mechanics*; Springer-Verlag: Berlin, 1978.
3. Chandler, D. *Introduction to Modern Statistical Mechanics*; Oxford University Press: New York, 1987, Chapter 8.
4. Mukamel, S. *Principles of Nonlinear Optical Spectroscopy*; Oxford University Press: New York, 1995.
5. Nitzan, A. *Chemical Dynamics in Condensed Phases*; Oxford University Press: Oxford, England, 2006.
6. May, V.; Kuhn, O. *Charge and Energy Transfer Dynamics in Molecular Systems*. Wiley-VCH: Berlin, 2000.
7. Micha, D. A. From Few-atom to Many-atom Quantum Dynamics. *Adv. Quantum Chem.* **2002**, *41*, 139.
8. Micha, D. A.; Thorndyke, B. The Quantum-Classical Density Operator for Electronically Excited Molecular Systems. *Adv. Quantum Chem.* **2004**, *47*, 294.
9. Reichl, L. E. *A Modern Course in Statistical Physics*; University of Texas Press: Austin, Texas, 1980.

10. Hanggi, P.; Thomas, H. Stochastic Processes, Time-Evolution, Symmetries and Linear Response. *Phys. Rep.* **1982**, *88*, 207.
11. Ruelle, D. General Linear Response Formula in Statistical Mechanics, and the Fluctuation-dissipation Theorem Far from Equilibrium. *Phys. Lett. A* **1998**, *245*, 220.
12. Speck, T.; Seifert, U. Restoring a Fluctuation-dissipation Theorem in a Nonequilibrium Steady State. *Europhys. Lett.* **2006**, *74*, 391.
13. Prost, J.; Joanny, J.-F.; Parrondo, J. M. R. Generalized Fluctuation-Dissipation Theorem for Steady State Systems. *Phys. Rev. Lett.* **2009**, *103*, 090601.
14. Esposito, M.; Harbola, U.; Mukamel, S. Nonequilibrium Fluctuations, Fluctuation Theorems, and Counting Statistics in Quantum Systems. *Rev. Mod. Phys.* **2009**, *81*, 1665, Erratum 2014, 86, 1125.
15. Yeganeh, S.; Ratner, M.; Mujica, V. Dynamics of Charge Transfer: Rate Processes Formulated with Non-Equilibrium Green's Functions. *J. Chem. Phys.* **2007**, *126*, 161103.
16. Galperin, M.; Tretiak, S. Linear Optical Response of Current Carrying Molecular Junctions: A Nonequilibrium Green's Function-Time Dependent Density Functional Theory Approach. *J. Chem. Phys.* **2008**, *128*, 124705.
17. Ness, H.; Dash, L. K. Nonequilibrium Fluctuation-dissipation Relations for One- and Two-particle Correlation Functions in Steady State Quantum Transport. *J. Chem. Phys.* **2014**, *140*, 144106.
18. Gross, E. K. U.; Dreizler, R. M. Eds. *Density Functional Theory*; Plenum Press: New York, 1995.
19. Zheng, X.; Cheng, G.-H.; Mo, Y.; Koo, S.-K.; Tian, H.; Yam, C.-Y.; Yan, Y.-J. Time Dependent Density Functional Theory for Quantum Transport. *J. Chem. Phys.* **2010**, *133*, 114101.
20. Tempel, D. G.; Watson, M. A.; Olivares-Amaya, R.; Aspuru-Guzik, A. Time Dependent Density Functional Theory of Open Quantum Systems in the Linear Response Regime. *J. Chem. Phys.* **2011**, *134*, 074116.
21. Casida, M. E.; Huix-Rotllant, M. Progress in Time Dependent Density Functional Theory. *Annu. Rev. Phys. Chem.* **2012**, *63*, 287.
22. Leathers, S.; Micha, D. A.; Kilin, D. S. Density Matrix Treatment of Combined Instantaneous and Delayed Dissipation for an Electronically Excited Adsorbate on a Solid Surface. *J. Chem. Phys.* **2009**, *131*, 144106-1.
23. Micha, D. A.; Leathers, A. S. Reduced Density Matrix Equations for Combined Instantaneous and Delayed Dissipation in Many-Atom Systems, and Their Numerical Treatment. In: *Energy Transfer Dynamics in Biomolecular Systems*; Burghardt, I., May, V., Micha, D. A., Bittner, E. R., Eds.; Springer-Verlag, 2009; p 363.
24. Kilin, D. S.; Micha, D. A. Surface Photovoltage at Nanostructures on Si Surfaces: Ab Initio Results. *J. Phys. Chem. C* **2009**, *113*, 3530.
25. Leathers, S.; Micha, D. A.; Kilin, D. S. Direct and Indirect Electron Transfer at a Semiconductor Surface with an Adsorbate: Theory and Application to $Ag_3Si(111)$:H. *J. Chem. Phys.* **2010**, *132*, 114702.
26. Kilin, D. S.; Micha, D. A. Relaxation of Photoexcited Electrons at a Nanostructured Si(111) Surface. *J. Phys. Chem. Lett.* **2010**, *1*, 1073.
27. Micha, D. A. Density Matrix Treatment of Non-Adiabatic Photoinduced Electron Transfer at a Semiconductor Surface. *J. Chem. Phys.* **2012**, *137*, 22A521.
28. Hembree, R. H.; Micha, D. A. Photoinduced Electron Transfer at a Si(111) Nanostructured Surface: Effect of Varying Light Wavelength, Temperature, and Structural Parameters. *J. Chem. Phys.* **2013**, *138*, 184708.
29. Vazhappilly, T.; Micha, D. A. Computational Modeling of the Dielectric Function of a Silicon Slab with Varying Thickness. *J. Phys. Chem. C* **2014**, *118*, 4429.
30. Han, Y.; Micha, D. A.; Kilin, D. S. Ab initio Study of the Photocurrent at the Au/Si Metal-Semiconductor Interface. *Mol. Phys.* **2015**, *113*, 327–335.

31. Micha, D. A. Density Matrix Treatment of Optical Properties in Photovoltaic Materials: Photoconductivity at a Semiconductor Surface. In: *Photoinduced Processes at Surfaces and in Nanomaterials*; Kilin, D. S., Ed.; ACS Book Series; American Chemical Society: Washington, DC, 2015, Accepted to appear.
32. Weinelt, M.; Kutschera, M.; Fauster, Th.; Rohlfing, M. Dynamics of Exciton Formation at the Si(100)c(4x2) Surface. *Phys. Rev. Lett.* **2004**, *92*, 126801-1.
33. Katz, A. *Principles of Statistical Mechanics: The Information Theory Approach*; W. H. Freeman: San Francisco, 1967.
34. Luzzi, R.; Vasconcellos, A. R.; Galvao Ramos, J. Irreversible Thermodynamics in a Non-Equilibrium Statistical Ensemble Formalism. *Riv. Nuovo Cimento* **2001**, *24*, 1.
35. Seifert, U. Stochastic Thermodynamics, Fluctuation Theorems and Molecular Machines. *Rep. Progr. Phys.* **2012**, *75*, 126001.

CHAPTER NINE

Frank Discussion of the Status of Ground-State Orbital-Free DFT

Valentin V. Karasiev*, Samuel B. Trickey*,[1]
*Quantum Theory Project, Departments of Physics and of Chemistry, University of Florida, Gainesville, Florida, USA
[1]Corresponding author: e-mail address: trickey@qtp.ufl.edu

Contents

Abstract

F.E. Harris has been a significant partner in our work on orbital-free density functional approximations for use in *ab initio* molecular dynamics. Here we mention briefly the essential progress in single-point functionals since our original paper (2006). Then we focus on the advantages and limitations of generalized gradient approximation (GGA) noninteracting kinetic energy (KE) functionals. We reconsider the constraints provided by near-origin conditions in atomic-like systems and their relationship to regularized versus physical external potentials. Then we seek the best empirical GGA for the noninteracting KE for a modest-sized set of molecules with well-defined near-origin behavior of their densities. The search is motivated by a desire for insight into GGA limitations and for a target for constraint-based development.

Advances in Quantum Chemistry, Volume 71
ISSN 0065-3276
http://dx.doi.org/10.1016/bs.aiq.2015.02.004

1. INTRODUCTION

1.1 Background

At the Harris Workshop (10–12 December 2014), the second author spoke about recent progress by the University of Florida orbital-free density functional theory (OFDFT) group of which Frank Harris is a member. Substantially all that work is reported in Refs. 1–3 and references therein. Earlier work and voluminous references for context are in two review articles from our group.[4,5]

Rather than recapitulate the talk and publications, here we provide a particular perspective on what has been learnt. The phrase "near origin" rather than "cusp condition" is a clue to the role the external potential plays in enforcing behavior upon generalized gradient approximations (GGAs) for the noninteracting kinetic energy (KE). We present some new results on near-origin conditions applied to GGAs. These extend work we did with Frank Harris.[6–8] Then we explore implications of a generic regularization of the usual external potential (from a nuclear array) by empirical determination of the most nearly optimal GGA for a set of molecular data. That continues the study of binding in simple molecules by non-self-consistent OFDFT with key ingredients of the methodology introduced in our previous publications.[6–8] Those ingredients, besides the near-origin analysis of the Pauli potential, include (i) the use of a set of nuclear spatial configurations for the same molecule; (ii) the use of Gaussian Kohn–Sham (KS) molecular densities as input; (iii) so-called ΔE criterion which enforces binding; and (iv) the E criterion which enforces correct absolute energies. See also the recent work of Borgoo et al.[9] in which the relationship between binding and the effective homogeneity of approximate noninteracting KE functional is considered.

Though our research agenda emphasizes functionals for free energy DFT[10] primarily for use in the warm dense matter regime, here we restrict attention to ground-state OFDFT. There are three reasons. First, ground-state OFDFT is a hard challenge (as history going all the way back to Thomas[11] and Fermi[12] demonstrates). That challenge is worsened by going to finite-T (one must devise an entropy functional and incorporate the intrinsic T-dependence of other functionals). Third, the ground-state approximations must be reliable and well-founded if there is to be a sensible $T = 0$ K limit for approximate free energy functionals.

1.2 Basics and Notation

For context and to set notation, the Levy–Lieb version of the Hohenberg–Kohn universal functional[13–15] is

$$\mathcal{E}[n] = T[n] + U_{ee}[n] \tag{1}$$

with $T[n]$, $U_{ee}[n]$, and $n(\mathbf{r})$ the total KE, total Coulomb energy (Hartree, exchange, and correlation), and the electron density at point $\mathbf{r}$, respectively ($\int d\mathbf{r} n(\mathbf{r}) = N_e$, with N_e the number of electrons). Assuming that the sum is bounded below, addition of an external potential energy $E_{ext}[n]$ gives the usual DFT variational principle,

$$\min_n \{\mathcal{E}[n] + E_{ext}[n]\} = E_0[n_0]. \tag{2}$$

Zero subscripts indicate ground-state values.

The KS[16] rearrangement of $\mathcal{E}$ invokes an auxiliary noninteracting Fermion system with the same density as the physical system. This raises so-called v-representability requirements which we assume to be satisfied. The KS system has KE and exchange (X) energies T_s, E_x, which enable the regrouping of (1) into

$$\mathcal{E}[n] = T_s[n] + E_H[n] + E_x[n] + E_c[n] \tag{3}$$

$$E_c[n] := U_{ee}[n] - E_H[n] - E_x[n] + T[n] - T_s[n] \tag{4}$$

$$E_H[n] := \frac{1}{2}\int d\mathbf{r}_1 d\mathbf{r}_2 \frac{n(\mathbf{r}_1)n(\mathbf{r}_2)}{|\mathbf{r}_1 - \mathbf{r}_2|} \tag{5}$$

$$E_x[n] := \langle \Phi_s[n] | \hat{V}_{ee} | \Phi_s[n] \rangle - E_H[n]. \tag{6}$$

E_c is the DFT correlation energy. It often is useful to write $E_{xc} = E_x + E_c$. $\hat{V}_{ee}$ is the electron–electron Coulomb interaction. The KS Slater determinant $\Phi_s[n]$ is comprised of orbitals from the KS system Euler equations,

$$h_{KS}[n]\varphi_i = \varepsilon_i \varphi_i, \quad n(\mathbf{r}) = \sum_i f_i |\varphi_i(\mathbf{r})|^2. \tag{7}$$

Here the $f_i = 0,1,2$ are occupation numbers in the non-spin-polarized case.[17,18] The KS potential is

$$\begin{aligned} v_{KS} &= v_H + v_{xc} + v_{ext} \\ v_H &= \int d\mathbf{r}_2 \frac{n(\mathbf{r}_2)}{|\mathbf{r} - \mathbf{r}_2|}, \quad v_{xc} = \frac{\delta E_{xc}}{\delta n}, \end{aligned} \tag{8}$$

For now, we leave $v_{\rm ext}$ unspecified. Finally, the KS KE is

$$T_{\rm s}[\{\varphi_i\}] = \frac{1}{2}\sum f_i \int d\mathbf{r}|\nabla\phi_i(\mathbf{r})|^2 := \int d\mathbf{r} t_{\rm orb}[n(\mathbf{r})] \tag{9}$$

in Hartree atomic units. This positive-definite integrand form of $T_{\rm s}$ is preferable for OFDFT because the integrand of the ordinary Laplacian form of $T_{\rm s}$ can have both signs. The two forms differ by a surface integral which is zero for physically significant systems.

1.3 Essential Challenge of OFDFT

Posed succinctly, the OFDFT opportunity is that the computational costs of direct minimization of Eq. (3) scale with system size. In contrast, solution of the KS equations, (7), has computational cost scaling as $N_{\rm e}^3$ or worse.

The OFDFT challenge can be stated succinctly too. $E_{\rm x}[n]$ is *defined* in terms of the KS orbitals and hence is known exactly only as an implicit functional of n. $T_{\rm s}[n]$ is an implicit functional as well. $E_{\rm c}[n]$ is *defined* in terms of those two. One might try reversion to $\mathcal{E}[n]$ for construction of approximations, but most rigorous knowledge about $\mathcal{E}$ (scaling, bounds, limits, etc.) is in terms of the KS rearrangement. Roughly a half century of effort has been devoted to finding good approximations to $E_{\rm x}$ and $E_{\rm xc}$. Abandoning the KS decomposition would discard that resource and, worse, disconnect the result from a huge literature of calculations with such functionals. And $T_{\rm s}$ has several rigorously demonstrable properties which serve as stringent constraints on approximations.[4,6–8]

In short, retention and use of the KS decomposition is practically inescapable. OFDFT thus aims at reliable approximations for KS DFT quantities without explicit dependence on the KS orbitals. The allowed variables therefore are n and its spatial derivatives. For $E_{\rm xc}$, the consequence is a restriction to the meta-generalized-gradient approximation (mGGA) rung of the widely quoted Perdew–Schmidt Jacobs' ladder of complexity.[19] mGGAs depend upon n, $|\nabla n|$, $\nabla^2 n$, and the KS KE density $t_{\rm orb}$. Immediately, the OFDFT challenge is in play: an explicit functional for $t_{\rm orb}$ is required.

In practice, the highest spatial derivative dependence that so far has been useful for $t_{\rm orb}$ is a GGA, to wit

$$\begin{aligned} T_{\rm s}^{\rm GGA}[n] &= c_{\rm TF}\int d\mathbf{r}\, n^{5/3}(\mathbf{r})F_{\rm t}(s(\mathbf{r})) \\ c_{\rm TF} &= \frac{3}{10}(3\pi)^{2/3}. \end{aligned} \tag{10}$$

F_t is called the enhancement factor. For $F_t = 1$, $T_s^{\mathrm{GGA}} = T_{\mathrm{TF}}$, the Thomas–Fermi functional. The dimensionless reduced density gradient is

$$s := \frac{1}{2(3\pi^2)^{1/3}} \frac{|\nabla n|}{n^{4/3}} \equiv \kappa \frac{|\nabla n|}{n^{4/3}}. \tag{11}$$

Remark: The s variable occurs in GGA X functionals also. They have the same form as Eq. (10) but with $n^{4/3}$ rather than $n^{5/3}$, F_x rather than F_t, and a different prefactor, $c_x = -(3/4)(3/\pi)^{1/3}$.

Equation (10) is a one-point GGA functional. In the OFKE literature, there is extensive work on two-point functionals, generically $\int d\mathbf{r}d\mathbf{r}' n^{\alpha}(\mathbf{r})K(\mathbf{r},\mathbf{r}')n^{\beta}(\mathbf{r}')$. See Section 2.3 of Ref. 4 for brief discussion and references. One readily can imagine constructing a GGA for two-point functionals, but we are unaware of effort along that line. Instead, the emphasis has been on constructing $K(\mathbf{r},\mathbf{r}')$ via constraints, mostly to match response properties of the weakly perturbed homogeneous electron gas.

Motivated to optimize computational performance, our group has focused on finding and exhausting the limits of single-point GGAs for T_s. The remaining discussion assesses what we have found, with a focus on the surprising nonuniversality of approximate OFKE functionals, implications for their common use with external potentials of regularized Coulomb form, and an empirical attempt to ascertain the limits of GGA performance for a particular kind of regularized potential.

2. QUALITATIVE DISTINCTIONS AMONG GGAs FOR t_{orb}

Our work began[6] by testing multiple published T_s^{GGA} functionals. A rough classification introduced then was standard GGA and modified-conjoint GGA (mcGGA). The latter term stems from conjoint functionals,[20] i.e., those for which $F_t \propto F_x$. Standard GGAs include the second-order gradient approximation (SGA)

$$T^{\mathrm{SGA}} = T_{\mathrm{TF}} + \frac{1}{9} T_{\mathrm{W}} \tag{12}$$

$$T_{\mathrm{W}}[n] := \frac{1}{8} \int d\mathbf{r} \frac{|\nabla n(\mathbf{r})|^2}{n(\mathbf{r})} \tag{13}$$

and the von Weizsäcker KE T_{W} itself, along with most of the GGAs of the modern era, e.g., that by Perdew,[21] the PW91 KE functional based on the Perdew–Wang X functional,[22] and those from DePristo and Kress,[23] Thakkar,[24] and Tran and Wesolowski,[25] and the APBEK functional based on the PBE X functional.[26] Modified conjoint GGAs arise from altering or refining the conjointness conjecture (which is not strictly correct; see Ref. 8). These include $T_{\mathrm{TF}} + T_{\mathrm{W}}$ and our PBE2, KST2,[6,8] and VT84F[3] functionals.

The two functional types have qualitatively different performance. Ordinary GGAs predict the KE order of magnitude correctly but fail to give binding for simple molecules and solids. There are some exceptions for solids in which a pseudo-density is used. mcGGAs do bind simple molecules as well as many solids at least semi-quantitatively, but they overestimate the KE strongly. As a consequence, the total energy also is strongly overestimated (too high). We also have found that these functionals exhibit peculiar sensitivity to the type of pseudo-potential used, behavior found by others as well.[27]

The main difference between ordinary GGAs and the mcGGAs is the enforcement of positivity constraints on the mcGGAs. Enforcement is via imposition of requirements upon the density in the case that the external potential is Coulombic,

$$v_{\mathrm{ext}}(\mathbf{r}) = -\sum_{\alpha} \frac{Z_\alpha}{|\mathbf{r} - \mathbf{R}_\alpha|} \tag{14}$$

with Z_α the atomic number of the nucleus at site $\mathbf{R}_\alpha$. Such approximate $T_{\mathrm{s}}^{\mathrm{GGA}}$ functionals therefore are not guaranteed to be universal, even though T_{s} is. Two questions then arise. Are both the overly large KE and sensitivity to pseudo-potentials of mcGGAs connected with this nonuniversality? Is there an example of a $T_{\mathrm{s}}^{\mathrm{GGA}}$ that has both the good KE magnitudes of an ordinary GGA and the good binding properties of an mcGGA? We address these two issues in the remainder of this chapter.

3. POSITIVITY AND NEAR-ORIGIN CONDITIONS

The Pauli term decomposition

$$T_{\mathrm{s}}[n] = T_{\mathrm{W}}[n] + T_\theta[n], \tag{15}$$

provides a rigorous bound,[28–32]

$$T_\theta[n] \geq 0, \tag{16}$$

because T_{W} is a lower bound to the KS KE.[33–36] T_{W} also is the exact T_{s} for one electron, a fact that will become useful shortly. (It also is exact for a two-electron singlet.) The Pauli term potential also is rigorously non-negative:

$$v_\theta(\mathbf{r}) := \frac{\delta T_\theta[n]}{\delta n(\mathbf{r})} \geq 0, \quad \forall \mathbf{r}. \tag{17}$$

These are universal properties of T_s. For a GGA, the Pauli term separation corresponds to a T_θ^{GGA} with energy density t_θ and enhancement factor

$$F_\theta(s) = F_t(s) - \frac{5}{3}s^2. \tag{18}$$

Though $T_\theta \geq 0$, it is not necessarily the case that the associated Pauli term energy density t_θ obeys the same positivity $t_\theta \geq 0$ because energy densities are defined only up to additive functions which integrate to zero. Refs. 31,36,37 chose the canonical form for t_θ (i.e., that which comes from the KS equation), which is positive semi-definite. We adopted that argument in Ref. 38. The consequence, to which we return in Section 4, is

$$F_\theta(s(\mathbf{r})) \geq 0, \quad \forall \mathbf{r}. \tag{19}$$

To have enough additional constraints to determine a useful approximate F_θ, we used[3,6,8,38] requisites of physical many-electron systems, i.e., those with an external potential given by Eq. (14). Nonuniversality enters.

The nuclear-cusp condition[39] density

$$n(r) \sim e^{-2Zr} = (1 - 2Zr) + O(r^2). \tag{20}$$

gives $v_\theta^{\mathrm{GGA}}(r) \sim a/r$, where a is a constant which depends on the specific enhancement factor.[8] So far as we know, the first mention of this consequence was by Levy and Ou-Yang (see the latter part of Section III of Ref. 31).

The one-electron character of the tail region of a many-electron atom[37] forces the approximate functional to go over to T_W in that region.[18] For a GGA therefore we may require,

$$\lim_{s \to \infty} F_\theta(s) = 0. \tag{21}$$

Kato cusp behavior Eq. (20) is not exhibited by any density that results from a regularized potential, e.g., a pseudo-potential. See, for example, Eq. (6) in Ref. 40 and associated discussion. Removing that cusp to allow use of compact basis sets (especially a plane-wave basis) is the motive for pseudo-potentials. Densities from expansion in a finite Gaussian-type basis set, even in all-electron calculations that use Eq. (14), also do not have Kato cusp behavior. Similarly, the proper tail behavior, also exponential, is not found in any finite Gaussian expansion density. Here we focus on the former issue, the near-origin behavior of atomic-like systems.

Consider one-center N_e electron densities of the flexible form

$$n_f(\mathbf{r}) := A_f \exp(-\lambda r^\gamma), \quad 1 \le \gamma \le 2 \tag{22}$$

$$A_f = \frac{N_e \gamma \lambda^{3/\gamma}}{4\pi\Gamma(3/\gamma)}, \tag{23}$$

The norm follows from Ref. 41, with the usual Γ function. With $\gamma = 1$, $\lambda = 2N_e$, $N_e = 1$, this is the H atom density in the central field approximation. For $\gamma = 2$, it is pure Gaussian. For use in what follows, the von Weizsäcker potential for densities of this form is

$$v_W = \frac{\delta T_W}{\delta n} = \frac{\lambda\gamma}{8} r^{\gamma-2}[2(\gamma+1) - \lambda\gamma r^\gamma] . \tag{24}$$

With densities of the form (22), we can explore two simple but illuminating issues. The first is to determine the external potential that corresponds to the given density for the case $N_e = 1$. Recall the bijectivity of the external potential and the density guaranteed by the first Hohenberg–Kohn theorem.[13] The central-field hydrogenic case is obvious but it is instructive to do it in the context of OFKE functionals. The Euler equation is

$$\frac{\delta(\mathcal{E} + E_{ext})}{\delta n} = v_W + v_\theta + v_H + v_{xc} + v_{ext} = \mu, \tag{25}$$

with μ the Lagrangian multiplier for charge normalization. T_W is exact for the one-electron case, so $v_\theta = 0$. Exact exchange cancels the Hartree self-interaction, so $v_H = -v_x$, and there is no correlation, $v_c = 0$. The von Weizsäcker potential (24) for the hydrogenic densities ($\gamma = 1$, $\lambda = 2N_e$) is

$$v_W = \frac{N_e}{r} - \frac{N_e^2}{2}. \tag{26}$$

For H, $\mu = -\frac{1}{2}$, $N_e = 1$, (25) gives the expected result:

$$0 = \frac{1}{r} - \frac{1}{2} + v_{ext}(r) - (-\frac{1}{2}) \Rightarrow v_{ext}(r) = -\frac{1}{r}. \tag{27}$$

Redoing the argument with $\gamma = 2$, $N_e = 1$ gives

$$v_W = \frac{\lambda}{2}(3 - \lambda r^2) \Rightarrow v_{ext} = \frac{1}{2}\lambda^2 r^2 + (\mu - \frac{3\lambda}{2}), \tag{28}$$

the expected quadratic dependence for v_{ext}.

This elementary exercise illustrates a significant point for approximate functionals. Repeat the argument for $\gamma = 2$ but now with the physically important external Coulomb potential imposed and with an approximate T_θ functional (not necessarily a GGA; for the moment, the discussion is general). Then the Euler equation becomes

$$\mu = -\frac{1}{r} - \frac{1}{2}\lambda^2 r^2 + \frac{3\lambda}{2} + v_\theta^{\text{approx}}(r). \tag{29}$$

The only way this can be satisfied is for there to be an *incorrect*, i.e., nonzero, v_θ^{approx} for the one-electron case.

In the case of pseudo-potentials, the argument runs in reverse. Suppose a pseudo-potential prescription to be used at the so-called one-electron level, i.e., one electron outside the core, and suppose it to deliver the form (28). Assume that one can contrive a satisfying approximate functional with the property that for $N_e = 1$, the approximate functional respects rigorous constraints for the corresponding pseudo-density. Now shift to an all-electron pseudo-potential and shrink the core toward the bare Coulomb potential. In an arbitrarily small region around the origin, the pseudo-density will remain harmonic but the pseudo-potential in almost all space will be essentially Coulombic, leading to the kind of mismatch given in Eq. (29). Even at this level (two pseudo-potentials with the same regularization procedure but significantly different core radii and populations), there is a lack of universality for the approximate T_θ.

For arbitrary γ dependence, $1 \leq \gamma \leq 2$, the imputed external potential is

$$v_{\text{ext}} = \mu - v_{\text{W}} = \mu + \frac{\lambda\gamma}{8} r^{\gamma-2}[\lambda\gamma r^\gamma - 2(\gamma+1)], \tag{30}$$

(with suitably adjusted μ of course). Similar mismatch difficulties will occur for all intermediate γ values, as will the singularities for $\gamma \neq 2$.

Now consider GGA functionals with arbitrary N_e. The GGA Pauli potential is[8]

$$v_\theta^{\text{GGA}}(s^2) = c_0 n^{2/3}\left\{\frac{5}{3}F_\theta(s^2) - \left(\frac{2}{3}s^2 + 2p\right)\frac{\partial F_\theta}{\partial(s^2)} + 4\left(\frac{4}{3}s^4 - q\right)\frac{\partial^2 F_\theta}{\partial(s^2)^2}\right\}, \tag{31}$$

with higher-order reduced density derivatives

$$p := \kappa^2 \frac{\nabla^2 n}{n^{5/3}}, \quad q := \kappa^4 \frac{\nabla n \cdot (\nabla\nabla n) \cdot \nabla n}{n^{13/3}}. \tag{32}$$

Evaluation with the flexible density (22) yields

$$s^2(r) = \kappa^2 \lambda^2 \gamma^2 r^{2(\gamma-1)} n_{\mathrm{f}}^{-2/3}(r), \tag{33}$$

$$p(r) = \kappa^2 \lambda \gamma r^{\gamma-2} [\lambda \gamma r^{\gamma} - (\gamma+1)] n_{\mathrm{f}}^{-2/3}(r), \tag{34}$$

and

$$q(r) = \kappa^4 \lambda^3 \gamma^3 r^{3\gamma-4} [\lambda \gamma r^{\gamma} - (\gamma-1)] n_{\mathrm{f}}^{-4/3}(r). \tag{35}$$

Except for a negative sign, the coefficient of $\partial F_\theta / \partial s^2$ in Eq. (31) is

$$\frac{2}{3} s^2 + 2p = 2\kappa^2 \lambda \gamma r^{\gamma-2} \left[\frac{4}{3} \lambda \gamma r^{\gamma} - (\gamma+1) \right] n_{\mathrm{f}}^{-2/3}(r). \tag{36}$$

Notice the singularity at the origin for $\gamma < 2$. Up to a factor of 4, the coefficient of $\partial^2 F_\theta / \partial (s^2)^2$ in Eq. (31) is

$$\left[\frac{4}{3} s^4 - q \right] = \kappa^4 \lambda^3 \gamma^3 r^{3\gamma-4} \left[\frac{\lambda \gamma}{3} r^{\gamma} + (\gamma-1) \right] n_{\mathrm{f}}^{-4/3}. \tag{37}$$

This is nonsingular only for $\gamma \geq 4/3$.

For small s, one usually enforces gradient expansion behavior on F_θ,

$$F_\theta = 1 + a s^2 \tag{38}$$

and only the first derivative term in v_θ, Eq. (31), is at issue. After a bit of manipulation,

$$p(r) = s^2 \left[1 - \frac{\gamma+1}{\lambda \gamma r^{\gamma}} \right] \Rightarrow \frac{2}{3} s^2 + 2p = 2s^2 \left[\frac{4}{3} - \frac{\gamma+1}{\lambda \gamma r^{\gamma}} \right]. \tag{39}$$

The singularity structure in v_θ then is evident. The general result is

$$v_\theta^{\mathrm{GGA}}[n_{\mathrm{f}}] = c_0 n_{\mathrm{f}}^{2/3} \left\{ \frac{5}{3} + a\kappa^2 \gamma^2 \lambda^2 r^{2(\gamma-1)} n_{\mathrm{f}}^{-2/3} \left[\frac{2(\gamma+1)}{\lambda \gamma r^{\gamma}} - 1 \right] \right\}. \tag{40}$$

For convenience, the two limiting cases are

$$v_\theta^{\mathrm{GGA}}[n_{\mathrm{f}}, \gamma = 1] = c_0 n_{\mathrm{f}}^{2/3} \left\{ \frac{5}{3} + a\kappa^2 \lambda^2 n_{\mathrm{f}}^{-2/3} \left[\frac{4}{\lambda r} - 1 \right] \right\} \tag{41}$$

and

$$v_{\theta}^{\mathrm{GGA}}[n_{\mathrm{f}},\gamma=2]=c_0 n_{\mathrm{f}}^{2/3}\left\{\frac{5}{3}+4a\kappa^2\lambda n_{\mathrm{f}}^{-2/3}\left[3-\lambda r^2\right]\right\}. \tag{42}$$

The takeaway point is that if one sets out to build an approximation constrained to behave properly for $\gamma = 1$ (the physical case), the singularity is inevitable and the near-origin positivity is determined by the sign of the gradient expansion coefficient a. Our mcGGAs are built to have $a > 0$. However, if the actual density is regularized and has Gaussian form near the origin, then if that density is "cuspy" enough, i.e., has large λ, even with $a > 0$ and positivity constraints enforced on building the approximation, there still can be small-r regions for which $v_{\theta}^{\mathrm{approx}} < 0$.

4. EMPIRICAL EXPLORATION OF THE LIMITS OF GGA KE

4.1 Methodology

Our approach to the development of GGA OFKE functionals has been to adopt some suitable analytical form for the KE enhancement factor F_{t} with a few parameters determined from imposing constraints (e.g., correct scaling if applicable, correct small-s and large-s behavior) and, if unavoidable, fitting to a small set of training data. Interpolation between small- and large-s is defined by the chosen analytical form for F_{t}. The analytical forms usually are relatively simple with deliberately limited flexibility to avoid introduction of nonphysical kinks or other artifacts in that interpolation. In this sense, the properties are analogous to those of standard finite basis sets (see, for example, Refs. 3,6,7,21,23–25).

The unwelcome effects of limited flexibility can be avoided, at least in principle, by use of a numerical enhancement factor given on a mesh $s_0 = 0, s_1, \ldots, s_n = s_{\max}$. There is a practical barrier however. To determine such a numerical F_{t} requires numerical integration in real space of the complicated $(n, |\nabla n|)$ dependence of the KE functional integrand t_{orb}, Eq. (9), evaluated on a numerical s-mesh. Experience demonstrates that the result is unphysical, noisy, numerically unstable. One can see the difficulty simply by considering the numerical integration of an s "density of states" on a mesh of points s_i:

$$\mathcal{D}(s_i) := \int d\mathbf{r}\,\delta(s_i - s(\mathbf{r})) \approx \sum_j w_j \delta_{s_i, s(\mathbf{r}_j)}, \tag{43}$$

with w_j the quadrature weights. Numerical experiment shows that a modest change in even a very fine **r** or s mesh (or both) leads to distinctly different results.

An effective alternative is Padé approximants[42] of high orders such as were used recently for analytical representation of common Fermi–Dirac integral combinations.[43] They provide the simultaneous flexibility and smoothness required by the numerical integration in **r**. Numerical exploration led to the Padé approximant

$$F_t(s) = \frac{1+\sum_{i=1}^{k} a_i s^i}{1+\sum_{i=1}^{l} b_i s^i}, \tag{44}$$

of order [9,10] in the variable s ($k = 9$, $l = 10$) as a workable compromise between flexibility and number of free parameters.

Only a few parameters in the approximant can be determined from imposition of constraints. The remainder must be obtained by fitting. For this study, the *only* constraint imposed on Eq. (44) is recovery of the correct second-order gradient expansion at small s,

$$F_t(s) \approx 1 + \frac{5}{27}s^2, \quad s << 1. \tag{45}$$

This is accomplished by setting $a_1 = b_1$, $a_2 = (5/27) + b_2$. To allow maximal freedom for the fitted F_t, we have not imposed the large-s von Weizsäcker limit given in Eq. (21). Studies of X GGA functionals[44] show that the distribution of s, Eq. (43), is negligible above about $s = 3$ for most systems of interest, thus suggesting that such large-s behavior constraints are not critically important, at least for fitting. Also note well that in what follows, we have *not* imposed *any* of the positivity constraints, Eqs. (16), (17), and (19). The motivation is to make the empirical fitting as unconstrained as possible.

The absolute KE versus binding energy dilemma posed at the end of Section 2 has implications for the fitting criteria to be used. The usual *KE* fitting criterion is equivalent to the *total energy* or *E* criterion,[6] namely to minimize the squared energy difference between nonself-consistent OFDFT and reference KS energies at system equilibrium geometries (from standard KS calculations). The procedure is nonself-consistent for the OFDFT calculations because KS densities are used as input. The obvious flaw in the *E* criterion (which was investigated in Ref. 8) is that it forces OFDFT total energies at KS equilibrium configurations to be as nearly correct as possible but ignores the shape of the KS binding energy curve near equilibrium.

In Refs. 6–8, Frank Harris and two of us introduced what was called the ΔE criterion. In it, the objective function to be optimized is formed from energy differences between a point away from equilibrium and the equilibrium point as predicted by the reference conventional KS calculation. The objective function has two of those energy differences, one from OFDFT and the other from the KS calculations. Obviously, the ΔE criterion enforces binding upon the OFDFT approximation while leaving the total energy uncontrolled. The result can be an excessively high total energy.

In this work, we address these two limitations by making a convex sum of average versions of the two criteria. The averages are calculated over all atoms and molecules (and their geometries) in a training set. To put the two criteria on the same scale, we use the mean absolute relative error (MARE) of energy differences rather than average absolute energy differences,

$$\omega_{\Delta E} = \frac{1}{N} \sum_{M,i \neq e} \frac{\left| \Delta E_{M,i}^{\mathrm{KS}} - \Delta E_{M,i}^{\mathrm{OF-DFT}} \right|}{\left| \Delta E_{M,i}^{\mathrm{KS}} \right|}. \tag{46}$$

Here, for the nuclear spatial configuration i of molecule M, $\Delta E_{M,i} = E_{M,i} - E_{M,e}$, with $E_{M,e}$ the energy associated with the equilibrium nuclear configuration as predicted from conventional KS computations, and N the total number of terms in the sum (Eq. 46). Similarly, the mean absolute relative error of the total energy is

$$\omega_E = \frac{1}{N} \sum_{M,i} \frac{\left| E_{M,i}^{\mathrm{KS}} - E_{M,i}^{\mathrm{OF-DFT}} \right|}{\left| E_{M,i}^{\mathrm{KS}} \right|}. \tag{47}$$

The objective function is a convex combination of both

$$\omega(\alpha) = \alpha \omega_E + (1 - \alpha) \omega_{\Delta E}, \tag{48}$$

with $\alpha \in [0,1]$. Minimization of $\omega(0)$ is essentially the ΔE criterion and conversely for $\omega(1)$. One expects, or at least hopes, that some intermediate α will provide a KE functional with both reasonable binding and reasonable absolute energy errors.

The training set we used includes nine molecules comprised of first- and second-row atoms and of diverse bonding types along with three closed shell atoms, $M = \{$LiH, CO, N_2, LiF, BF, NaF, SiO, H_4SiO, H_4SiO$_4$, Be, Ne, Ar$\}$. A set of six bond lengths was used for each molecule. Molecular geometries were changed by varying the single bond length in the diatomics, the central bond length R(Si–O) in H_4SiO, and by varying R(Si–O$_i$) in H_4SiO_4 deformed in the T_d mode. This set is small by comparison with the training sets used in the Minnesota series of XC functionals[45] because our purpose is different. We do not seek a broadly useful empirical functional. The issue here is narrower, namely whether there exists an OFDFT GGA functional which does well both on absolute energies and binding even on a small sample of systems.

One other technical point is that the enhancement factor F_t which results from fitting is checked for poles on the interval $s \in [0,1000]$. If the denominator of Eq. (44) has a root on that interval, the corresponding set of parameters is rejected.

All reference KS calculations were done in the local density approximation (LDA) for XC (see Refs. 16,46–52) using a triple-zeta Gaussian-type basis with polarization functions (TZVP).[53–55] Orbital-free KE integrals were calculated by numerical quadrature, as in our previous work.[6] Weight functions, $w_I(\mathbf{r})$, localized near each center with the properties that $w_I(\mathbf{r}) \geq 0$ and $\sum_I w_I(\mathbf{r}) = 1$ are used to represent the multicenter integrals exactly as a sum of atom-centered contributions[56]

$$T_s^{\mathrm{GGA}}[n] = \sum_{I=1}^{N_{\mathrm{atoms}}} c_{\mathrm{TF}} \int d\mathbf{r} w_I(\mathbf{r}) n^{5/3}(\mathbf{r}) F_t(s(\mathbf{r})). \tag{49}$$

Radial integration of the resulting single-center forms was via a Gauss–Legendre procedure, while integration over the angular variables used high-order quadrature formulae.[57] A dense mesh consisting of 150 radial and 434 angular grid points was used to calculate atom-centered integrals. These computations used routines developed by Salvador and Mayer[58] and included in their code Fuzzy.

Before proceeding to results, one should note the implications of these numerical procedures. The finite Gaussian-type basis inexorably yields Gaussian near-origin behavior of the density. Yet the calculations are all-electron in the bare Coulomb external potential (14). This is precisely the inconsistency between external potential and near-origin density behavior discussed in Section 3.

4.2 Results

There are 17 independent parameters left in Eq. (44) after constraining to the second-order gradient expansion. Those were optimized to minimize the objective function $\omega(\alpha)$. Figure 1 shows the ω_E and $\omega_{\Delta E}$ MAREs,

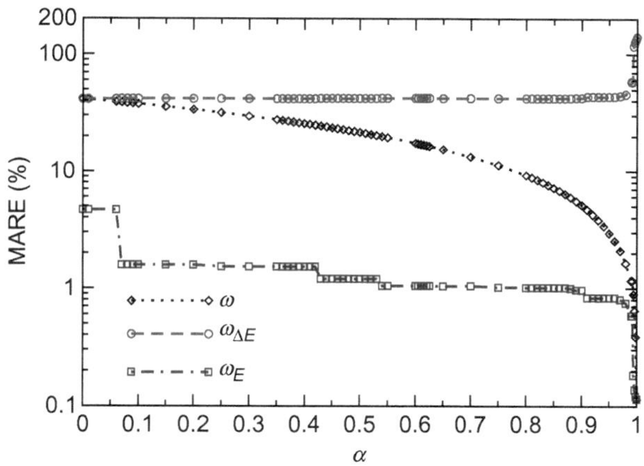

Figure 1 Minimum $\omega(\alpha)$ and corresponding $\omega_{\Delta E}$ and ω_E values as functions of α.

Eqs. (47) and (46), respectively, as functions of α. The minimum $\omega(\alpha)$ value also is shown. It decreases monotonically from 41% to 0.12%. Up to about $\alpha = 0.97$, ω_E decreases slowly with a few jumps (from about 5% to 0.8%), while $\omega_{\Delta E}$ is almost flat (from 41% to 44%). Unsurprisingly, they diverge as $\alpha \rightarrow 1$ ($\omega_{\Delta E} = 140\%$, $\omega = 0.12\%$), an illustration of the absolute energy versus binding energy dilemma.

Figure 2 shows the fitted F_t and F_θ for selected α values. Notice the violation of $F_\theta \geq 0$, a consequence of the unconstrained fitting. Notice also the structure in F_t for $\alpha < 1.0$.

Figure 2A clearly shows the separation of enhancement factors into two groups corresponding to $\alpha \leq 0.99$ and $\alpha = 1.0$. The $\alpha = 1.0$ curve is almost a line (F_θ is shown as a function of s^2) and is practically indistinguishable from the SGA curve (shown for comparison) for $s < 1.5$. F_θ (and F_t) has some structure (oscillations around the SGA curve) for $s \leq 1.5$ and $\alpha \leq 0.99$, though that may change upon changing the training set and/or the analytical form for the enhancement factor. Also, during the optimization process we noted the existence of many significantly different enhancement factors which cannot be discriminated clearly by the objective function $\omega(\alpha)$.

Any choice of $0.1 \leq \alpha \leq 0.98$ corresponds to the required KE functional, namely one which provides semi-quantitative binding, $\omega_{\Delta E} \approx 44\%$, and reasonable absolute energy error ($\omega_E \approx 1\%$). Table 1 lists the ω_E and $\omega_{\Delta E}$ values for the systems from the training set corresponding to the optimized $\omega(\alpha = 0.95)$. The highest $\omega_{\Delta E} = 85\%$ corresponds to LiH. Values for the optimized $\omega(\alpha = 1.0)$ are also shown for comparison. All $\omega_{\Delta E}$ values for $\alpha = 1.0$

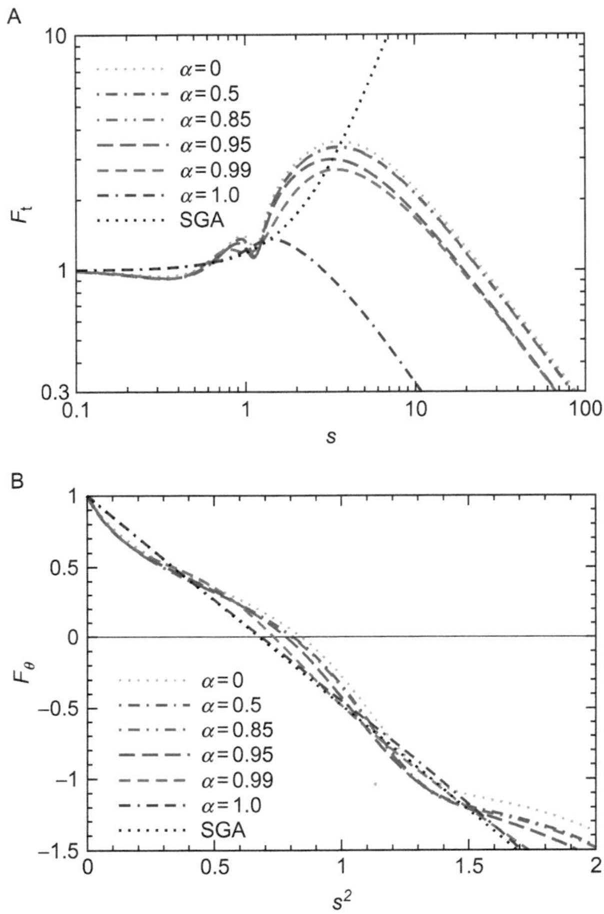

Figure 2 Noninteracting kinetic energy (A) and Pauli term (B) enhancement factors, F_t and F_θ, as functions of s and s^2, respectively.

(except the LiH molecule) are between 110% and 200%, signifying no binding.

The quality of energy curves for the CO and H_4SiO molecules corresponding to the $\alpha = 0.95$ functional is shown in Fig. 3. The $\alpha = 1$ curves (functional parameters fitted to optimize $\omega(\alpha = 1)$, i.e., pure E criterion) are shown for comparison. They have no minima. For the CO molecule, the minimum in the $\alpha = 0.95$ curve is too shallow compared to the reference KS result. In contrast, for H_4SiO the agreement between the $\alpha = 0.95$ optimized functional and KS results is excellent. Figure 4 shows a similar comparison for two molecules not in the training set. For the simplest, H_2, both

Table 1 MARE Values ω_E and $\omega_{\Delta E}$ (in %) Calculated Individually for Each System, with Parameters from Minimization of $\omega(\alpha = 0.95)$ and $\omega(\alpha = 1.0)$

	$\alpha = 0.95$		$\alpha = 1.0$	
System	ω_E	$\omega_{\Delta E}$	ω_E	$\omega_{\Delta E}$
LiH	4.2	85	0.2	85
CO	0.2	28	0.3	150
N_2	0.2	35	0.3	150
LiF	0.1	80	0.1	200
BF	0.1	15	0.1	160
NaF	0.1	25	0.03	180
SiO	0.02	32	0.04	120
H_4SiO_4	0.3	41	0.04	120
H_4SiO	0.02	50	0.6	110
Be	4.0	–	0.2	–
Ne	0.1	–	0.02	–
Ar	0.7	–	0.1	–
Average	0.83	44	0.12	140

functionals ($\alpha = 0.95$ and $\alpha = 1.0$) provide qualitatively correct binding, but the minimum is too shallow and there are large discrepancies at the dissociation limit relative to the KS result. Single-bond stretching in H_2O is described qualitatively roughly correctly only by the $\alpha = 0.95$ functional. As expected, the $\alpha = 1.0$ functional fails to give binding.

Table 2 lists the set of coefficients in the KE enhancement factor (Eq. 44) for the optimized $\omega(\alpha = 0.95)$.

5. SUMMARY DISCUSSION

A GGA OFKE form is the simplest one-point functional which explicitly includes effects of electron density inhomogeneity. The GGA form, with parameters determined by constraints, has been very successful for approximate XC functionals, to the point that such functionals dominate in practical calculations. A crucial distinction with respect to OFKE is the order of magnitude of the XC energy, about 10% of the total energy.

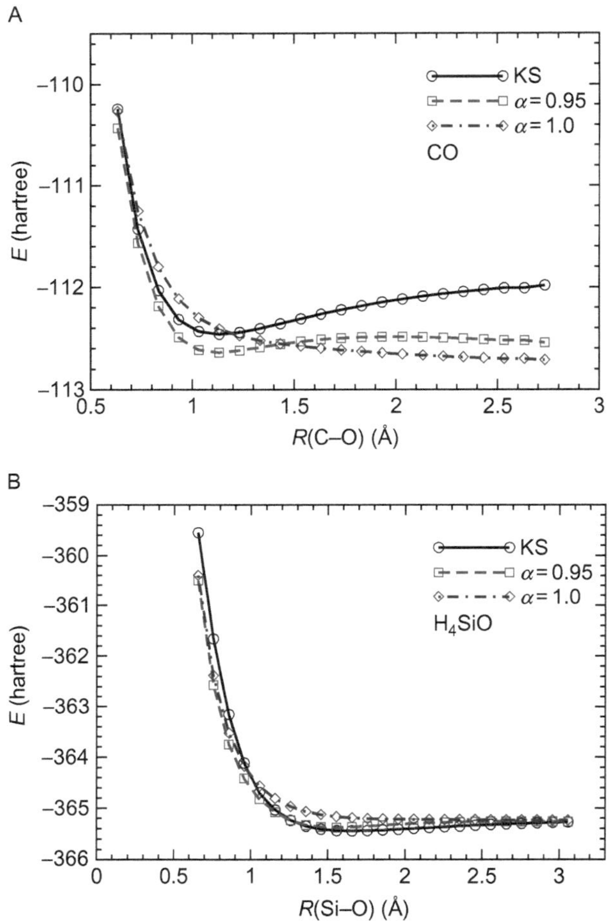

Figure 3 Total energy as a function of bond distance for the CO (A) and H_4SiO_4 (B) molecules from the training set obtained from a KS LDA calculation and from the post-KS orbital-free calculation with approximate GGA functionals.

The ground-state KE, however, has the same order of magnitude as the total energy and the KS KE is a large fraction of the total KE. Hence, a relative error of even a few percent in an approximate OFKE functional will have much bigger impact on calculated properties than would the same relative error in an approximate XC functional.

This distinction has significant implications for the constraint-based development of GGA OFKE functionals. There is a related, but perhaps less-obvious distinction. Because exchange in physical systems is purely Coulombic and correlation (as defined in DFT; recall Eq. 4) is mostly

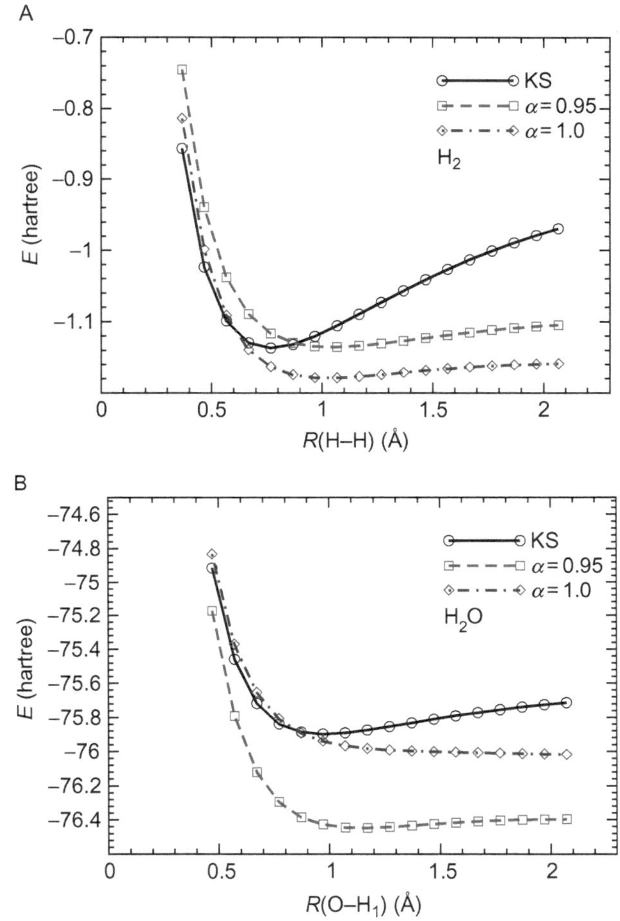

Figure 4 Total energy as a function of bond distance for the H_2 (A) and H_2O (B) molecules (neither in training set) obtained from a KS LDA calculation and from the post-KS orbital-free calculation with approximate GGA functionals.

Coulombic (the KE contribution is small), it is eminently sensible to impose Coulombic constraints on an approximate E_{xc}. The resulting functional should be applicable to a broad range of v_{ext}, if not truly universal. Experience shows that to be the case. Good GGA E_{xc} functionals deliver variations in MARE over classes of properties and types of bonding but they are broadly applicable.

Generating useful constraints on a GGA OFKE functional that preserve universality has not been as straightforward so far as for the XC functionals. One way to see the underlying difficulty is that T_s is a noninteracting system

Table 2 Coefficients in Eq. (44) from Optimizing $\omega(\alpha = 0.95)$

Coefficient	**Value**
$a_1 = b_1$	12.100994770272
$a_2 = 5/27 + b_2$	10.829496969896
a_3	− 27.327919841144
a_4	73.841590552393
a_5	25.096089580269
a_6	− 45.306369888376
a_7	− 77.901835391837
a_8	− 20.862438996553
a_9	67.083330246208
b_1	12.100994770272
b_2	10.644311784711
b_3	14.896876304511
b_4	5.5830951758904
b_5	− 24.558524755221
b_6	− 31.914940553009
b_7	4.3293607988211
b_8	17.169012815532
b_9	2.4210601059537
b_{10}	3.2527234245842

quantity, whereas E_{xc} is, as just remarked, predominantly a Coulombic quantity. This means a lack of specificity for T_s compared to E_{xc}. Surmounting that lack is what we have done by abandoning universality and imposing conditions that follow from Coulombic v_{ext}. Precisely because they are nonuniversal conditions, use of pseudo-potentials or Gaussian-type basis sets immediately introduces inconsistencies. In Section 3, we have shown how the consequences can be delineated clearly with simple one-center densities.

Section 4 then considered whether any GGA can have acceptable errors in both total energies and binding energies. The result is not as encouraging as we would like. For a small training set, the best empirical OFKE GGA we have been able to develop so far provides a relatively small MARE for the total energy ($\omega_E \approx 0.8\%$) but only semi-quantitative binding ($\omega_{\Delta E} \approx 44\%$). The use of mixed $E - \Delta E$ criteria is essential to get both correct total energies and roughly reasonable binding simultaneously.

The correct second-order gradient expansion was the only constraint imposed on the KE enhancement factor. The result is violation of a constraint, $F_\theta \geq 0$, (Eq. 19) which depends on a particular choice of t_θ. We have not checked whether $v_\theta \geq 0$ is violated for the empirical GGA, but are certain that it will be violated because of the negative slope of $F_\theta^{\text{empirical}}$ for $s^2 \leq 1$. One might hope that incorporation of more constraints should make the functional better or, at least, more nearly universal in the limited sense of improving its transferability to different systems and/or conditions. The counterargument is that, except for the high-order Padé form itself, the empirical functional was not restricted in any other way, a fact which should facilitate optimization (even at the cost of realism or transferability).

A weak caveat is that we have not yet tested the empirical functionals in SCF calculations. This may be important because the large-s behavior of the empirical functionals for Padé approximants of different orders might be very different, while the ω_E and $\omega_{\Delta E}$ errors are very similar. That difference in large-s behavior might be important for SCF calculations, but not for post-KS calculations. However, the somewhat disappointing performance on post-KS binding energy curves makes this, in our judgment, a somewhat problematic conjecture.

During minimization of $\omega(\alpha)$ with respect to the independent parameters of the Padé enhancement factor (at fixed α of course), we encountered three main difficulties. First is the precision of numerical integration required (very dense radial and angular meshes) to handle roughness in the KE enhancement factor. Second is that, for any given α, the optimization process very frequently sticks in local minima. Optimizations therefore were run for multiple values of α. Those results were analyzed to find one or a few superior parameter sets (based on values of ω_E and $\omega_{\Delta E}$). Then those parameter sets were used as initial ones to start a repeat optimization for all α. Eventually, this "by hand" procedure yielded optimal sets of parameters for every α. Third, the existence of multiple enhancement factors which deliver very similar values of the objective function $\omega(\alpha)$ despite being very different functions of s makes the final functional for each α sensitive to the choice of the training set. Increasing the size of the training set might help to overcome that difficulty. But we note again that our purpose here is not to attempt a general, widely applicable empirical GGA for OFKE but simply to find the best for a modest selection of molecules. Even with that narrow goal, the outcome seems to be that there are significant limits on what can be expected of a GGA OFKE functional.

ACKNOWLEDGMENTS

We thank the University of Florida High-Performance Computing Center for computational resources and technical support. We thank Keith Runge for help with the title. This work was supported by U.S. Department of Energy grant DE-SC0002139.

REFERENCES

1. Karasiev, V. V.; Sjostrom, T.; Dufty, J.; Trickey, S. B. Accurate Homogeneous Electron Gas Exchange-Correlation Free Energy for Local Spin-Density Calculations. *Phys. Rev. Lett.* **2014**, *112,* 076403(1–5).
2. Karasiev, V. V.; Sjostrom, T.; Trickey, S. Finite-Temperature Orbital-Free DFT Molecular Dynamics: Coupling PROFESS and Quantum Espresso. *Comput. Phys. Commun.* **2014**, *185*, 3240–3249.
3. Karasiev, V. V.; Chakraborty, D.; Shukruto, O. A.; Trickey, S. B. Nonempirical Generalized Gradient Approximation Free-Energy Functional for Orbital-Free Simulations. *Phys. Rev. B* **2013**, *88*, 161108(R)(1–5).
4. Karasiev, V. V.; Chakraborty, D.; Trickey, S. B. Progress on New Approaches to Old Ideas: Orbital-Free Density Functionals. In: Delle Site, L.; Bach, V. Eds.; *Many-Electron Approaches in Physics, Chemistry, and Mathematics: A Multidisciplinary View*; Springer: Heidelberg, 2014; pp 113–134.
5. Karasiev, V. V.; Chakraborty, D.; Dufty, J. W.; Harris, F. E.; Runge, K.; Trickey, S. B. Innovations in Finite-Temperature Density Functionals. In: Graziani, F. et al., Eds.; *Frontiers and Challenges in Warm Dense Matter*; Springer Verlag: Heidelberg, 2014; pp 61–85.
6. Karasiev, V. V.; Trickey, S. B.; Harris, F. E. Born-Oppenheimer Interatomic Forces from Simple, Local Kinetic Energy Density Functionals. *J. Comput. Aided Mater. Des.* **2006**, *13*, 111–129.
7. Karasiev, V.; Jones, R. S.; Trickey, S. B.; Harris, F. E. Recent Advances in Developing Orbital-Free Kinetic Energy Functionals. In: Paz, J. L.; Hernández, A. J. Eds.; *New Developments in Quantum Chemistry*; Transworld Research Network: Kerala, India, 2009; pp 25–54.
8. Karasiev, V. V.; Jones, R. S.; Trickey, S. B.; Harris, F. E. Properties of Constraint-Based Single-Point Approximate Kinetic Energy Functionals. *Phys. Rev. B* **2009**, *80,* 245120 (1–17); Erratum *ibid.* **2013**, *87*, 239902.
9. Borgoo, A.; Green, J. A.; Tozer, D. J. Molecular Binding in Post-Kohn-Sham Orbital-Free DFT. *J. Chem. Theory Comput.* **2014**, *10*, 5338–5345.
10. Mermin, N. D. Thermal Properties of Inhomogeneous Electron Gas. *Phys. Rev.* **1965**, *137*, A1441–A1443.
11. Thomas, L. The Calculation of Atomic Fields. *Proc. Camb. Philos. Soc.* **1927**, *23*, 542–548.
12. Fermi, E. Un metodo statistico per la determinazione di alcune proprietá dell'atomo. *Atti Accad. Nazl. Lincei* **1927**, *6*, 602–607.
13. Hohenberg, P.; Kohn, W. Inhomogeneous Electron Gas. *Phys. Rev.* **1964**, *136*, B864–B871.

14. Levy, M. Universal Variational Functionals of Electron Densities, First-Order Density Matrices, and Natural Spin-Orbitals, and Solution to the v-Representability Problem. *Proc. Natl. Acad. Sci. U.S.A.* **1979**, *76*, 6062–6065.
15. Lieb, E. H. Density Functionals for Coulomb Systems. *Int. J. Quantum Chem.* **1983**, *24*, 243–277.
16. Kohn, W.; Sham, L. Self-Consistent Equations Including Exchange and Correlation Effects. *Phys. Rev.* **1965**, *140*, A1133–A1138.
17. Parr, R.; Yang, W. *Density Functional Theory of Atoms and Molecules*; Oxford: New York, 1989.
18. Dreizler, R.; Gross, E. *Density Functional Theory*; Springer-Verlag: Berlin, 1990.
19. Perdew, J. P.; Schmidt, K. Jacob's Ladder of Density Functional Approximations for the Exchange-Correlation Energy. *A.I.P. Conf. Proc.* **2001**, *577*, 1–20.
20. Lee, H.; Lee, C.; Parr, R. G. Conjoint Gradient Correction to the Hartree-Fock Kinetic- and Exchange-Energy Density Functionals. *Phys. Rev. A* **1991**, *44*, 768–771.
21. Perdew, J. P. Generalized Gradient Approximation for the Fermion Kinetic Energy as a Functional of the Density. *Phys. Lett. A* **1992**, *165*, 79–82.
22. Lacks, D. J.; Gordon, R. G. Tests of Nonlocal Kinetic Energy Functionals. *J. Chem. Phys.* **1994**, *100*, 4446–4452.
23. DePristo, A. E.; Kress, J. D. Kinetic-Energy Functionals via Padé Approximations. *Phys. Rev. A* **1987**, *35*, 438–441.
24. Thakkar, A. J. Comparison of Kinetic-Energy Density Functionals. *Phys. Rev. A* **1992**, *46*, 6920–6924.
25. Tran, F.; Wesolowski, T. A. Link Between the Kinetic- and Exchange-Energy Functionals in the Generalized Gradient Approximation. *Int. J. Quantum Chem.* **2002**, *89*, 441–446.
26. Constantin, L.; Fabiano, E.; Laricchia, S.; Sala, F. D. Semiclassical Neutral Atom as a Reference System in Density Functional Theory. *Phys. Rev. Lett.* **2011**, *106,* 186406 (1–4).
27. Lehtomäki, J.; Makkonen, I.; Caro, M. A.; Lopez-Acevedo, O. Orbital-Free Density Functional Theory Implementation with the Projector Augmented-Wave Method. *J. Chem. Phys.* **2014**, *141,* 234102(1–7).
28. Tal, Y.; Bader, R. F. W. Studies of the Energy Density Functional Approach. I. Kinetic Energy. *Int. J. Quantum Chem.* **1978**, *S12*, 153–168.
29. Bartolotti, L.; Acharya, P. K. On the Functional Derivative of the Kinetic Energy Density Functional. *J. Chem. Phys.* **1982**, *77*, 4576–4585.
30. Harriman, J. The Interface Between Reduced Density Matrices and Density Functional Theory. In: Erdahl, R.; Smith, J. V. Eds.; *Density Matrices and Density Functionals*; D. Reidel: Dordrecht, 1987; pp 359–373.
31. Levy, M.; Ou-Yang, H. Exact Properties of the Pauli Potential for the Square Root of the Electron Density and the Kinetic Energy Functional. *Phys. Rev. A* **1988**, *38*, 625–629.
32. Baltin, R. The 3-Dimensional Kinetic-Energy Density Functional Compatible with the Exact Differential-Equation for Its Associated Tensor. *J. Chem. Phys.* **1987**, *86*, 947–952.
33. Hoffmann-Ostenhof, M.; Hoffmann-Ostenhof, T. Schrödinger Inequalities" and Asymptotic Behavior of the Electron Density of Atoms and Molecules. *Phys. Rev. A* **1977**, *16*, 1782–1785, see Eq. (2.20). See also; Romera, E.; Dehesa, J. S. Weizsacker Energy of Many-Electron Systems. *Phys. Rev. A* **1994**, *50*, 256–266.
34. Sears, S. B.; Parr, R. G.; Dinur, U. On the Quantum-Mechanical Kinetic Energy as a Measure of the Information in a Distribution. *Isr. J. Chem.* **1980**, *19*, 165–173.
35. Harriman, J. E. A Kinetic Energy Density Functional. *J. Chem. Phys.* **1985**, *83*, 6283–6287.

36. Herring, C. Explicit Estimation of Ground-State Kinetic Energies from Electron Densities. *Phys. Rev. A* **1986**, *34*, 2614–2631.
37. Levy, M.; Perdew, J. P.; Sahni, V. Exact Differential Equation for the Density and Ionization Energy of a Many-Particle System. *Phys. Rev. A* **1984**, *30*, 2745–2748.
38. Trickey, S. B.; Karasiev, V. V.; Jones, R. S. Conditions on the Kohn-Sham Kinetic Energy and Associated Density. *Int. J. Quantum Chem.* **2009**, *109*, 2943–2952.
39. Kato, T. On the Eigenfunctions of Many-Particle Systems in Quantum Mechanics. *Commun. Pure Appl. Math.* **1957**, *10*, 151–177.
40. Lambert, F.; Clérouin, J.; Zérah, G. Very-High-Temperature Molecular Dynamics. *Phys. Rev. E* **2006**, *73*, 016403(1–5).
41. Gradshteyn, I.; Ryzik, I. *Table of Integrals, Series, and Products*. Academic Press: New York, 1980, Corrected and enlarged edition by A. Jeffrey (entry 3.478).
42. Baker, G., Jr.; Gammel, J. L. *The Padé Approximant in Theoretical Physics*; Academic Press: New York, 1970.
43. Karasiev, V. V.; Chakraborty, D.; Trickey, S. B. Improved Analytical Representation of Combinations of Fermi-Dirac Integrals for Finite-Temperature Density Functional Calculations. *Comput. Phys. Commun.* **2015**, *192*, 114–123. URL: http://dx.doi.org/10.1016/j.cpc.2015.03.002.
44. Gázquez, J.; del Campo, J.; Trickey, S.; Alvarez-Mendez, R.; Vela, A. Analysis of the Generalized Gradient Approximation for the Exchange Energy. In: Ghosh, S. K.; Chattaraj, P. K. Eds.; *Concepts and Methods in Modern Theoretical Chemistry*; 1, CRC Press: Boca Raton, FL, 2013; pp 295–311(In honor of Prof. B. M. Deb).
45. Peverati, R.; Truhlar, D. G. Quest for a Universal Density Functional the Accuracy of Density Functionals Across a Broad Spectrum of Databases in Chemistry and Physics. Philos. Trans. R. Soc. A, 372(20120476).
46. Slater, J. C. A Simplification of the Hartree-Fock Method. *Phys. Rev.* **1951**, *81*, 385–390.
47. Slater, J. C. Magnetic Effects and the Hartree-Fock Equation. *Phys. Rev.* **1951**, *82*, 538–541.
48. Slater, J. C. Suggestions from Solid-State Theory Regarding Molecular Calculations. *J. Chem. Phys.* **1965**, *43*, S228.
49. Gáspár, R. Über eine approximation des Hartree-Fockschen potentials durch eine universelle potentialfunktion (On an Approximation of Hartree-Fock Potentials Through a Universal Potential Function). *Acta Phys. Hung.* **1954**, *3*, 263–286.
50. Tong, B.; Sham, L. Application of a Self-Consistent Scheme Including Exchange and Correlation Effects to Atoms. *Phys. Rev.* **1966**, *144*, 1–4.
51. Vosko, S. H.; Wilk, L.; Nusair, M. Accurate Spin-Dependent Electron Liquid Correlation Energies for Local Spin Density Calculations: A Critical Analysis. *Can. J. Phys.* **1980**, *58*, 1200–1211.
52. Ceperley, D. M.; Alder, B. J. Ground State of the Electron Gas by Stochastic Method. *Phys. Rev. Lett.* **1980**, *45*, 566–569.
53. Schäfer, A.; Horn, H.; Ahlrichs, R. Fully Optimized Contracted Gaussian Basis Sets for Atoms Li to Kr. *J. Chem. Phys.* **1992**, *97*, 2571–2577.
54. Schäfer, A.; Huber, C.; Ahlrichs, R. Fully Optimized Contracted Gaussian Basis Sets of Triple Zeta Valence Quality for Atoms Li to Kr. *J. Chem. Phys.* **1994**, *100*, 5829–5835.
55. Taken from the Extensible Computational Chemistry Environment Basis Set Database, Version 02/25/04, Molecular Science Computing Facility, Environmental and Molecular Sciences Laboratory, Pacific Northwest Laboratory, P.O. Box 999, Richland, Washington 99352, USA, funded by the U.S. Department of Energy (contract DE-AC06-76RLO). See URL: http://www.emsl.pnl.gov/forms/basisform.html.

56. Becke, A. D. A Multicenter Numerical Integration Scheme for Polyatomic Molecules. *J. Chem. Phys.* **1988**, *88*, 2547–2553.
57. Lebedev, V. I.; Laikov, D. N. Quadrature Formula for the Sphere of 131st Algebraic Order of Accuracy. *Dokl. Akad. Nauk* **1999**, *366*, 741–745.
58. Salvador, P.; Mayer, I. Energy Partitioning for "Fuzzy" Atoms. *J. Chem. Phys.* **2004**, *120*, 5046–5052.

CHAPTER TEN

Statistical Inference with Minimum Relative Entropy: A Robust Numerical Algorithm Employing Sinc Quadrature

Vasilios G. Koures[1]
IISAM L3C, Cheyenne, WY, USA
[1]Corresponding author: e-mail address: koures@iisam.com

Contents

Abstract

Given partial information (i.e., constraints) about a probability distribution, the distribution that maximizes the entropy with respect to the constraints is the one that is least prejudiced about the missing information. As new information arrives, the prior distribution must change. To maintain maximum uncertainty given new information, we must minimize the relative entropy (the Kullback–Leibler distance) between the prior (p_0) and posterior (p) distributions. Robust constrained optimization algorithms to implement this minimization have proven difficult to find. However, such an algorithm, which incorporates Sinc quadrature and is applied to option market fitting, is presented here. The quick convergence and robust behavior of this algorithm may open up statistical inference with minimum relative entropy to a wider range of scientific applications.

Advances in Quantum Chemistry, Volume 71
ISSN 0065-3276
http://dx.doi.org/10.1016/bs.aiq.2015.03.005

1. INTRODUCTION

Many processes in physics, chemistry, and economics are modeled probabilistically. Ideally, we would infer a probability distribution to model a stochastic process by a method that is least prejudiced with respect to missing information. For example, as observational data of a chemical process arrives, we would update our underlying probability distribution in a least biased way. To do this, we would have to include the new information in a minimal way so that information that is still missing is maximized.

Since entropy is a measure of missing information,[1,2] the maximum entropy distribution, constrained by incoming information, maximizes the missing information. It is thus the least prejudiced distribution. From a statistical inference point of view, there is no reason to prefer any other distribution.

Nevertheless, the maximum likelihood estimation (MLE) method[3] is commonly used because, although it introduces bias into the statistical inference, it is easier to model mathematically. So, unlike the maximum entropy method, MLE is not built by maximizing the missing information, constrained only by incoming data. Therefore, even though it is suboptimal for statistical inference, researchers often have little choice but to go with MLE.

Recently, however, a generalization of the maximum entropy principle (MEP), known as minimum relative entropy (MRE), has increasingly been deployed in the investigation of interesting problems in physics and chemistry. One of the most interesting of these applications in chemical physics is the use of MRE to provide a multiscale analysis by measuring the overlap between two molecular-thermodynamic ensembles.[4,5]

In this chapter, we introduce a robust numerical algorithm that may open up the MRE method to a wider range of applications in chemistry and physics. This method employs Sinc quadrature, which is introduced in Section 5. Sinc quadrature is a powerful numerical integration technique that seamlessly handles sharp boundaries and enables our optimization routine, sinc-MRE, to converge even for constraints that are highly correlated (see Sections 6 and 7).

2. MAXIMUM ENTROPY PRINCIPLE

The information function, $I(p)$, decreases monotonically with probability:

$$I(p) = -\log p. \tag{1}$$

More probable events thus provide less information; because of this property, the information function is also known as the *surprisal.*[6]

Entropy is defined as the expected surprisal and it is denoted by the letter H:

$$H = -\sum_{i=1}^{N} p_i \log p_i, \tag{2}$$

where the set of positive numbers, $\{p_1, p_2, \ldots, p_N\}$ whose sum equals one, represents the probabilities for a discrete set of N events.

As a measure of missing information, entropy may also be viewed as measuring the uncertainty associated with a probability distribution. Therefore, since a uniform distribution corresponds to maximum uncertainty, its entropy is at a maximum. However, a certain event gives minimum (zero) entropy.[7]

Given partial information (i.e., constraints) about a probability distribution, we can maximize the entropy with respect to the constraints and obtain the distribution that is least prejudiced about the missing information. The MEP may thus be employed via constrained optimization.

For example, simple applications of the MEP lead to the exponential and normal distributions.[7] That is, given a mean on $(0, \infty)$, the maximum entropy distribution is exponential. On the other hand, given a mean and variance on $(-\infty, \infty)$, then the maximum entropy distribution is normal.

2.1 Example: MEP Applied to Option Markets

It is well known that market prices of stocks, commodities, bonds, and other financial assets are modeled as stochastic processes.[8] Options market prices are then modeled as derivative prices depending on the underlying asset prices.[9] Statistical inference may be employed to obtain an underlying probability distribution for prices, and this distribution may be used to price and risk manage options.

In fact, it was originally suggested long ago that options prices themselves should contain useful information about the underlying asset price distributions.[10] This insight suggests that the MEP may be employed to extract an asset price probability distribution from the partial information available in options markets.[11]

We will analyze this problem here to demonstrate the robustness of our Sinc quadrature implementation of the MEP. The options markets have provided a formidable challenge to the implementation of MEP because options market prices tend to be clustered around the at-the-money price.

This clustering may produce highly correlated option prices that tend to destabilize the optimization algorithm.[11]

Our Sinc quadrature method allows for a stable, robust algorithm that converges even for the challenging options markets. Therefore, we would expect that this algorithm would be robust across many applications in chemistry and physics and we will report on such applications in the future.

Now, if we denote the price of a stock by x, we can express the continuum version of Eq. (2) for the entropy as follows[12]:

$$H(p) = -\int_0^\infty p(x)\log[p(x)]\,dx. \tag{3}$$

Note that the quantity $p(x)$ is really a probability density function (pdf). It is not a probability distribution although it is often referred to as such. We will refer to it simply as a distribution so that we can be in line with the language of functional analysis.

We optimize expression (3) subject to the constraints,

$$\int_0^\infty p(x)\,dx = 1 \tag{4}$$

and

$$E[c_i(X)] = \int_0^\infty c_i(x)\,p(x)\,dx = d_i,\ i = 1, 2, \ldots, n \tag{5}$$

where the data set $\{d_1, d_2, \ldots, d_n\}$ is the set of option prices obtained from the market. At maturity, T, the payoffs for *European* call options are[9]

$$c_i(x_T) = \max(x_T - K_i, 0). \tag{6}$$

For *European* put options, the payoffs are[9]

$$c_i(x_T) = \max(K_i - x_T, 0). \tag{7}$$

The quantity, K_i, is the strike price for the ith option, which is, the price that determines whether the holder will exercise his option or not. If $K_i = x$, the option is said to be *at the money*; if the payoff would be positive, the option is *in the money*; otherwise, the option is *out of the money*.[9]

Equations (3)–(5) constitute the MEP. Equation (3) is the entropy and it is the functional that must be maximized. Constraint equation (4) ensures that the distribution we obtain is normalized. Equation (5) constrains this distribution to fit the observed market prices to the expected option payoffs.

The observed market prices thus provide partial information about the distribution, but the full distribution is obtained by maximizing the entropy. We now maximize the uncertainty associated with the distribution subject to the partial information provided by the market prices.

2.2 MEP: Solving for *p*(*x*) Using Lagrange Multipliers

We solve for the distribution $p(x)$ that maximizes $H(p)$ in (3), subject to constraints (4) and (5), by using the method of Lagrange multipliers.[7,11] We write

$$H(p) = -\int_0^\infty p(x)\log[p(x)]\,\mathrm{d}x + (1+\lambda_0)\int_0^\infty p(x)\,\mathrm{d}x + \sum_{i=1}^{n}\lambda_i\int_0^\infty c_i(x)p(x)\,\mathrm{d}x \tag{8}$$

where $(\lambda_0, \lambda_1, \ldots, \lambda_n)$ is the set of Lagrange multipliers. To maximize $H(p)$, we take its functional derivative with respect to $p(x)$ and set it to zero:

$$\frac{\delta H}{\delta p(x)} = -\log[p(x)] + \lambda_0 + \sum_{i=1}^{n}\lambda_i c_i(x) = 0. \tag{9}$$

Solving (9) for $p(x)$, we obtain

$$p(x) = \frac{1}{Z}\exp\left(\sum_{i=1}^{n}\lambda_i c_i(x)\right), \tag{10}$$

where Z is the partition function[7]:

$$Z = \mathrm{e}^{-\lambda_0} = \int_0^\infty \exp\left(\sum_{i=1}^{n}\lambda_i c_i(x)\right)\mathrm{d}x. \tag{11}$$

It is easy to see that (10) maximizes the entropy because the second functional derivative of $H(p)$ is strictly negative.

Here are some things to note about the maximizing distribution in (10):

- It is the formal solution to *any* MEP problem where the constraints can be expressed as expectation values, as in (5) above.
- We use it to obtain the optimal Lagrange multipliers by calculating the expectation values of the constraints and setting them equal to the observed data.
- Since (9) is nonlinear, the Lagrange multipliers must usually be obtained numerically.

- We will soon see that the optimal set of Lagrange multipliers may in principle always be found by optimization.
- Given only a mean, it is now clear why $p(x)$ is the exponential distribution. Also, given a mean and variance, it is now clear why $p(x)$ is the normal distribution.

It is also worth noting that the MEP problem using information from option markets is difficult by conventional numerical methods because the payoff functions, (6) and (7), each exhibit a sharp, nondifferentiable corner. However, we will soon see that numerical Sinc methods allow nonuniform sampling in the neighborhood of singular boundaries and this contributes to their robust convergence properties.

3. MINIMUM RELATIVE ENTROPY

Now, assume we are already working with a prior distribution. How does that distribution change upon arrival of new information (e.g., option market prices)? In that case, we would like to minimize the expected difference between information functions,

$$S(p, p_0) = E_p[\log p - \log p_0] = \sum_{i=1}^{N} p_i \log \frac{p_i}{(p_0)_i}, \tag{12}$$

where the prior distribution is p_0 and the posterior distribution is p. The expectation value is computed relative to the posterior distribution where, as discussed in relation to Eq. (2), the set of positive numbers, $\{p_1, p_2, \cdots, p_N\}$, represents the probabilities for a discrete set of N events.

The function $S(p, p_0)$ is known as the relative entropy.[12] It is strictly nonnegative and it is zero if and only if p is identically equal to p_0. The relative entropy is thought of as a sort of distance between distributions and it is called the Kullback–Leibler distance.[12] However, it is not a true distance in the mathematical sense because $S(p, p_0)$ is not symmetric in its arguments and it does not satisfy the triangle inequality.

Now, to maintain maximum uncertainty given new information, we must minimize the relative entropy between the prior and posterior distributions. Similar to the case of the MEP, we can go to the continuum and obtain the MRE by minimizing the Kullback–Leibler distance,

$$S(p, p_0) = \int_0^{\infty} p(x) \log[p(x)/p_0(x)] \, dx, \tag{13}$$

subject to constraints equivalent in form to those in (4) and (5). As before, we use the method of Lagrange multipliers to enforce the constraints:

$$S(p,p_0) = \int_0^\infty p(x)\log[p(x)/p_0(x)]\,\mathrm{d}x - (1+\lambda_0)\int_0^\infty p(x)\,\mathrm{d}x \\ -\sum_{i=1}^{n}\lambda_i\int_0^\infty c_i(x)p(x)\,\mathrm{d}x. \tag{14}$$

As in the case of MEP, we take the functional derivative of S with respect to $p(x)$ and set it equal to zero:

$$\frac{\delta S(p,p_0)}{\delta p(x)} = \log[p(x)/p_0(x)] - \lambda_0 - \sum_{i=1}^{n}\lambda_i c_i(x) = 0. \tag{15}$$

Solving (15) for $p(x)$, we obtain

$$p(x) = \frac{p_0(x)}{\zeta}\exp\left(\sum_{i=1}^{n}\lambda_i c_i(x)\right), \tag{16}$$

where ζ is seen as the grand partition function with "fugacity" $p_0(x)$:

$$\zeta = \mathrm{e}^{-\lambda_0} = \int_0^\infty p_0(x)\exp\left(\sum_{i=1}^{n}\lambda_i c_i(x)\right)\mathrm{d}x. \tag{17}$$

Note that if the prior distribution is uniform (i.e., constant), then the MRE distribution in (16) reduces to the MEP distribution in (10). So, while still satisfying the constraints, the MEP distribution is the closest (in entropic distance) to the uninformative uniform distribution.[7,11] If there were no constraints at all, then the MEP distribution itself would be a uniform distribution, as we would expect.[7,11]

Since the MRE distribution includes the MEP distribution as a special case, for the rest of this chapter we will consider only the MRE method. Any results we derive for MRE may readily be specialized to MEP.

4. SOLVING FOR THE MRE DISTRIBUTION

This method of solution applies to any statistical inference problem where the observed data quantities are expressed as expectation values, as in (5).

If we substitute the MRE distribution (16) into (13), we get after some algebra:

$$-S_{\max} = \log\zeta - \sum_{i=1}^{n} \lambda_i d_i \equiv f(\lambda_1, \lambda_2, \ldots, \lambda_n), \tag{18}$$

where, as in Eq. (5), the set $\{d_1, d_2, \ldots, d_n\}$ is the observed data that is theoretically expressed as a set of expectation values.

What remains now is to solve for the Lagrange multipliers. We do this by finding the minimum of $f(\vec{\lambda})$. To see this, note that setting the gradient of $f(\vec{\lambda})$ to zero ensures that the constraints in Eq. (5) are satisfied:

$$0 = \partial_i f = E[c_i(X)] - d_i, \tag{19}$$

where $\partial_i \equiv \partial / \partial\lambda_i$.

Proof. From (18), we see that

$$\partial_i f = \frac{1}{\zeta}\frac{\partial\zeta}{\partial\lambda_i} - d_i = E[c_i(X)] - d_i,$$

where the second equality comes from (16) and (17). ■

Next, we prove that the extremum of $f(\vec{\lambda})$ is indeed a minimum by showing that its Hessian is equal to the covariance of the constraints; hence, the Hessian matrix is positive definite:

$$\partial_i\partial_j f = \mathrm{cov}\left[c_i(X), c_j(X)\right]. \tag{20}$$

Proof. From (18), we get

$$\begin{aligned}\partial_i\partial_j f = \partial_i\partial_j \log\zeta &= \frac{1}{\zeta}\partial_i\partial_j\zeta - \left(\frac{1}{\zeta}\partial_i\zeta\right)\left(\frac{1}{\zeta}\partial_j\zeta\right) \\ &= E\left[c_i(X)\cdot c_j(X)\right] - E[c_i(X)]\cdot E\left[c_j(X)\right] \\ &= \mathrm{cov}\left[c_i(X), c_j(X)\right].\end{aligned}$$

The equality in the second line comes from (16) and (17). ■

We have thus proven that the Lagrange multipliers are obtained by minimizing the objective function on the right-hand side of (18). Since we can directly compute the gradient and Hessian, we have, in principle, a menu of choices of optimization algorithms. Therefore, the remaining challenge is to implement a numerical integration scheme that is robust and accurate. Sinc quadrature offers a fast, accurate integration algorithm that is well suited to the task.

5. SINC QUADRATURE

To minimize (18), we must be able to numerically integrate the grand partition function, ζ, in (17), along with the expectation and covariance values of the constraints in (19) and (20), respectively. These numerical integrals are used as iterative inputs to an optimization algorithm. It is therefore imperative that the computed integrals be accurate enough so that numerical errors do not propagate and cause the optimization algorithm to diverge.

This is particularly challenging in the options market-fitting problem because the constraints (6) and (7) are not differentiable at the strike price, K_i. Furthermore, options market data tend to be clustered close to the strike price and this may make the constraints highly correlated, possibly rendering an ill-conditioned Hessian matrix in (20).

It is thus evident that the option market-fitting problem provides a good test for our optimization with Sinc quadrature. That is, if we can get our algorithm to converge for this problem, we can be confident that it will converge for a wider class of problems in chemistry and physics.

5.1 Sinc Quadrature in a Nutshell

Numerical Sinc methods provide a powerful suite of computational tools that may be applied to a wide range of problems, including interpolation, differentiation, integration, and partial differential equations.[13–15]

The incentive to develop Sinc methods started with the realization that, for the Paley–Wiener class of functions,[13,15] interpolation and quadrature are exact. For example, if a function $f(x)$ belongs to the Paley–Wiener class, then[13,15]

$$f(x) = \sum_{k=-\infty}^{\infty} f(kh)\,\mathrm{sinc}\left(\frac{x-kh}{h}\right) \tag{21}$$

and

$$f(kh) = \frac{1}{h}\int_{-\infty}^{\infty} f(x)\,\mathrm{sinc}\left(\frac{x-kh}{h}\right)\mathrm{d}x, \tag{22}$$

where

$$\mathrm{sinc}\left(\frac{x-kh}{h}\right) \equiv \begin{cases} \dfrac{\sin[\pi(x-kh)/h]}{\pi(x-kh)/h}, & x \neq kh \\ 1, & x = kh \end{cases}. \tag{23}$$

When the series in (21) converges, it is called the *cardinal function.*[13,15]

Note that if $f(x)$ belongs to the Paley–Wiener class of functions, then the infinite sum (21) converges for any value of h.[13,15] That is, if we sample the function uniformly at points, kh for any given h, then the cardinal function on the right-hand side of (21) converges to $f(x)$ for any value x in the domain of this function.[13,15]

The results (21) and (22) motivated a search for a more expansive class of functions for which a truncated cardinal function may converge exponentially. However, it was found that exponential convergence is spoiled by truncation errors.[13,15]

The requirement that functions be entire was then relaxed and a class of functions that were analytic on an infinite strip, centered on the real line, was considered. It turned out that compatibility between exponential convergence and truncation errors was recovered for this class.[13,15]

Conformal maps from general arcs in the complex plane to the infinite strip about the real axis further expanded the class of functions for Sinc methods that converge exponentially.[13–15] In addition to exponential convergence, Sinc methods also seamlessly handle singular boundary conditions. These results, together with ongoing developments,[14] continue to expand the realm of applications of numerical Sinc methods.

6. EMPLOYING SINC QUADRATURE TO FIND MRE

Recalling (18)–(20), our aim is to numerically integrate the grand partition function and its derivatives. We input these integrals into an optimization algorithm and solve for the Lagrange multipliers that produce the MRE distribution.

Let us denote the integrand in (17) by $F(x)$:

$$F(x) = p_0(x)\exp\left(\sum_{i=1}^{n}\lambda_i c_i(x)\right). \tag{24}$$

Since this is a finite distribution, we can find positive constants α, β, and C such that

$$|F(x)| \leq C\begin{cases} x^{\alpha-1}, & x \in \left(0, \log\left(1+\sqrt{2}\right)\right) \\ \exp(-\beta x), & x \in \left[\log\left(1+\sqrt{2}\right), \infty\right) \end{cases}. \tag{25}$$

We then choose these constant parameters so that[15]

$$N = \left\lfloor \frac{\alpha}{\beta} M + 1 \right\rfloor \tag{26}$$

and

$$h = \left(\frac{2\pi d}{\alpha M}\right)^{1/2} \le \frac{2\pi d}{\log(2)}, \tag{27}$$

where $2d$ is the width of the strip about the real axis.[13,15] The absolute error between the integral (17) and its Sinc quadrature is exponentially damped[13,15]:

$$\left| \int_0^\infty F(x)\,\mathrm{d}x - h \sum_{k=-M}^{N} w_k F(x_k) \right| = \mathrm{O}\left(\exp\left[-(2\pi d\alpha M)^{1/2}\right]\right). \tag{28}$$

The nodes are given by

$$x_k = \log\left[\mathrm{e}^{kh} + \sqrt{\mathrm{e}^{2kh}+1}\right] \tag{29}$$

and the weights are

$$w_k = \frac{\mathrm{e}^{kh}}{\sqrt{\mathrm{e}^{2kh}+1}}. \tag{30}$$

The nodes are more densely distributed near the origin and their distribution becomes uniform as k goes to infinity. This nonuniform sampling property plays a significant role in the ability of Sinc methods to handle singular boundary conditions since it produces denser sampling near fast-changing regions.

A similar result has been worked out for Sinc interpolation[13,15]:

$$\left| F(x) - \sum_{k=-M}^{N} F(x_k)\,\mathrm{sinc}\left(\frac{\log(x/(1-x)) - kh}{h}\right) \right| = \mathrm{O}\left(\sqrt{M}\exp\left[-\pi\sqrt{\alpha M/2}\right]\right). \tag{31}$$

The integers N and M are again chosen according to (26), but the constants α and β are based on an assumed asymptotic behavior that is slightly different than above:

$$|F(x)| \le C \begin{cases} x^{\alpha}, & x \in (0, 1/2) \\ (1-x)^{\beta}, & x \in [1/2, 1) \end{cases}. \tag{32}$$

The interpolation nodes are

$$x_k = \frac{e^{kh}}{e^{2kh}+1} \tag{33}$$

and the interpolation step size is

$$h = \frac{\pi}{\sqrt{2\alpha M}} \leq \frac{\pi^2}{\log(2)}. \tag{34}$$

Detailed discussions of Eqs. (28)–(34), together with rigorous proofs of (28) and (31), may be found in Refs. 13,15.

We now make use of the results (28) and (31) to obtain the MRE. That is, we use (28) to calculate the grand partition function (17) and then input it into (18) to find the Lagrange multipliers via a minimization algorithm. We then use these Lagrange multipliers to obtain the MRE posterior distribution. The Sinc interpolation (31) is used to obtain the value of the MRE distribution at any point on the real line.

7. COMPUTATION RESULTS

The minimization of (18) was obtained by deploying a line search method with quasi-Newton directions.[16] The parameter set that gave us fast and accurate convergence is

$$\alpha = 2, \beta = 1, d = \pi/2, M = 11. \tag{35}$$

However, a wide set of parameter values, particularly larger values of M, also led to convergence.

The market fit, shown in Fig. 1, was performed with normalized historical S&P 500 index options data. Our prior distribution was a lognormal distribution[9] with implied at-the-money volatility (sigma) and time to maturity in years (T), as shown in the figure. The interest rate (r_0) and dividend rate (r_s) are also shown. Note that an implied forward market price was also calculated from the rates and it was fit to the market by treating it as a zero-strike call option. Put and call option market prices are denoted with a P and C, respectively. Column 3 shows the computed prices from the MRE distribution, column 4 shows the percent error of the fit, and column 5 shows the Lagrange multipliers. It is evident that this was a very successful fit of the options market data.

r_0 = 0.0628
r_S = 0.04943
Sigma = 0.23043
T = 0.249315

Forward price:
101.2399888

Market	Strike	Fit	% Error of fit	Lambda
0.22P	70	0.22	−0.0245	0.3207
0.66P	80	0.66	−0.1669	0.1748
1.65P	90	1.65	0.0916	0.0633
4.15P	100	4.15	0.0804	0.0313
1.05C	110	1.05	0.0315	−0.1337
0.05C	120	0.05	−0.0034	−0.2544
Forward	0	101.24	−0.0011	−0.3149

Figure 1 Market fit of option prices computed from MRE distribution. The Lagrange multipliers are listed in the fifth column from the left.

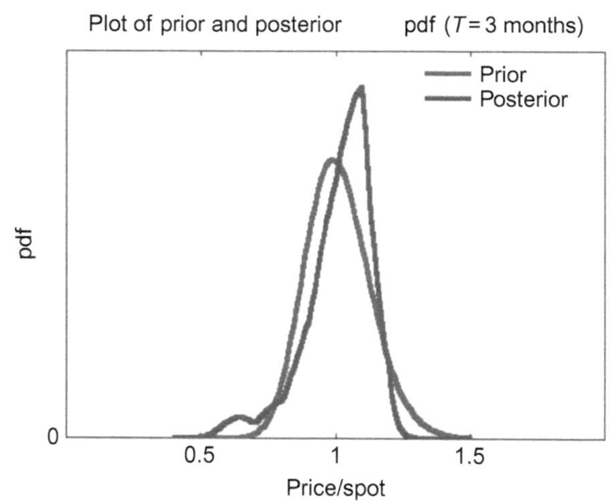

Figure 2 Plot of prior and posterior probability density functions.

We can also use Sinc interpolation (31) to plot the prior and posterior distributions. They are shown in Fig. 2.

Note that the posterior distribution appears to be bimodal. This is indicative of price inefficiency in the options markets. We can try to smooth the posterior distribution out by inputting a prior distribution that is skewed to the left relative to the lognormal distribution.

In Fig. 3, a skewed distribution is plotted on top of the distributions as shown in Fig. 2. Since the skewed distribution fits the posterior distribution

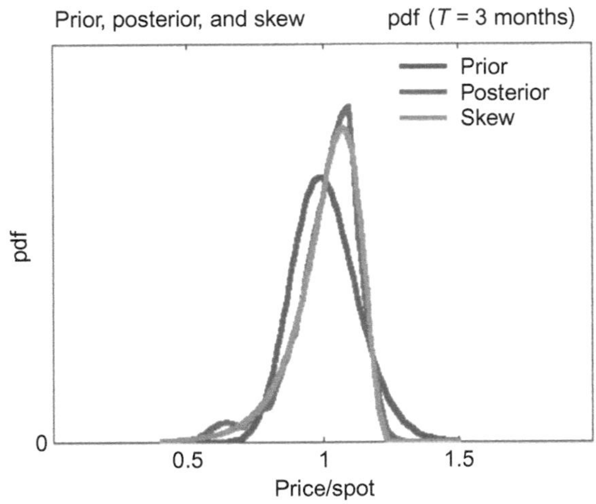

Figure 3 Skewed pdf plotted on top of prior and posterior pdf's in Fig. 2.

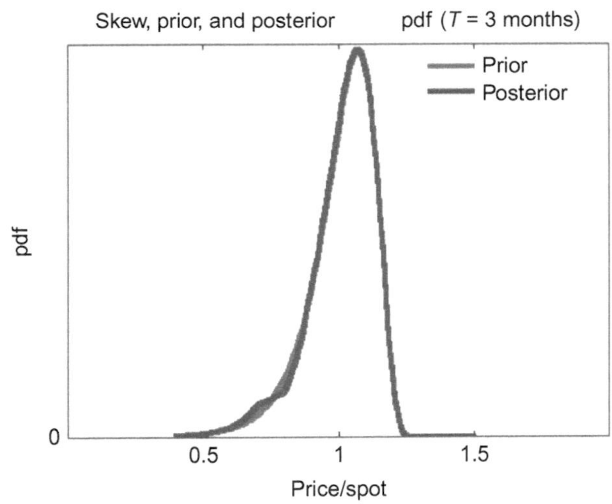

Figure 4 Skewed prior pdf produces posterior with single mode. The red (light gray in the print version) pdf in this figure is the same as the green (light gray in the print version) pdf in Fig. 3.

from Fig. 2 reasonably well, we may use the skewed distribution as a prior in hopes that it would lead to a posterior distribution that is not bimodal.

In Fig. 4, the results of fitting a skewed prior distribution to the market are shown; the MRE distribution is no longer bimodal. However, the bumpiness of the distribution at low strikes may still indicate price inefficiency in the markets.

The skewed prior distribution is closer in entropic distance to the new posterior distribution, but both posterior distributions produced equally good fits to the market prices. However, the skewed distribution incorporated more *a priori* knowledge about the markets.

The new, better-behaved posterior distribution would thus be more useful to financial engineers as a tool for pricing and risk managing off-market instruments.

8. CONCLUSION

We have shown that statistical inference with MRE can be implemented robustly with Sinc quadrature. The power of sinc-MRE optimization is rooted in the exponentially damped error of Sinc quadrature and in the nonuniform sampling property of Sinc expansions.

The exponential damping of absolute errors allows the optimization to converge as it computes integrals iteratively without the growth and propagation of round-off errors. The nonuniform sampling of Sinc expansions enables a seamless handling of singular boundaries and fast-changing regions.[14]

The sinc-MRE method, applied to a real options market-fitting problem, provided a robust field test. It was robust in terms of choice of parameters since a wide variety of parameter values, particularly larger values of M, led to convergence. It was also robust in terms of the choice of prior distribution, and it was robust in terms of real market inputs. That is, the real market data set that we used was typical because it consisted of option prices that were clustered around the at-the-money strike price.

This clustering property of real market data leads to highly correlated prices, which subsequently lead to a nearly singular Hessian matrix that can destabilize the minimization algorithm.[11] In fact, the best that could be done with conventional quadrature techniques in Ref. 11 was to apply MRE minimization to synthetic data so that convergence may at least be demonstrated in principle.

In our case, we demonstrated robust convergence in practice by using real market data and implementing the sinc-MRE minimization algorithm. In spite of the fact that convergence is difficult to obtain with highly correlated market prices that also have sharp payoff boundaries, we nevertheless obtained fast and successful convergence.

The successful convergence of this problem provides incentive for us to apply sinc-MRE optimization to other outstanding problems in chemistry and physics. Sinc methods have already been reformulated in the Dirac bra-ket notation.[17] They may thus be adapted readily to many-particle problems in quantum mechanics and quantum field theory. Other related work, including another paper submitted to these proceedings,[18] has been done that deploys Sinc methods to the solution of Schrödinger problems.[17,19–21]

Work along these lines originated as part of an extended effort, by Frank Harris and myself, to apply localized basis sets to problems in quantum field theory. Our first papers in this effort presented a Gaussian basis representation of quantum electrodynamics.[22,23]

We will be reporting on more applications of Sinc methods to problems in quantum field theory in the future.

ACKNOWLEDGMENTS

This chapter was written in honor of Frank E. Harris' 85th birthday. Frank helped spawn my career as an interdisciplinary scientist as he has always thrived in and encouraged work that crossed traditional disciplinary boundaries. It has been a great pleasure to work with Frank and to develop a strong friendship with him that has lasted for years.

This work was supported by IISAM, L3C (Interdisciplinary Institute for Science and Applied Mathematics).

REFERENCES

1. Ben-Naim, A. *A Farewell to Entropy: Statistical Thermodynamics Based on Information*; World Scientific: Singapore, 2008.
2. Lemons, D. S. *A Student's Guide to Entropy*; Cambridge University Press: New York, 2013.
3. Johnson, R. A.; Wichern, D. W. *Applied Multivariate Statistical Analysis*, 6th ed.; Pearson: Upper Saddle River, NJ, 2007.
4. Shell, M. S. The Relative Entropy Is Fundamental to Multiscale and Inverse Thermodynamic Problems. *J. Chem. Phys.* **2008**, *129*, 144108.
5. Carmichael, S. P.; Shell, M. S. A New Multiscale Algorithm and Its Application to Coarse-Grained Peptide Models for Self-Assembly. *J. Phys. Chem. B* **2012**, *116*, 8383–8393.
6. Tribus, M. *Thermodynamics and Thermostatics: An Introduction to Energy, Information and States of Matter, with Engineering Applications*; D Van Nostrand Company: Princeton, NJ, 1961.
7. Papoulis, A.; Pillai, S. U. *Probability, Random Variables, and Stochastic Processes*, 4th ed.; McGraw-Hill: New York, 2001.
8. Franke, J.; Hardle, W. K.; Hafner, C. M. *Statistics of Financial Markets: An Introduction*, 4th ed.; Springer: New York, 2015.
9. Hull, J. C. *Options, Futures, and Other Derivatives*, 9th ed.; Prentice Hall: Upper Saddle River, NJ, 2014.
10. Breeden, D. T.; Litzenberger, R. H. Price of State-Contingent Claims Implicit in Option Prices. *J. Financ.* **1978**, *34*, 621–651.
11. Buchen, P. W.; Kelly, M. The Maximum Entropy Distribution of an Asset Inferred from Option Prices. *J. Financ. Quant. Anal.* **1996**, *31*, 143–159.

12. Cover, T. M.; Thomas, J. A. *Elements of Information Theory*, 2nd ed.; Wiley: Hoboken, NJ, 2006.
13. Stenger, F. *Numerical Methods Based on Sinc and Analytic Functions*; Springer-Verlag: New York, 1993.
14. Stenger, F. *Handbook of Sinc Numerical Methods*; CRC Press: Taylor & Francis Group, Boca Raton, FL, 2011.
15. Lund, J.; Bowers, K. L. *Sinc Methods for Quadrature and Differential Equations*; SIAM: Philadelphia, PA, 1992.
16. Quarteroni, A.; Saleri, F.; Gervasio, P. *Scientific Computing with Matlab and Octave*, 4th ed.; Springer-Verlag: Heidelberg, 2014.
17. Koures, V. G.; Harris, F. E. Sinc Collocation in Quantum Chemistry: Solving the Planar Coulomb Schrödinger Equation. *Int. J. Quantum. Chem. Symp.* **1996**, *30*, 1311–1318.
18. Stenger, F.; Baumann, G.; Koures, V. G. Computational Methods for Chemistry and Physics, and Schrödinger in 3+1. *Adv. Quantum. Chem.* **2015**, this volume.
19. Koures, V. G. Solving the Coulomb Schrödinger Equation in d=2+1 via Sinc Collocation. *J. Comput. Phys.* **1996**, *128*, 1–5.
20. Amore, P. A Variational Sinc Collocation Method for Strong-Coupling Problems. *J. Phys. A* **2006**, *38*, L349–L355.
21. Jerke, J. L.; Lee, Y.; Tymczak, C. I. *Solving One-Electron System in a Novel Gaussian-Sinc Mixed Basis Set.* **2014**, arXiv: 1405.5073v2 [physics.comp-ph].
22. Koures, V. G. Gaussian Basis Expansion of the QED Light Cone Hamiltonian. *Phys. Lett. B* **1995**, *348*, 170–177.
23. Koures, V. G.; Harris, F. E. Light Cone Hamiltonian in Quantum Chemistry: Gaussian Basis Representation for Quantum Electrodynamics. *Int. J. Quantum. Chem. Symp.* **1995**, *29*, 277–288.

CHAPTER ELEVEN

Computational Methods for Chemistry and Physics, and Schrödinger in 3 + 1[1]

Frank Stenger[*,1], **Gerd Baumann**[†,‡], **Vasilios G. Koures**[§]
[*]SINC, LLC, School of Computing, University of Utah, Salt Lake City, Utah, USA
[†]Mathematics Department, German University in Cairo, New Cairo City, Egypt
[‡]University of Ulm, Ulm, Germany
[§]IISAM L3C, Cheyenne, WY, USA
[1]Corresponding author: e-mail address: stenger@cs.utah.edu

Contents

Abstract

Many authors have written papers about solving Schrödinger's equation. Some of these, unaware of the existence of approximate methods for solving problems over unbounded regions, have introduced the use of *absorbing boundary conditions*. Our motivation for writing this chapter is to acknowledge the many contributions that

[1] This chapter is written in honor of Frank Harris' 85th birthday.

Advances in Quantum Chemistry, Volume 71
ISSN 0065-3276
http://dx.doi.org/10.1016/bs.aiq.2015.02.005

Frank Harris has made to physics and chemistry, including his excellent contribution to the Schrödinger problem, and also, to introduce physics and chemistry researchers to a family of new and powerful computational methods that readers of these areas are largely unaware of. Included in this chapter is a procedure for computing an accurate approximation of Schrödinger's equation over $\mathbb{R}^3 \times (0,T)$, where T can be, infinity, depending on the problem.

1. INTRODUCTION AND SUMMARY

One of us (Stenger) was particularly impressed with the Koures–Harris Schrödinger paper of 1996,[1] with their results, with how well it was written, and with their extension of the Sinc notation to the physics *Sinc kets* and *Sinc bras*. Sinc methods have also been used by other authors for computing solutions to Schrödinger and other difficult to solve partial differential equations (PDE). In Ref. 2, Amore developed an efficient method for solving Schrödinger's PDE in 1 + 1 ; in Ref. 3 Ley-Koo used elliptic functions to compute eigenstates at energy levels. This method of approximation is also effective; indeed Sinc methods were first derived using elliptic functions—see Ref. 4. In Ref. 5 Schulte used a product of trigonometric (for time) and Gaussian decreasing exponentials (for spacial) approximations to compute solutions to 2 + 1 Schrödinger PDE problems. Such bases are similar to the ones we are using in this chapter to solve 3 + 1 problems; indeed, it may be shown that in one dimension, such bases have similar accuracy (see Ref. 6, section 1.4). But finite difference methods were used to approximate derivatives of the PDE in his approximating scheme, which leads to large matrices and *a fortiori* results that are less accurate; we avoided the pitfalls of such methods in the present chapter via our use of multi-dimensional indefinite convolution,[7] Chapters 3 and 4. The authors Jerke, Lee, and Tymczak obtained excellent results via use of Sinc methods in Ref. 8; they were furthermore impressed with the effectiveness of Sinc. Evidently, they were unaware of the many already established computational procedures of such methods of Refs. 4,6,7 which would have both simplified, clarified and increased the efficiency of their computations.

Evidently, many others who have made contributions to solving Schrödinger's equation have been unaware of the Koures–Harris effective method for solving such problems. In place, they have introduced *absorbing boundary* methods to enable them to solve this $\mathbb{R}^3 \times (0, T)$ problem over a finite region, since they were evidently unaware of the existence of methods for solving problems over infinite regions in space and/or time.

We thus felt the need to include in our presentation relatively unknown, but effective methods that are useful to computational researchers of chemistry and physics. To this end, we have presented in this chapter, simple methods of interpolation, quadrature, indefinite integration, definite and indefinite convolution, inversion of Laplace transforms, and methods for solving ordinary and partial differential equations.

Regarding Section 2 of this chapter, the polynomial approximation part, the ordinary differential equation part, the Burgers equation part, and all of the examples are new.

In Section 3 of the chapter we examine Schrödinger's equation in 3 + 1 dimensions. We first convert the PDE to an integral equation (IE). We then describe our method of obtaining a solution to this IE based on a Sinc—exponential approximation and multidimensional indefinite convolution. We obtain the four-dimensional Laplace transform of the kernel of the IE, and we then describe our method of solving the IE based on successive approximation. This procedure enables the evaluation of the right hand side of the IE via use of a finite number of one-dimensional matrix multiplications, i.e., we do not require the use of large matrices. This enables high accuracy for both space and time, and circumvents finite difference methods which have limited accuracy. We have included several examples of solutions and illustrations of our computations in Section 5, for: the harmonic oscillator, over $\mathbb{R}^3 \times (0, T)$, the modified Pöscher–Teller potential over $\mathbb{R}^3 \times (0, \infty)$, and the Paul trap potential over $\mathbb{R}^3 \times (0, T)$ and over $\mathbb{R}^3 \times (0, \infty)$.

These three Schrödinger equation examples demonstrate that Sinc convolution methods are straight forward and effective in applications. The examples also demonstrate that there are new types of solutions for Schrödinger's equation. The solutions form localized structures in spatial and temporal coordinates. Such localized structures are consistent with the probability interpretation of the density function $\Psi(\bar{r}, t)$. Thus the spatial temporal solution derived via Sinc convolution open a new field in applications which now need no longer be based on standard quantum mechanical perturbation methods.

All of the computations for our illustrations were done using *Mathematica*.

2. COMPUTATIONAL METHODS IN 1 D

In this section we shall discuss various computational methods, classical and new.

Popular methods currently in use also have problems, for example:

1. One-step methods for solving time problems are either accurate, but have stability problems, or else they have limited accuracy. This is particularly the case of partial differential equations (PDE), such as Schrödinger's equation, for which the PDE has both bounded and unbounded solutions. This phenomenon can at times be eliminated by converting the PDE to an integral equation (IE) which eliminates the unbounded solution. But we shall illustrate here that we can make such PDE to IE conversions even more effective.
2. Currently, most methods for solving PDE , such as finite element and finite difference methods are based on approximating the highest derivative of the PDE. This leads to large matrices having very large condition numbers. We describe in this chapter methods which circumvent such difficulties.
3. PDE and IE frequently have singularities which impede the rate of convergence to the solution. The methods we shall describe in this chapter are not plagued by such phenomena.
4. Fourier transform methods (FFT) are also very popular for solving PDE and IE, but here, too, one frequently requires an approximation of very high order, due to, e.g., the presence of singularities of the solution, or due to the slow decrease of the approximated function. For example, the use of Laplace transforms could be used to circumvent such difficulties, but Laplace transforms are seldom used, due to difficulties for determining the inverse transforms. Here, too, we introduce a novel method for Laplace transform inversion, which is very efficient and easy to apply, and which does not have the shortcomings of FFT.
5. Indefinite integrals are difficult to deal with. They are frequently evaluated by one-step methods which are not very accurate, and which are often unstable. The method of approximating indefinite integrals which we describe here plays a central role in the other methods which we present in this chapter—methods for approximating derivatives, methods for approximating Hilbert transforms, methods for solving ordinary differential equations (ODE), and methods for solving PDE and IE. In particular, the resulting methods for solving elliptic, parabolic and hyperbolic PDE are not appreciably different, contrary to other methods that are currently in use.
6. Finally, all of the methods of this chapter stem from interpolation. Our methods compute function values at interpolation points; these points are nearly always sparsely spaced, since relatively few are required.

Our methods of interpolation can then be used to e.g., plot a smooth solution, and furthermore, this computed smooth solution is uniformly accurate throughout the domain of the solution.

2.1 Interpolation Methods on $(a,b) \subseteq \mathbb{R}$

Consider an interpolation procedure on a finite interval (a, b), of the form

$$f(x) \approx P(x) = \sum_{k=-M}^{N} b_k(x) f(x_k), \tag{1}$$

where the $a < x_{-M} < x_{-M+1} < \ldots < x_N < b$, and with $(a, b) \subseteq \mathbb{R}$. The b_k are basis functions, which can be polynomials, rational functions, Sinc functions, wavelets, splines, etc.

For example:

i. For polynomial interpolation, on a finite interval, we have $-\infty < a < b < \infty$, and

$$b_k(x) = \prod_{\ell=-M}^{N} \prod_{\ell \neq k} \frac{x - x_\ell}{x_k - x_\ell}. \tag{2}$$

The use of Lagrange polynomials is frequently used for computing in Chemistry and Physics. A similar formula results for interpolation on $\mathbb{R}$ using Hermite polynomials on $\mathbb{R}$ or Laguerre polynomials on $(0, \infty)$.

ii. For *Sinc interpolation*, we can have $-\infty \leq a < b \leq \infty$. We use a one-to-one transformation $\varphi : (a, b) \rightarrow \mathbb{R}$ to get an approximation of the form (1) on (a,b), with

$$\begin{aligned} b_k(x) &= \frac{\sin\left\{\frac{\pi}{h}(\varphi(x) - k\,h)\right\}}{\frac{\pi}{h}(\varphi(x) - k\,h)}, \quad \text{or} \\ b_k(x) &= \exp\left(-(\varphi(x) - k\,h)/h)^2\right) \frac{\sin\left\{\frac{\pi}{h}(\varphi(x) - k\,h)\right\}}{\frac{\pi}{h}(\varphi(x) - k\,h)}. \end{aligned} \tag{3}$$

Corresponding to such a scheme, we define a row vector $\mathbf{b}$ of basis functions and an operator V that maps a function f into a column vector of order $m = M + N + 1$ by

$$\begin{aligned} \mathbf{b} &= (b_{-M}, \ldots, b_N), \\ Vf &= (f(x_{-M}), \ \ldots, \ f(x_N))^T. \end{aligned} \tag{4}$$

This notation enables us to write the above interpolation scheme in simple operator form, as

$$f \approx P = \mathbf{b}\, Vf. \tag{5}$$

Remark 2.1. Five or six significant figure accuracy will suffice for most problems of applications, and to this end, with reference to Equation (1), our methods of computation require relatively small values of M and N. Also, our methods compute solutions at interpolation points, and we usually do not interpolate our computed solutions. However, with proper selection of a method of computation, the underlying computed solution is uniformly accurate, and it is nearly always possible to obtain a smooth solution via our methods of interpolation. On rare occasions we may require both large values of M and/or N, and to also use interpolation. If in such circumstances the user is restricted to a limited precision of arithmetic, we recommend evaluation of the interplations based on the barycentric method of J.-P. Berrut.[9]

2.2 Quadrature

We omit the details of this section, since these can be easily found in other textbooks. For example descriptions of Gaussian quadrature can be found in most texts on numerical analysis, while the texts[6,7] contain descriptions of Sinc quadratures over finite, semi-infinite, infinite intervals and contours. Indeed, the text[7] contains a CD of computer programs for nearly all of the methods presented in this section.

2.3 Indefinite Integration

Given b_k and x_k as above for $k = -M, \ldots, N$, define numbers $\beta^{\pm}_{jk}$ and $m \times m$ ($m = M + N + 1$) matrices $B^{\pm}$ by

$$\begin{aligned} &\beta^{+}_{jk} = \int_a^{x_j} b_k(x)\, \mathrm{d}x, \quad \beta^{-}_{jk} = \int_{x_j}^{b} b_k(x)\, \mathrm{d}x, \\ &B^{\pm} = [\beta^{\pm}_{jk}]. \end{aligned} \tag{6}$$

Note. If we form $B^{\pm}$ for some finite interval (c, d) then the matrix $(B^{\pm})'$ for any other finite interval (c', d') is just $(B^{\pm})' = (d' - c')/(d - c)\, B^{\pm}$. The eigenvalues of these matrices obey the same proportionality relations. Their eigenvectors are independent of the length of the interval. Similar simple transformations apply also for infinite intervals. This enables efficient storage of a small number of such matrices that a user requires.

Let us also introduce definitions of operators $\mathcal{J}^+$, $\mathcal{J}^-$, and their corresponding approximations $\mathcal{J}_m^+$, $\mathcal{J}_m^-$ that are defined in terms of the above $m \times m$ matrices B^+ and B^-,

$$\begin{aligned}(\mathcal{J}^+ f)(x) &= \int_a^x f(t)\,\mathrm{d}t,\\ (\mathcal{J}^- f)(x) &= \int_x^b f(t)\,\mathrm{d}t,\\ (\mathcal{J}_m^+ f)(x) &= \mathbf{b}(x)\,B^+\,Vf,\\ (\mathcal{J}_m^- f)(x) &= \mathbf{b}(x)\,B^-\,Vf.\end{aligned} \tag{7}$$

We thus get the approximations (Ref. 6, section 4.5)

$$\mathcal{J}^\pm f \approx \mathcal{J}_m^\pm f. \tag{8}$$

The matrices $B^\pm$ can be explicitly expressed for Sinc methods (Ref. 6, sections 3.6, and 4.5).

20-Year-Old Conjecture, worth \$300, recently solved: If λ is an eigenvalue of $B^\pm$ for Sinc indefinite integration then $\Re\lambda > 0$. This conjecture was recently solved by Han and Xu.[10]

Remark 2.2. We have computed the eigenvalues the indefinite integration matrices $W^\pm$ of Legendre, Hermite, Chebyshev, Hermite Laguerre as well as for the trigonometric polynomials derived in Ref. 7, section 1.4 to find that all of the eigenvalues of these matrices are located on the open right half of the complex plane. That this is true for all matrices $\mathbf{W}^\pm$ defined as above has also not yet been proved. In addition, we do not include polynomials that interpolate at end-points of an interval. Also, this eigenvalue property does not hold for all polynomials, for example, it does not hold for those Newton–Cotes polynomials on (a, b) having the interpolation points $x_j = a + (j + M + 1/2)(b - a)/(m + 1), j = -M,\ldots,N)$, $(m = M + N + 1)$ do not have this positivity property for every degree whenever these polynomials have negative quadrature weights, e.g., for degrees greater than 5.

2.4 A New Conjecture

We here describe a new conjecture, this time for polynomial, instead of for Sinc methods.

Conjecture. *Let ω denote a function that is positive a.e. on an interval $(a, b) \subseteq \mathbb{R}$, and assume that the moments $\int_a^b \omega(x)\, x^k \, dx$ exist and are finite for all nonnegative integers k. Let $\{p_m\}_0^\infty$ denote the sequence of orthogonal polynomials, with $p_m(x)$ of degree m in x, i.e., with*

$$\int_a^b w(x) p_m(x)\, p_n(x)\, dx = c_m\, \delta_{m,n},$$

where $c_m > 0$ and where $\delta_{m,n}$ denotes the Kronecker delta. If for $m \geq 1$, $p_m(x) = 0$ for $x = x_{-M} < x_{-M+1}, < \ldots < x_N$, with $m = M + N + 1$, and if the numbers $\beta_{j,k}^{\pm}$ are defined by

$$\begin{aligned} \beta_{j,k}^{+} &= \int_a^{x_j} \omega(x) \frac{p_m(x)}{(x - x_k)\, p_m'(x_k)}\, dx \\ \beta_{j,k}^{-} &= \int_{x_j}^b \omega(x) \frac{p_m(x)}{(x - x_k)\, p_m'(x_k)}\, dx \end{aligned} \tag{9}$$

then every eigenvalue of $B^{\pm} = \left[\beta_{j,k}^{\pm}\right]$ lies on the open right half of the complex plane.

Stenger offers \$100 for proof or disproof of this conjecture.

2.5 Approximation of Derivatives

The matrices $\pm(B^{\pm})^{-1}$ yield accurate approximation of derivatives, i.e., their inverses exist, since the real parts of their eigenvalues are on the right half plane, and so

$$V f' \approx \pm (B^{\pm})^{-1}\, V f. \tag{10}$$

2.6 New methods for ODE

We first convert the initial value problem

$$y' = f(x, y), \quad x \in (a, b), \quad y(a) = c \tag{11}$$

to a Volterra IE, and then apply our indefinite integration method, to obtain the approximation

$$\mathbf{y} = B^{+}\, \mathbf{f} + \mathbf{c}, \tag{12}$$

where B^{+} is defined as in (6), $\mathbf{y} = (y_{-M}, \ldots, y_N)^T$, and where, with $f_j = f(x_j, y_j)$, with $\mathbf{f} = (f_{-M}, \ldots, f_N)^T$, and with $\mathbf{c} = (c, \ldots, c)^T$ ($M + N + 1$ "c's").

The system of equations (for **y**) can now be solved by successive approximation; since the eigenvalues of $B^{\pm}$ are proportional to $b - a$, we achieve convergence for all $b - a$ sufficiently small. Moreover, this approach for solving ODE initial value problems is stable. Because of the positivity of the eigenvalues, we can frequently solve a linear ODE initial or boundary value problem via use of just one matrix–vector multiplication, e.g., it is frequently possible to deduce by inspection that the requisite matrix inversion exists.

Remark 2.3. Consider, for example, the ODE problem

$$y' = f(x)\,y + g(x), \quad x \in (0, c) \quad y(0) = a.$$

If e.g., $f(x) < 0$ on $(0, c)$, then, denoting by B^{+} the indefinite integration matrix of Section 2.2 for the interval $(0, c)$, and converting this DE to the IE

$$y(x) - \int_0^x f(t)\,y(t)\mathrm{d}t = \int_0^x g(t)\mathrm{d}t + a,$$

we get, using the notations $\mathbf{y} = (y_{-M}, \ldots, y_n)^T$, (a column vector) $D = diag(f(x_{-M}), \ldots, f(x_N))$ (a diagonal matrix), $\mathbf{g} = (g(x_{-M}), \ldots, g(x_N))^T$, and $\mathbf{a} = (a, a, \ldots, a)^T$ ($m = M + N + 1$ entries of a) we get the approximate solution $y(x_j) \approx y_j$ at all points x_j, $j = -M, \ldots, N$ on $(0, c)$ from the matrix evaluation

$$\mathbf{y} = (I - B^{+}\,D)^{-1}(B^{+}\,\mathbf{g} + \mathbf{a})$$

where in this case we can tell by inspection that the matrix $(I - B^{+}\,D)$ is nonsingular. Similarly, if $f(x) > 0$, we can solve the above ODE by first converting to an IE as above, but by replacement of $\int_0^x$ with $-\int_x^c$, and then using B^{-}.

Example 2.1. Consider the ordinary differential equation problem

$$\frac{d^2 y}{(d\,x)^2} = x\,y, \quad y(0) = y_0 = \frac{1}{3^{2/3}\,\Gamma(2/3)}, \quad y'(0) = 0. \tag{13}$$

whose solution is the Airy function, $Ai(x)$. The value $y_0 = 1/(3^{2/3}\,\Gamma(2/3))$ is known to be the correct normalized value of $Ai(0)$. The Airy function is the solution to Schrödinger's equation for a particle confined within a triangular potential well and for a particle in a one-dimensional constant force field. $Ai(x)$ is frequently computed using a power series for x on a finite interval, and by the WKB method for x outside of this interval. Using our above ODE procedure, we can compute it for all $x \in (0, \infty)$, as follows.

We first write equation (13) as a system of first order equations,

$$\begin{aligned} y' &= v \\ v' &= x\,y \end{aligned} \tag{14}$$

which we solve subject to the initial condition $(y(0), v(0)) = (y_0, 0)$. We can thus write the system (14) as a system of integral equations in the form

$$\begin{aligned} y(x) &= y_0 + \int_0^x v(\eta)\, \mathrm{d}\eta \\ v(x) &= -\int_x^\infty \eta\, y(\eta)\, \mathrm{d}\eta \end{aligned} \tag{15}$$

We again use a Sinc indefinite integration procedure over $(0, \infty)$ (via the $\log(sinh)$ transformation) combined with the above described indefinite integration method to approximate each of the equations in (15). Thus, denoting by x_k the interpolation points, we set $\mathbf{y} = (y_{-N}, \ldots, y_N)^T$, with the y_k being approximations of $y(x_k)$, $\mathbf{v} = (v_{-N}, \ldots, v_N)^T$, we form a diagonal matrix $D = diag(x_{-N}, \ldots, x_N)$, we let B^+ (rsp. B^-) denote the indefinite integration matrices (gotten from the *CD* of Ref. 7) for approximating the integrals $\int_0^x$ (rsp. $\int_x^\infty$), and we let $\mathbf{1}$ denote the column vector of $m = M + N + 1 = 2\,N + 1$ ones, so that we can replace (15) with the approximating system,

$$\begin{aligned} \mathbf{y} &= y_0\, \mathbf{1} + B^+\ \mathbf{v} \\ \mathbf{v} &= -B^-\, D\mathbf{y}. \end{aligned} \tag{16}$$

Eliminating $\mathbf{v}$, we arrive at the expression

$$\mathbf{y} = y_0\, (I + \mathbf{C})^{-1}\, \mathbf{1}, \tag{17}$$

in which $C = B^+\, B^-\, D$. Note that the matrix $(I + C)$ is nonsingular, because of the above cited eigenvalue properties of $B^{\pm}$. By taking $M = N = 10$, and $h = \pi/(3\,N)^{1/2}$ we get the plot of Fig. 1.

Remark 2.4. We wish to point out here, that (17) enables a solution to *arbitrary precision*, and while asymptotic methods can only achieve limited accuracy, we could e.g., get 100 places of accuracy, provided that such precision is available, and provided that the entries of the matrix C are computed in such precision. This remark applies to all of the algorithms of this chapter[1].

[1] Bailey and Borwein used Algorithm 1.5.18 of Ref. 7 to check their procedure for computing π to 20 billion places.

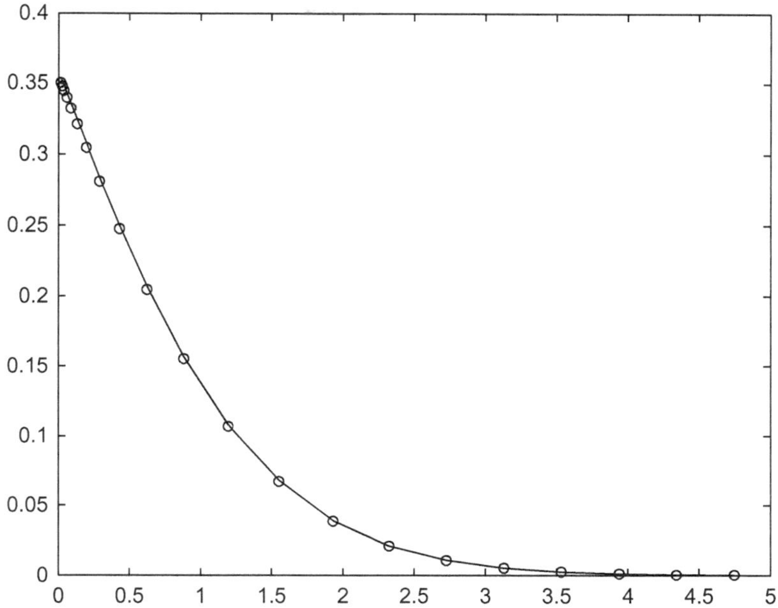

Figure 1 Exact (—) and approximate (o–o) airy function.

2.7 Indefinite Convolution

Our model integrals over $x \in (0,a)$ take the form

$$\begin{aligned} p(x) &= \int_0^x f(x-t)\, g(t)\, \mathrm{d}t, \\ q(x) &= \int_x^a f(x-t)\, g(t)\, \mathrm{d}t. \end{aligned} \tag{18}$$

These two equations describe the most often solved IE in real life !
Define the "Laplace transform" by

$$\mathcal{F}_{\pm}(s) = \int_0^c \mathrm{e}^{-t/s} f(\pm t)\, \mathrm{d}t, \qquad c \geq (b-a). \tag{19}$$

Remark 2.5. The transformation $s \to 1/s$ is a conformal map of the right half plane to itself, so that, if the usual Laplace transform of f is known, we can get the "Laplace transform" from it just by replacing s with $1/s$.

We assume that $\mathcal{F}_{\pm}(s)$ exists for all $\Re s > 0$.

Then, it can be shown that (Ref. 6, section 4.6; Ref. 7, section 1.5.9)

Theorem 1.

$$p=\mathcal{F}_+(\mathcal{J}^+)g, \quad q=\mathcal{F}_-(\mathcal{J}^-)g. \tag{20}$$

The above results are esoteric. However consider the following.
Approximations. Let $\mathbf{b}$ be defined as in (4), let B^+ be defined as in (6) above, and set $B^+ = X\,S\,X^{-1}$, where S is a diagonal matrix. Then

$$\begin{aligned} p &= \mathcal{F}_+(\mathcal{J}^+)g \\ &\approx \mathcal{F}_+\left(\mathcal{J}_m^+\right)g \\ &= \mathbf{b}\,\mathcal{F}_+(B^+)\,V g \\ &= \mathbf{b}\,X\,\mathcal{F}_+(S)\,X^{-1}\,V g, \end{aligned} \tag{21}$$

i.e., vector-matrix form

$$V\,p \approx X\,\mathcal{F}_+(S)\,X^{-1}\,V g.$$

Similarly for q.
Remark 2.6.

- The matrices B^+ and B^- have the same eigenvalues.
- The numbers $\mathcal{F}_\pm(s_j)$ denote the (diagonal) elements of the matrices $\mathcal{F}_\pm$. If the transforms $\mathcal{F}_\pm$ are not explicitly known, then the values $\mathcal{F}_\pm(s_j)$ can be approximated to arbitrary accuracy via use of Sinc or Gaussian quadrature applied to the "Laplace transform" integral.
- The operators $\mathcal{J}^\pm$ and the above matrices $\mathbf{B}^\pm$ are translation invariant, i.e., they are the same for $(a+c, b+c)$ for any $c \in \mathbb{R}$. The bases $\mathbf{w}$ will, of course, have to be modified, i.e., on $(a+c, b+c)$ the basis is $\mathbf{v} = (v_{-M}, \ldots, v_N)$, with $v_j(x) = w_j(x-c)$.
- Recall, we assumed that $\Re s > 0$ for the existence of the Laplace transform. We thus require that all eigenvalues of $\mathbf{B}^\pm$ have positive real parts, in order for the approximations in (21) to be well defined. The above polynomial conjecture and its recent proof for Sinc approximation enable this well defined property.

Using the above operator identities for p and q, we can readily derive the other formulas which follow.

2.8 Laplace Transform Inversion

Let the "Laplace transform" $\mathcal{F}_+$ be defined as in (13). If $\mathcal{J}^+$ denotes the indefinite integral operator defined on $(0,a) \subseteq (0,c) \subseteq (0,\infty)$, then[6,11]

$$f = (\mathcal{J}^+)^{-1} \mathcal{F}_+ (\mathcal{J}^+)\, 1, \tag{22}$$

where "1" denotes the function that is identically 1 on $(0, a)$. Hence, with $\mathcal{J}_m^+ \approx \mathbf{w}\, B^+\; V$, with $B^+ = X\, S\, X^{-1}$, $S = \mathrm{diag}(s_{-M}, \ldots, s_N)$, **we can proceed as follows to compute the values** $f_j \approx f(x_j)$ of f:

i. Assume that the matrix X and vector $\mathbf{s} = (s_{-M}, \ldots, s_N)^T$ have already been stored for some interval (α, β), corresponding to a matix B^+ of Section 2.2, make the replacement $\mathbf{s} \rightarrow a/(\beta - \alpha)\, \mathbf{s}$, and compute the column vector $\mathbf{v} = (v_{-M}, \ldots, v_N)^T = X^{-1}\, \mathbf{1}$, where $\mathbf{1}$ is a vector of $M + N + 1$ ones;

ii. Compute

$$\mathbf{g} \equiv (\mathcal{F}_+(s_{-M})/s_{-M}, \ldots, \mathcal{F}_+(s_N)/s_N)$$

$$\mathbf{z} \equiv \mathbf{g} * \mathbf{v} \text{ with "*" a Hadamard product}$$

$$\mathbf{f} = (f_{-M}, \ldots, f_N)^T = X\, \mathbf{z}.$$

iii. Note: If B^+ is an indefinite integration matrix defined as above for a finite interval (a, b), then all of the $M = M + N + 1$ values $f_j \approx f(x_j)$ are computed for $x_j - a$ with

$$0 < x_{-M} - a < x_{-M+1} < \ldots < x_N < b - a.$$

All operations of this evaluation take a trivial amount of time, except for last matrix-vector evaluation. However, the size of these matrices is nearly always much smaller than the size of the DFT matrices for solving similar problems via FFT.

Then we have the approximation $f(x_j) \approx f_j$, which we can combine, if necessary, with our interpolation formula (1) to get a continuous approximation of f on $(0, a)$.

Example 2.2. Let us approximate f, given the Laplace transform

$$\mathcal{F} + (s) = \frac{2\sqrt{\pi D s^3}}{d}\left((s+2)(D+s+2) - 4s^2\right)$$

in which $D = \sqrt{(s+2)^2 + 4s^2}$, $d = \sqrt{(D+s+2)^2 + 4s^2}$, and for which the exact solution is $f(t) = t^{1/2}\, e^{-t/2} \cos(t)$.

Notice that the singularities of $\mathcal{F}_+(s)$ at $s = 0$, and of $f(t)$ at $t = 0$, i.e., $\mathcal{F}_+(s)/s \sim (s\, \pi/2)^{1/2}$, $s \rightarrow 0$ and $f(t) \sim t^{1/2}$, $t \rightarrow 0$ are of the same order.

Figure 2 illustrates the plots of the exact and approximate solutions. The approximate solution was computed using the Sinc transformation

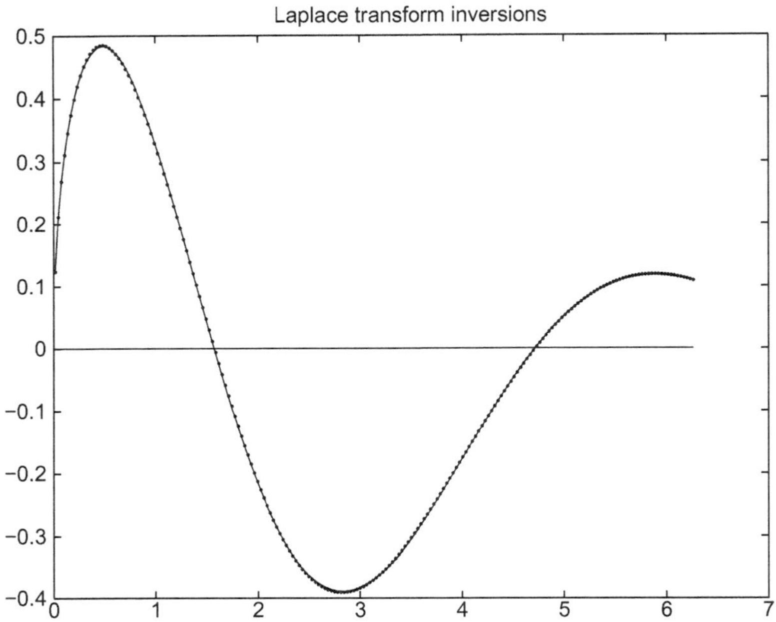

Figure 2 Exact (—) and approximate (···) laplace transform inversion for $t^{1/2}\exp(-t/2)\cos(t)$

(see Ref. 5, section 1.5.3) $\varphi(z) = \log(\sinh(z))$, with $M = N = 10$, i.e., with $M + N + 1 = 21$ points, and with $h = \pi/\sqrt{N}$. It can be shown that the error for any N is of the order of $\exp\left(-(\pi/2)\, N^{1/2}\right)$. Although a small number of points were used based on the procedure of Section 2.8, intermediate values for this plot were then computed via Sinc interpolation.

Remark 2.7. In this example we have computed the approximations $f_j \approx f(x_j)$ over $(0, \infty)$ via the $\log(\sinh)(\cdot)$ transformation. However, there are several options for computing approximations over a subinterval[2] of $(0, \infty)$:

i. Using the interpolation formula, based on the $\log(\sinh)(\cdot)$ transformation (Ref. 7, Equation (1.5.24) and Example 1.5.7). This enables evaluation over an arbitrary subinterval of $(0, \infty)$;

ii. Using an indefinite integration matrix over $(0, c)$, for some finite $c > 0$, based on the transformation $\varphi(x) = c\exp(x)/(1 + \exp(x))$ (Ref. 7, Equation (1.5.24) and Example 1.5.5). This approach enables close examination of $f(t)$ in a neighborhood of $t = 0$. Indeed, this approach also enables an approximation to within a relative error, even when $|f(t)| \to \infty$ as $t \to 0$ (see Ref. 7, section 3.6); or

[2] Although our discussion of this remark is based on the interval $(0, \infty)$, it also applies, e.g., to approximation of f over the real line $\mathbb{R} = (-\infty, \infty)$.

iii. To obtain an accurate picture of what happens in a neighborhood of e.g., $t=\infty$, one can substitute $-\mathcal{F}_{-}(C^{-})$ for $\mathcal{F}_{+}(B^{+})$ in the above described algorithm, where C^{-} is the indefinite integration matrix for (c, ∞) for some positive number c. The matrix C^{-} is easily obtainable from B^{-}, where B^{-} is the usual indefinite integration matrix for $(0, \infty)$.

2.9 Hilbert Transforms

Presently methods based on Hilbert transforms are used extensively in the area of geophysics, in Australia. See [Ref. 25] for applications.

Using the above operator identities for p and q, we can derive the identity

$$\begin{aligned}(\mathcal{H}g)(x) &= P.V. \int_a^b \frac{g(t)}{t-x}\,\mathrm{d}t \\ &= \left(\log\left(\mathcal{J}^{+}\right)-\log\left(\mathcal{J}^{-}\right)\right)g.\end{aligned} \tag{23}$$

That is, for numerical approximation,

$$V\,\mathcal{H}\,g \approx \left(\log\left(B^{+}\right)-\log\left(B^{-}\right)\right)\,V\,g. \tag{24}$$

2.10 Convergence of Approximations

Sinc methods nearly always converge at an exponential rate if the functions they approximate are analytic in the interior of the interval of approximation. End-point singularities are also allowed, provided that they are integrable. Indeed, Sinc interpolation is effective even if functions "blow up" at end-points, in which case convergence occurs in a relative sense. Generally, the rate of convergence of using m points on an interval is of the order of $\exp(-c\,m^{1/2})$ where c is a positive constant. Indeed, this rate is the same whether or not the function to be approximated has singularities at end-points[3]. Similarly, for approximating solutions to multidimensional problems, it is best to keep singularities on boundaries of regions of approximation, and to this end, it is often advisable to subdivide the region of approximation to achieve this setting.

Sinc methods even work for curvilinear regions (Ref. 7, sections 2.5.4, and 4.2.5).

Further details are found in Refs. 4,6,7.

[3] m-point polynomial approximation is more accurate than Sinc if the function to be approximated is analytic in a region containing the interval of approximation. On the other hand, the error of polynomial methods is much worse that of Sinc in the presence of singularities at end-points.

3. SOLVING PARTIAL DIFFERENTIAL EQUATIONS

Consider the solution of a partial differential equation (PDE) having one of the forms

$$\mathbf{L}\,u = \begin{cases} -\Delta\, u, \\ \dfrac{\partial u}{\partial t} - \Delta\, u, \quad \text{or} \\ \dfrac{\partial^2 u}{\partial t^2} - \Delta\, u. \end{cases} \tag{25}$$

All such PDE can be reduced to the solution of integral equations (IE). All these IE can then be solved by essentially the same procedure, based on one dimensional matrix multiplication using one-dimensional, small-sized matrices. That is, we do not use the large sized matrices that are required by finite difference, finite element, or Galerkin methods; we furthermore get exponential convergence, even in the presence of unknown-type singularities.

Example 3.1. Consider the solution of *Burgers' equation*,

$$\begin{aligned} &\frac{\partial}{\partial t}u(x,t) - \varepsilon\,\frac{\partial^2}{\partial x^2}u(x,t) \\ &\quad = -\frac{1}{2}\frac{\partial}{\partial x}u^2(x,t), \quad x \in \mathbb{R}, \qquad t > 0, \\ &u(x,0) = u_0(x). \end{aligned} \tag{26}$$

To transform this to an IE, we need the *heat equation kernel*,

$$\mathcal{G}(x,t) = (4\,\pi\,\varepsilon\,t)^{-1/2}\exp\left(-\frac{x^2}{4\,\varepsilon\,t}\right), \tag{27}$$

as well as its derivative,

$$\mathcal{G}_x(x,t) = -\pi\,\frac{x}{(4\pi\varepsilon t)^{3/2}}\exp\left(-\frac{x^2}{4\varepsilon t}\right). \tag{28}$$

By means of these, we transform the PDE (26) into the IE

$$\begin{aligned} &u(x,t) = v(x,t) \\ &-\pi\int_0^t\int_{\mathcal{R}} G_x(x-\xi,t-t')\frac{x-\xi}{(4\pi\varepsilon(t-t'))^{3/2}}\;u^2(\xi,\tau)\,\mathrm{d}\xi\,\mathrm{d}t', \end{aligned} \tag{29}$$

where

$$v(x,t) = \frac{1}{(4\pi\varepsilon t)^{1/2}} \int_{\mathcal{R}} \exp\left\{ -\frac{(x-\xi)^2}{4\varepsilon t} \right\} u_0(\xi)\, d\xi \tag{30}$$

We shall evaluate the two integrals on the right hand side of (27) via indefinite convolution. For the first integral, we require the "Laplace transform"

$$\begin{aligned} G_x(s,\tau) &= \int_0^\infty \int_0^\infty \exp\left\{ -\frac{x}{s} - \frac{t}{\tau} \right\} \mathcal{G}_x(x,t)\, dx\, dt \\ &= -\frac{1}{4\,\varepsilon^{1/2}} \frac{s\,\tau^{1/2}}{s + \varepsilon^{1/2}\,\tau^{1/2}}. \end{aligned} \tag{31}$$

To evaluate the second integral on the right hand side of (27) we would like to know the "Laplace transform",

$$\hat{G}(s,t) = \int_0^\infty e^{-x/s} \mathcal{G}(x,t)\, dx \tag{32}$$

Unfortunately this integral cannot be explicitly so expressed. However, upon noting that

$$\frac{\partial}{\partial t} \int_0^t \mathcal{G}(x, t-t')\, dt' = \frac{\partial}{\partial t} \int_0^t \mathcal{G}(x,t')\, dt' \tag{33}$$

it follows that

$$v(x,t) = \frac{\partial}{\partial t} \int_0^t \int_{\mathbb{R}} G(x-\xi, t-t')\, u_0(\xi)\, d\xi\, dt'. \tag{34}$$

The "Laplace transform" of the convolution kernel of this of this equation *can* be explicitly expressed, i.e., we have

$$\begin{aligned} G_t(s,\tau) &= \int_0^\infty \int_0^\infty \exp\left\{ -\frac{x}{s} - \frac{t}{\tau} \right\} \mathcal{G}_t(x,t)\, dx\, dt \\ &= \frac{s}{2\,(s + \sqrt{\varepsilon\,\tau})}. \end{aligned} \tag{35}$$

The simultaneous approximation $v(x_j, t_k) \approx V_{j,k}$ at all points (x_j, t_k) is then done by means of the following two-line algorithm, in which $U = [U_{j,k}] = [u_0(x_j,\ t_k)]$, $B^+ = X^+ S\ (X^+)^{-1}$, $B^- = X^-\ S\ (X^-)^{-1}$, are indefinite integration matrices for approximating $\int_0^x$ and $\int_x^\infty$ on $(0, \infty)$, $C = Y\ P\ Y^{-1}$ is an indefinite integration matrix for approximating an integral over $(0,t)$

on $(0,T)$, $S = \mathrm{diag}(s_{-N},\ldots,s_N)$, $P = \mathrm{diag}(\tau_{-N},\ldots,\tau_N)$ denoting diagonal matrices, $\mathbf{G}_t = [G_t(s_j,\tau_k)]$, and "*" denotes a "Hadamard product",

$$\begin{aligned} V &= X^+ \left(\mathbf{G}_t * \left((X^+)^{-1}\, U\, (Y^{-1})^T\right)\right) Y^T, \\ V &= V + X^- \left(\mathbf{G}_t * \left((X^-)^{-1}\, U\, (Y^{-1})^T\right)\right) Y^T. \end{aligned} \tag{36}$$

The matrix V only needs to be evaluated once.

Next, since the above IE is nonlinear, the right hand side of the IE must be evaluated repeatedly until convergence is achieved, so that the integral on the extreme right hand side of (29) must be evaluated repeatedly. This evaluation can be carried out by means of the following algorithm, in which we first set $U = V$, with V already computed in (34), and then carrying out the steps

$$\begin{aligned} Z &= U * U \\ U &= X^+ \left(\mathbf{G}_x * \left((X^+)^{-1}\, Z\, (Y^{-1})^T\right)\right) Y^T, \\ U &= U - X^- \left(\mathbf{G}_x * \left((X^-)^{-1}\, Z\, (Y^{-1})^T\right)\right) Y^T, \end{aligned} \tag{37}$$

where the matrices $X^{\pm}$ and Y are the same as in (36), and where $\mathbf{G}_x = [G_x(s_j,\tau_k)]$. Note the negative sign on the right hand side of the last line of (37), due to the fact that $\mathcal{G}_x(x,t)$ is an odd function of x.

Remark 3.1. The error of approximation of u^2 is of the same order (as a function of the number of points) as that of u, since the solution may be shown to be analytic (Ref. 7, chapter 2).

For obtaining Fig. 3, we took $u_0(x) = \sin^2(\pi x)\exp(-x^2/2)$, and $\varepsilon = 1/2$, which enabled convergence of the iterative process for a solution over $(x,t) \in \mathbb{R} \times (0,\infty)$, by using the approximation

$$u(x,t) = \sum_{j=-M}^{M} \sum_{k=-N}^{N} u(j\,h_1, k\,h_2)(x)\, w_j^1(x)\, w_k^2(t) \tag{38}$$

in which we used $w^1(x) = \mathrm{sinc}((x - j\,h_1)/h_1)$, $w^2(t) = \mathrm{sinc}((\varphi(t) - k\,h_2)/h_2)$, $M = 18$, $N = 18$, $h_1 = \pi/\sqrt{(M)}$, $h_2 = 2/\sqrt{(N)}$, and $\varphi(t) = \log(\sinh(t))$.

4. SPACE–TIME SCHRÖDINGER'S PDE IN 3 + 1

Several authors, including F. Harris and V. Koures [Ref. 1], Koures,[12] Amore,[2] and Jerke, Lee, and Tymczak[8] have used Sinc methods to solve Schrödinger problems. Many more have first applied absorbing boundary conditions in order to be able to use methods that work only for bounded domains (see e.g., Arnold[13]). We now present our approach.

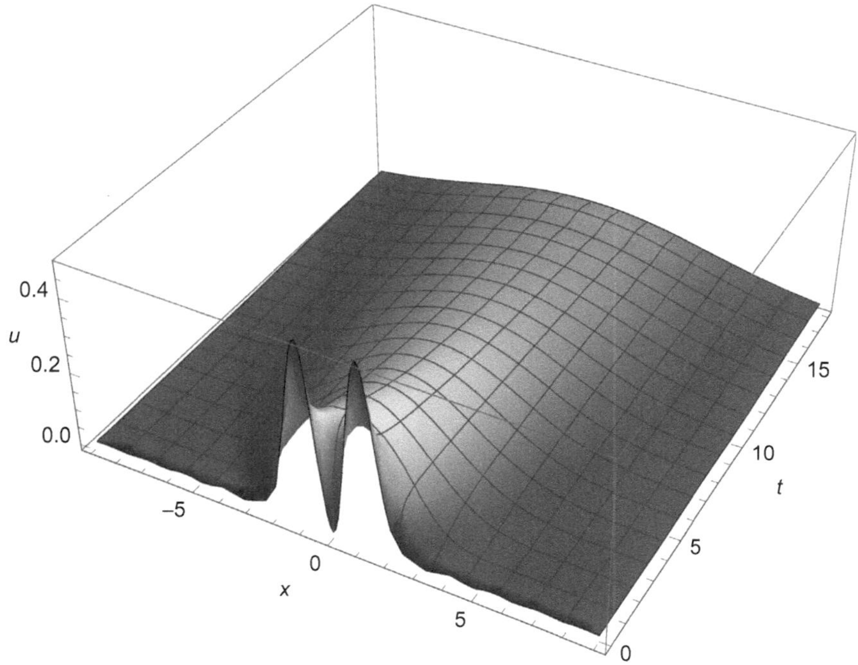

Figure 3 Plot of computed solution of Burger's equation.

1. We shall use multidimensional indefinite convolution to solve PDE the problem

$$i\hbar\,\frac{\partial \Psi}{\partial t} + \frac{\hbar^2}{2m}\,\nabla^2\,\Psi = V\,\Psi, \quad \Psi = \Psi(\bar{r},t), \quad (\bar{r},t) \in \mathbb{R}^3 \times (0,T) \qquad (39)$$
$$\Psi(\bar{r},0) = \Psi^0(\bar{r}).$$

Here, $\hbar$ is Planck's constant, m is the mass of an electron, V is an electrostatic potential, and $|\Psi|^2$ is the probability density i.e., $|\Psi(\bar{r},t)|^2\,\mathrm{d}\bar{r}$ is the probability that the electron is located in a neighborhood of $\bar{r} \in \mathbb{R}^3$ at time t, and T is either finite or infinite.

2. We first convert this PDE to the IE:

$$\begin{aligned}\Psi(\bar{r},t) &= \int_0^t \int_{\bar{R}^3} \mathcal{G}(\bar{r}-\bar{r}',t-t'), V(\bar{r}',t')\,\Psi(\bar{r}',t')\,\mathrm{d}\bar{r}'\,\mathrm{d}t' \\ &+ \int_{\mathbb{R}^3} \mathcal{G}(\bar{r}-\bar{r}',t)\,\Psi^0(\bar{r}')\,\mathrm{d}\bar{r}',\end{aligned} \qquad (40)$$

where

$$\mathcal{G}(\bar{r},t) = \mathrm{e}^{-3\,\pi\,i\,/4}\left(\tfrac{m}{2\,\pi\,\hbar\,t}\right)^{3/2} \exp\left(\frac{m\,i\,r^2}{2\,\hbar\,t}\right). \qquad (41)$$

3. We shall use Sinc approximation for obtaining a solution. This approximation takes the form

$$\Psi(\bar{r},t) \approx \sum_{j=-N}^{N} \sum_{k=-N}^{N} \sum_{\ell=-N}^{N} \sum_{n=-M}^{M} \Psi(\bar{r}_{j,k,\ell},t_n)\, U_j(x)\, U_k(y)\, U_\ell(z)\, V_n(t), \quad (42)$$

where

$$\begin{aligned} U_j(\xi) &= \exp(-(\xi - j\,h_1)^2/h_1^2)\; S(j,h_1) \circ (\xi) \\ V_n(t) &= S(n,h_2) \circ \phi(t), \end{aligned} \quad (43)$$

where h_1 and h_2 are constants to be determined, and where $\phi(t) = \log(\sinh(t))$ for approximation on $\mathbb{R}^3 \times (0,\infty)$, while $\phi(t) = \log(t/(T-t))$ for approximation on $\mathbb{R}^3 \times (0,T)$, for some finite T. Also, $\bar{r}_{j,k,\ell} = (j\,h_1, k\,h_1, \ell\,h_1)$ and $t_n = \phi^{-1}(n\ h_2)$. We will solve the above PDE using successive approximation, and to this end, we shall initially attempt a solution over $\mathbb{R}^3 \times (0,\infty)$. However, the thus computed approximations do not converge, then we shall solve the equation over $\mathbb{R}^3 \times (0,T)$. Convergence always occurs for all T sufficiently small—see Section 2.3 above.

4. The four-dimensional "Laplace transform" of $\mathcal{G}(\bar{r},t)$ is defined by

$$G(\bar{\Lambda},\tau) = \int_0^\infty \int_{\mathbb{R}^3_+} \mathcal{G}(\bar{r},t)\exp(-\bar{r}\cdot\bar{\Lambda} - t/\tau)\, d\bar{r}\, dt, \quad (44)$$

where $\bar{\Lambda} = (1/\lambda_1, 1/\lambda_2, 1/\lambda_3)$, and where we set

$$\Lambda = \sqrt{1/\lambda_1^2 + 1/\lambda_2^2 + 1/\lambda_3^2}.$$

The transform $G(\bar{\Lambda},\tau)$ is readily obtained from Ref. 7, sections 2.4.2, and 2.4.5; we get

$$\begin{aligned} &G(\bar{\Lambda},\tau) \\ &= -\frac{2\,m\,i\,\tau}{\hbar}\left(1 - \frac{i\,h\,\tau}{2\,m}\Lambda^2\right)^{-1} \cdot \\ &\cdot \left(\frac{1}{8} - H(\lambda_1,\lambda_2,\lambda_3,\tau) - H((\lambda_2,\lambda_3,\lambda_1,\tau) - H(\lambda_3,\lambda_1,\lambda_2,\tau)\right) \end{aligned} \quad (45)$$

where, the functions H are defined, as follows. Setting

$$
\begin{aligned}
z_{+} &= -\frac{\dfrac{1}{\lambda_2} - \dfrac{i}{\lambda_3}}{\left(\dfrac{2\,m}{\hbar\,\tau\,i}\right)^{1/2}\left(1+\sqrt{1+\dfrac{\hbar\,\tau\,i}{2\,m}\left(\dfrac{1}{\lambda_2^2}+\dfrac{1}{\lambda_3^2}\right)}\right)}, \\
z_{-} &= -\frac{\left(\dfrac{2\,m}{\hbar\,\tau\,i}\right)^{1/2}\left(1+\sqrt{1+\dfrac{\hbar\,\tau\,i}{2\,m}\left(\dfrac{1}{\lambda_2^2}+\dfrac{1}{\lambda_3^2}\right)}\right)}{\dfrac{1}{\lambda_2} + \dfrac{i}{\lambda_3}},
\end{aligned}
\tag{46}
$$

we have

$$
H(\lambda_1, \lambda_2, \lambda_3, \tau) = \frac{\left(\dfrac{\hbar\,\tau}{2\,m\,i}\right)^{1/2}}{4\,\pi\,\lambda_1\sqrt{1+\dfrac{\hbar\,\tau\,i}{2\,m}\left(\dfrac{1}{\lambda_2^2}+\dfrac{1}{\lambda_3^2}\right)}}\,(Q(z_{+}) - Q(z_{-})), \tag{47}
$$

and where, with $a \in \mathbb{C}$, and with $\mathcal{C} = \{z = \exp(i\,\theta) \in \mathbb{C} : 0 < \theta < \pi/2\}$,

$$
Q(a) = \int_{\mathcal{C}} \frac{\mathrm{d}z}{z - a}. \tag{48}
$$

The correct evaluation of this integral is given in Theorem 2.4.1 of Ref. 7.

5. The four-dimensional "Laplace transform" $G = G(\bar{\Lambda}, \tau)$ will be used to compute an approximation of the first integral on the right hand side of (40), via use of one four-dimensional indefinite convolution. This requires that we first write this integral in the form

$$
\begin{aligned}
&\int_0^t \int_{\mathbb{R}^3_+} \mathcal{G}(\bar{r} - \bar{r}', t - t')\, V(\bar{r}'\, t')\, \Psi(\bar{r}', t')\, \mathrm{d}\bar{r}'\, \mathrm{d}t' \\
&= \int_0^t \left(\sum_{j,\,k,\,\ell=1}^{2} \int_{\mathbb{R}_{j,k,\ell}} \mathcal{G}(\bar{r} - \bar{r}', t - t')\, V(\bar{r}'\, t')\, \Psi(\bar{r}', t')\, \mathrm{d}\bar{r}' \right) \mathrm{d}t',
\end{aligned}
\tag{49}
$$

and then approximating each of these integrals using the indefinite convolution procedure of (Ref. 7, section 2.5), which enables getting an accurate solution at all points $(\bar{r}, t) = (\bar{r}_{j,k,\ell}, t_n)$ via one-dimensional matrix multiplications. In (49), $\mathbb{R}_{j,k,\ell} = \mathbb{R}_j \times \mathbb{R}_k \times \mathbb{R}_\ell$, with $\mathbb{R}_1 = R_+$, and with $\mathbb{R}_2 = \mathbb{R} \setminus \mathbb{R}_+$.

6. In this notation, the second integral on the right hand side of (40) may be written in the computable form similar to (49) above, namely,

$$\begin{aligned}&\int_{\mathbb{R}^3} \mathcal{G}(\bar{r}-\bar{r}',t)\,\Psi^0(\bar{r}',t)\,\mathrm{d}\bar{r}'\\&=\sum_{j,k,\ell=1}^{2}\int_{\mathbb{R}_{j,k,\ell}} \mathcal{G}(\bar{r}-\bar{r}',t)\,\Psi^0(\bar{r}')\,\mathrm{d}\bar{r}'.\end{aligned} \tag{50}$$

Unfortunately, the 3-d spacial "Laplace transform"

$$\int_{\mathbb{R}^3_+} \mathcal{G}(\bar{r},t)\exp(-\bar{r}\cdot\bar{\Lambda})\,\mathrm{d}\bar{r}$$

cannot be explicitly expressed, and so we must proceed as in (25) above.

By way of simplifying the problem, suppose that we have an explicit expression for the "Laplace transform" $\hat{w}$ of $w(t)=\int_0^t v(t-t')\,\mathrm{d}t'$, and that we also want an explicit expression for the "Laplace transform" $\hat{v}$ of v. Then, noting that w has the alternate integral expression $w(t)=\int_0^t v(t')\,\mathrm{d}t'$, and performing integration by parts on the integral $\hat{w}(\tau)=\int_0^\infty \mathrm{e}^{-t/\tau}\left(\int_0^t v(t')\,\mathrm{d}t'\right)\mathrm{d}t$, we get $\hat{v}(\tau)=\hat{w}(\tau)/\tau$.

Hence setting $K(\bar{\Lambda},\tau)=G(\bar{\Lambda},\tau)/\tau$, with $G(\bar{\Lambda},\tau)$ as in (45), it follows that we can use the transform K to approximate the second integral on the right hand side of (40) via essentially the same algorithmic steps as required for approximating the first integral on the right hand side of (40). To this end, since the array $\psi^0_{j,k,\ell}\equiv\Psi^0(j\,h_1,k\,h_1,\ell\,h_1)$ is three dimensional, we first convert this to a four-dimensional one, i.e., $\Phi^0_{j,k,\ell,n}\equiv\psi^0_{j,k,\ell}$, and we can then proceed to compute a four-dimensional array approximation of the first integral on the right of (40).

7. The solution to the IE (40) will then be carried out via successive approximation, in the form

$$\begin{aligned}&\Psi_{n+1}(\bar{r},t)\\&=\int_0^t\int_{\bar{R}^3} \mathcal{G}(\bar{r}-\bar{r}',t-t'),\,V(\bar{r}',t')\,\Psi_n(\bar{r}',t')\,\mathrm{d}\bar{r}'\,\mathrm{d}t'\\&\quad+\int_{\mathbb{R}^3} \mathcal{G}(\bar{r}-\bar{r}',t)\,\Psi^0(\bar{r}')\,\mathrm{d}\bar{r}',\end{aligned} \tag{51}$$

starting with

$$\Psi_0(\bar{r},t)=\int_{\mathbb{R}^3} \mathcal{G}(\bar{r}-\bar{r}',t)\,\Psi^0(\bar{r}')\,\mathrm{d}\bar{r}'. \tag{52}$$

8. Note, the integral in (52) has to be computed only once. Note, also, as for the case of Burgers' equation above, the four dimensional arrays of the transforms of G and K need only be evaluated once.
9. Convergence occurs for an approximation over $\mathbb{R}^3 \times (0, \infty)$ in essence, under the assumption of "sup-norm", if

$$\int_0^\infty \left(\int_{\mathbb{R}^3} |V(\bar{r},t)|^2 \, \mathrm{d}\bar{r} \right)^{1/2} \mathrm{d}t < 1.$$

If this criterion isn't satisfied, and/or convergence does not occur, then we can compute a solution over $\mathbb{R}^3 \times (0,T)$, for a finite T, in which case we get convergence if

$$\int_0^T \left(\int_{\mathbb{R}^3} |V(\bar{r},t)|^2 \, \mathrm{d}\bar{r} \right) \mathrm{d}t < 1.$$

However, these are theoretical criteria, wheras we can get convergence even when they are not satisfied. Our approach is to first try successive approximation with $(\bar{r},t)$ on $\mathbb{R}^3 \times (0,\infty)$, and if convergence fails, then to try it for $(\bar{r},t)$ on $\mathbb{R}^3 \times (0,T)$, for several values of T, if necessary, to determin a value of T for which convergence occurs. Convergence will then occur for all values of T', i.e., for $(\bar{r},t) \in \mathbb{R}^3 \times (0,T')$ with $T' < T$. One can prove that if convergence occurs for Sinc approximation which are accurate due to known properties of the potential, then there exists a unique solution to the PDE (see Ref. 7, section 2.6).
10. The approximate solution to the problem (39) is then given at all points $(\bar{r}_{j,k,\ell}, t_n)$, where $\bar{r}_{j,k,\ell} = (j, h_1, k\, h_1, \ell\, h_1)$, and where for the case of t-integration over $\mathbb{R}_+$,

$$t_n = \log\left(\mathrm{e}^{n\, h2} + \sqrt{1 + \mathrm{e}^{2\, n\, h_2}} \right),$$

whereas for t-integration over a finite interval $(0,T)$,

$$t_n = \frac{T \exp(n\, h_2)}{1 + \exp(n\, h_2)},$$

by proceeding as in section 4.4.2 of Ref. 7.

5. EXAMPLES OF COMPUTATIONS

This section presents a few examples for the 3 + 1 Schödinger equation. The derived approximations are based on Sinc convolution calculations of the Schrödinger integral equation. The IE (40) can be written in a short hand operator notation as

$$\Psi = \boldsymbol{N}\Psi + \Phi \tag{53}$$

using the abbreviations $\Psi = \Psi(\bar{r}, t)$ and

$$\Phi = \int_{\mathbb{R}^3} \mathcal{G}(\bar{r} - \bar{r}', t)\, \Psi^0(\bar{r}')\, \mathrm{d}\bar{r}', \tag{54}$$

as well as

$$N\Psi = \int_0^t \int_{\bar{R}^3} \mathcal{G}(\bar{r} - \bar{r}', t - t')\, V(\bar{r}', t')\, \Psi(\bar{r}', t')\, \mathrm{d}\bar{r}'\, \mathrm{d}t'. \tag{55}$$

The operator representation put us in position to write the integral equation in a Neumann iteration scheme as follows

$$\Psi_{n+1} = \boldsymbol{N}\Psi_n + \Phi. \tag{56}$$

The discrete representation is implemented for each coordinate using the relation given in (21). This implementation uses an extension of the convolution given in Algorithm 4.6.7 of Ref. 6 to 4 dimensions. In addition to the convolution algorithm the Laplace transform is used in the four dimensional representation. The final approximation of the Schrödinger equation uses relations (42) and (43). Details of the basis representation are as follows.

As introduced by Schmeisser and Stenger a Sinc approximation of a function f can be given in connection with a Gaussian multiplier in the following representation

$$C_{h,M,N}[f](z) = \sum_{k=-M}^{N} f(z_k)\, \mathrm{Sinc}\left(\frac{\phi(z)}{h} - k\right) \exp\left(-c\left(\frac{\phi(z)}{h} - k\right)^2\right). \tag{57}$$

with c a constant and ϕ denoting a conformal map. This type of approximation allows to represent a function $f(z)$ on an arc Γ with an exponential decaying accuracy.[14] As discussed the approximation works effective for analytic functions. The definition of the Sinc-Gauss basis by

$$\begin{aligned} B_{S/G} &= \left\{ S(k,h) \circ \phi(z) \exp\left(-c \left(\frac{\phi(z) - kh}{h} \right)^2 \right) \right\}_{k=-M}^{N} \\ &= \{ \mathfrak{S}(k,h,c) \circ \phi(z) \}_{k=-M}^{N} \end{aligned} \tag{58}$$

with $\mathfrak{S}(k,h,c)\circ\phi(x) = \mathrm{Sinc}\left(\frac{\phi(x)}{h} - k\right) \mathrm{e}^{-c\left(\frac{\phi(x)}{h} - k\right)^2}$ generated by a product of a Sinc function and a Gaussian allows us to write the approximation of a single variable function f in a compact form as

$$C_{h,c,M,N}[f](z) = \sum_{k=-M}^{N} f(z_k) \mathfrak{S}(k,h,c) \circ \phi(z), \tag{59}$$

where $z_k = \phi^{-1}(kh)$ are the discrete Sinc points kh on $\mathbb{R}$. To represent the basic matrices A_m and B_m of the convolution representation in $4d$, we need to calculate the Töplitz matrix σ_k^c representing a generalization of the corresponding Sinc approximation.[7] Let sinc(x) be given by (3) and σ_k^c and e_k^c be defined as

$$\sigma_k^c = \int_0^k \mathrm{sinc}(x) \exp\left(-cx^2\right) \mathrm{d}x. \tag{60}$$

This put us into position to write

$$e_k^c = \frac{1}{2} \mathrm{erf}\left(\frac{\pi}{2\sqrt{c}} \right) + \mathrm{sng}(j-k) \sigma_{|j-k|}^c, \ k \in \mathbb{Z}. \tag{61}$$

Thus the spectral properties of the approximation are included in the matrices $A_m = \mathrm{h}\, I^{(-1)} D(1/\phi')$ with $I_{j,k}^{(-1)} = e_{k-j}^c$ for a single coordinate. The details of the specific discrete and algorithmic representation will be discussed in a forthcoming paper. Instead of going into the details of the algorithmic implementation we will present a few results demonstrating the effective application of our approach.

5.1 The Harmonic Oscillator in 3 + 1

A well known model in quantum mechanics possessing analytic solutions is the spherical harmonic oscillator.[15] We demonstrate with this example how the method selects stable solutions starting from an initial state which is not an eigenstate of the time independent problem. It is known that for the stationary problem there exists an infinite set of eigenfunctions corresponding each to a specific eigenvalue. To test the convergence of the method to a stable state we examined the problem where the potential $V(\bar{r})$ is given by a three dimensional harmonic well[15,16]

$$V(x,y,z)=\frac{1}{2}\left(x^2+y^2+z^2\right). \tag{62}$$

As initial condition $\Psi^0(\bar{r})$ we use a Gaussian of the following form

$$\Psi^0(x,y,z)=a\,e^{-(x^2+y^2+z^2)} \tag{63}$$

where the amplitude a is a real parameter. The change of the initial condition will mimic the transition from an eigenstate to a dispersing state for $t\in(0,T)$. The boundary conditions on the domain $\mathbb{R}^3\times(0,T)$ are Dirichlet boundary conditions. Note that in our approximation there is no finite limit in the spatial coordinates as frequently used in approximations in quantum mechanics.[17,5] Because a Gaussian is part of the corresponding eigenfunctions of the spherical harmonic oscillator, we expect a convergence to a stable state (eigenstate). In Fig. 4, we present results for such kind of problem as spatial temporal plots and surfaces of constant probability (isoprob surfaces). The observation in Fig. 4 is that there is a symmetric probability distribution in the spatial coordinates. In addition the probability piles up along the temporal axis and approaches the initial conditions at the boundaries $(0,T)$ as expected. The interpretation of such structure is that a particle will be stable with a maximal probability. The plots of Fig. 4 resemble to a stable particle. However, in Fig. 5 where the initial state is not related to the eigenstates; i.e. $\Psi^0=1\cosh^2(r^2)$, of the spherical oscillator a decay of the probability density sets in. The derived solutions demonstrate that they are consistent with the known quantum mechanical behavior such as stability and dispersion of wave packages.

The isoprob surfaces (bottom right panel Fig. 4) show elliptically shaped surfaces of the probability for a fixed value $|\Psi(\bar{r},1/4)|=$ const.. The isoprob surfaces are the surfaces described by Schrödinger in his 1926 paper.[18]

In a second calculation using the harmonic trap potential, we used an initial condition which is not part of the basis functions of the spherical oscillator. However, the used initial is also decaying for large arguments. Specifically we use

$$\Psi^0(x,y,z)=\frac{1}{\cosh^2(x^2+y^2+z^2)}. \tag{64}$$

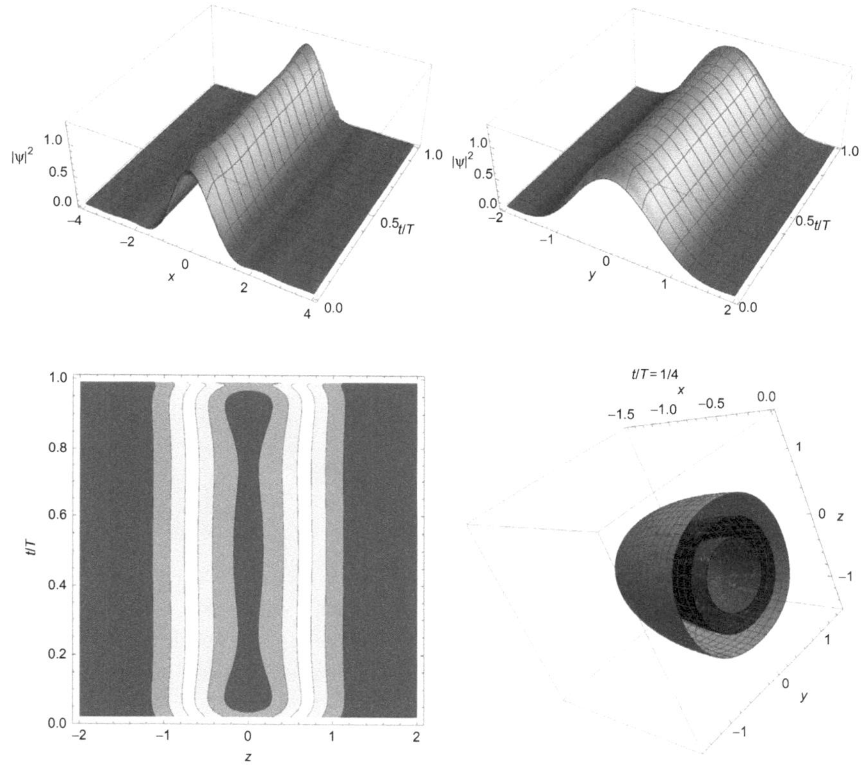

Figure 4 Probability distribution $|\Psi(\bar{r},t)|^2$ in different sections of space and the temporal coordinate. The complementary coordinates not present in the plots are set to zero. The isoprobability surfaces (isoprob surfaces) are calculated at $t/T = 1/4$. These surfaces, here elliptic spheres, were already discussed by Schrödinger in 1926.[18] The calculation uses a factor $a = 5(2/\pi)^{1/4}$ in the initial Gaussian. The number of Sinc points $N = 4$ for temporal and spatial coordinates. The factor c of the basis function was set to $c = 1/100$. The step length is $h = \pi/\sqrt{N}$ in spatial and temporal directions.

The initial function is shown in Fig. 5 for $z = 0$ in the top left panel. As a result of the time evolution of a hyperbolic initial condition, we gain again localized probability densities along the spatial directions. In all directions the probability density first increases in time and then decays on the long run. Such a behavior has to be expected because Schrödinger's equation is dispersive.

This first example demonstrates that spatial temporal solutions for the 3 + 1 Schrödinger equation are directly accessible by applying the Sinc convolution approximation. It becomes apparent that initial conditions which

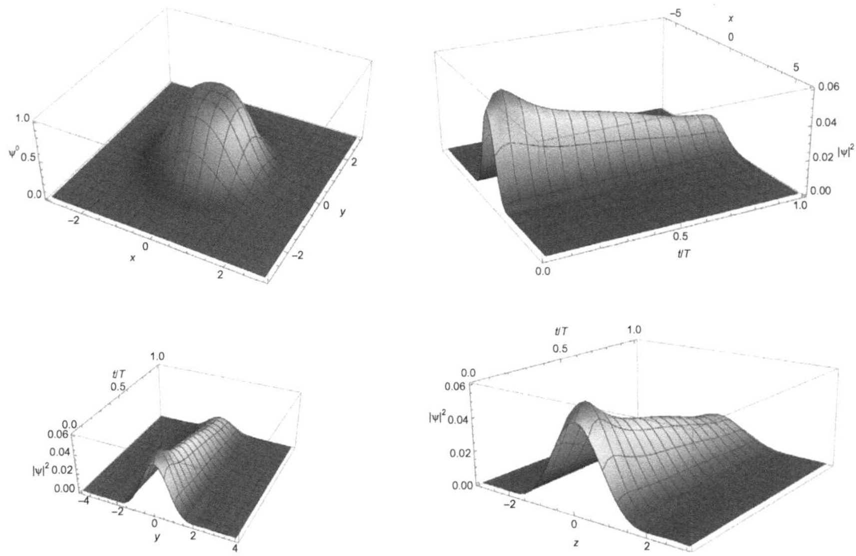

Figure 5 Dispersive behavior of the probability distribution $|\Psi(\bar{r},t)|^2$ in different sections of space along the scaled temporal coordinate. The potential $V(\bar{r}) = (x^2 + y^2 + z^2)/2$ is a harmonic trap. As initial state, a reciprocal hyperbolic function $\Psi^0 = 1/\cosh^2(x^2 + y^2 + z^2)$ is used (top row left panel). The complementary coordinates not present in the plots are set to zero. The number of Sinc points is set to $N = 4$ for temporal and spatial coordinates on $\Omega = \mathbb{R}^3 \times (0,1)$. The approximation was generated with a Gaussian–Sinc basis in all coordinates. The factor c of the basis function was set to $c = 1/100$. The step length is $h = \pi/\sqrt{N}$ in spatial and temporal directions.

are part of the basis functions of the spherical oscillator are stable while other initials disperse. Another example also known as a classical example in a one dimensional setup for the Schrödinger equation is the so called Pöschel-Teller model.

5.2 The Modified Pöschel–Teller Potential in 3 + 1

Pöschel and Teller introduced a potential function using sums of reciprocal squares of trigonometric functions to describe anharmonic oscillators.[19] The interesting point of this model is that it allows an analytic solution and the eigenvalues are related to the harmonic oscillator. Today a modified version of their potential is used in different applications in nonlinear dynamical systems to solve the inverse scatting problem to derive so called soliton solutions. To our best knowledge the higher dimensional problem was not dealt with in quantum mechanics. We will use the modified Pöschel-Teller potential as toy potential in three spatial dimensions. The problem is solved on the domain $\Omega = \mathbb{R}^3 \times (0, \infty)$. The one dimensional problem is discussed at full length by Flügge as a problem possessing analytic solutions.[16,19] The potential V we use is given by the following relation

$$V(x, y, z) = -\frac{\rho^2\lambda(\lambda - 1)}{\cosh^2(\rho(x^2 + y^2 + z^2))}. \tag{65}$$

We again take for our initial condition $\Psi^0(\bar{r})$ a Gaussian of the form

$$\Psi^0(x, y, z) = a(\bar{r})e^{-(x^2 + y^2 + z^2)} \tag{66}$$

where the amplitude $a(\bar{r}) = \cos^2(x^2 + 2(y - 1/2)^2)10$ is chosen to be a real valued function in $\mathbb{R}$. The results for this choice is shown in Fig. 6. The four

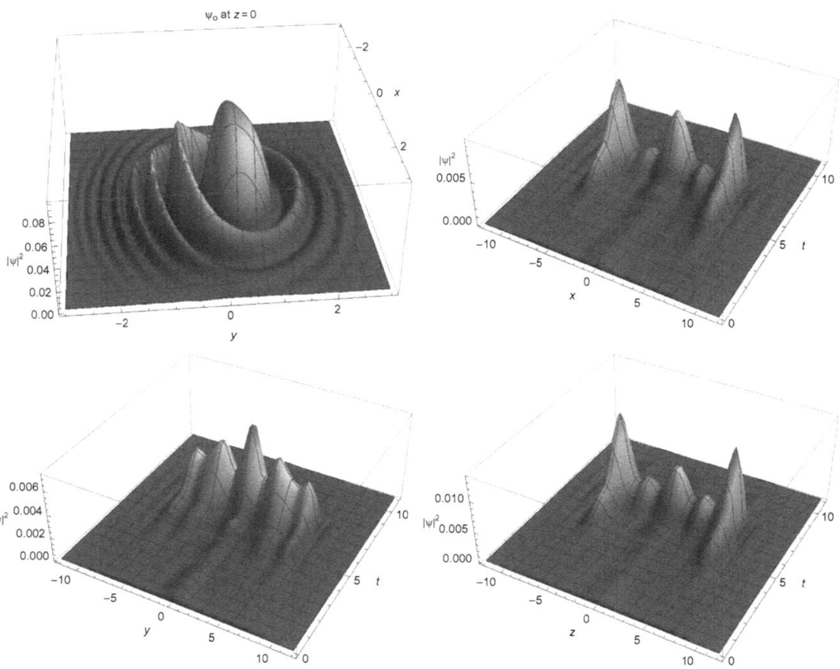

Figure 6 Probability distribution $|\Psi(\bar{r}, t)|^2$ trapped in a Pöschel–Teller potential at different sections of space. The calculations were performed on the domain $\Omega = \mathbb{R}^3 \times (0, \infty)$, only a finite temporal interval is displayed. The top left panel shows the initial wave function $\Psi^0 = \cos^2\left(x^2 + 2(y - 1/2)^2\right)\exp(-(x^2 + y^2 + z^2))/10$. The complementary coordinates not present in the plots are set to zero. The number of Sinc points $N = 4$ for temporal and spatial coordinates. The factor c of the basis function was set to $c = 1/100$, the step length is $h = \pi/\sqrt{N}$ in spatial and temporal directions. The potential parameters were set to $\rho = 1$ and $\lambda = 2$.

panels show the initial distribution (top left panel) and the temporal spatial sections of the probability density on the time interval $(0,T) = (0,11)$. In all density plots the missing coordinates were set to zero. The probability densities are again localized in spatial directions. In all three spatial sections we observe a trace along the time axis at the origin of the spatial coordinate. The density fans out to a set of peaks after a certain time. The peaks look like to drag a part of the probability in the past. The structure disappears completely after a finite life time.

5.3 The Paul Trap a Potential in 3 + 1

The Paul trap originally invented by Paul and coworkers[20–22] is an electric trap to keep single atoms at a fixed location. Paul used that trap primarily for spectroscopic examinations. Nowadays Paul traps are used to trap single atoms and cool them down to zero temperature. An review on Paul traps mainly concerned with the quantum mechanics and the application to different fields in physics is given by Leibfried et al.[23,24] The current example demonstrates the application of the Sinc convolution approximation to a spatial-temporal potential of a Paul trap as introduced by Paul 1958.[22] Paul used his trap to store ions and atoms in a time varying electric field to examine the spectroscopic properties at that time. The time dependent potential Paul used in his experiments is given in scaled coordinates as follows

$$V(x,y,z,t) = \frac{U_0 + V_0 \cos(\Omega\, t)}{r_0^2 + 2z_0^2}\left(x^2 + y^2 - 2z^2\right), \tag{67}$$

where U_0 and V_0 are the dc and ac parts of the trap potential, r_0 is the radius of the ring electrode, and z_0 is half of the distance between the end caps of the trap, and Ω is the ac frequency of the electric field. Introducing dimensionless constants

$$A = \frac{8eU_0}{m\Omega^2\left(r_0^2 + 2z_0^2\right)}, \qquad Q = \frac{4eV_0}{m\Omega^2\left(r_0^2 + 2z_0^2\right)}, \text{and } \tau = \Omega\, t/2, \tag{68}$$

the potential can be written as

$$V(x,y,z,\tau) = \frac{1}{2}(A + 2\, Q\cos(2\,\tau))\left(x^2 + y^2 - 2z^2\right). \tag{69}$$

Use of this type of potential in connection with the initial condition $\Psi^0(\bar{r})$ allows to predict the solutions of the Schrödinger equation. Here Ψ^0 is related to the probability density of a particle of mass m and charge e.

The specific form of the initial condition is taken here as a Gaussian multiplied by a function $a(\bar{r})$ that is given by

$$\Psi^0(x, y, z) = a(\bar{r})e^{-(x^2+y^2+z^2)} \tag{70}$$

where the amplitude $a(\bar{r}) = 4\cos^2(x^2 - 2y^2)$ is chosen to be a real valued function. The Initial condition of this type is shown as an isoprob surface in Fig. 7 for a configuration where the particle is distributed over a single location.

The initial condition Ψ^0 and the Paul potential $v(\bar{r}, \tau)$ is used in the Sinc convolution approximation of the IE (54). The resulting spatial temporal probability distributions for a finite time interval $(0, T)$ are shown in Fig. 8. The probability distribution $|\Psi(\bar{r}, t)|^2$ shows that the particle is localized in all spatial directions. In the time domain the density $|\Psi(\bar{r}, t)|^2$ is nearly constant and representing approximately the initial condition. There is only a slight increase and decrease of the probability density at the beginning and at the end of the time domain. The structure shows that a stable particle exists for the time domain examined.

In another calculation we examined the quantum mechanical behavior of the Paul trap for a parameter set which is unstable in the corresponding mechanical model (see Fig. 9). The expectation is that the solution of Schrödinger's equation will become somehow singular. The calculations demonstrate that starting with a very tiny initial condition of the same type as above given as

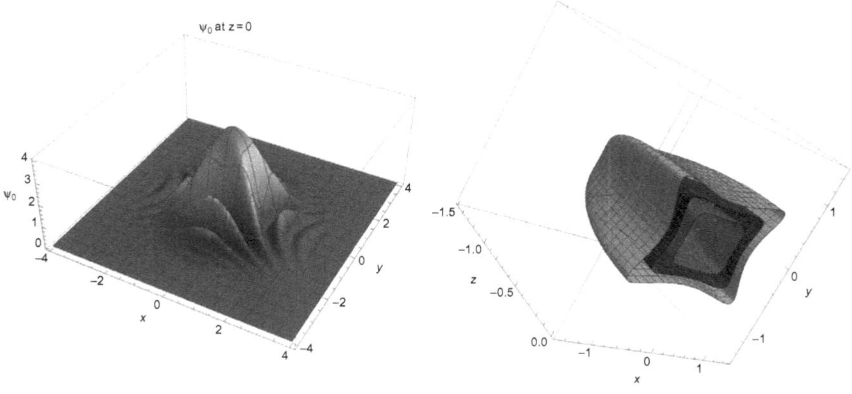

Figure 7 Initial probability distribution $\Psi(\bar{r}, t=0) = \Psi^0(\bar{r}) = a(\bar{r})e^{-(x^2+y^2+z^2)}$ with $a(\bar{r}) = 4\cos^2(x^2 - 2y^2)/10$ (left panel). The right panel shows one half of the total configuration of a single particle of the function Ψ^0 as isoprob surfaces. The "particle" is localized initially.

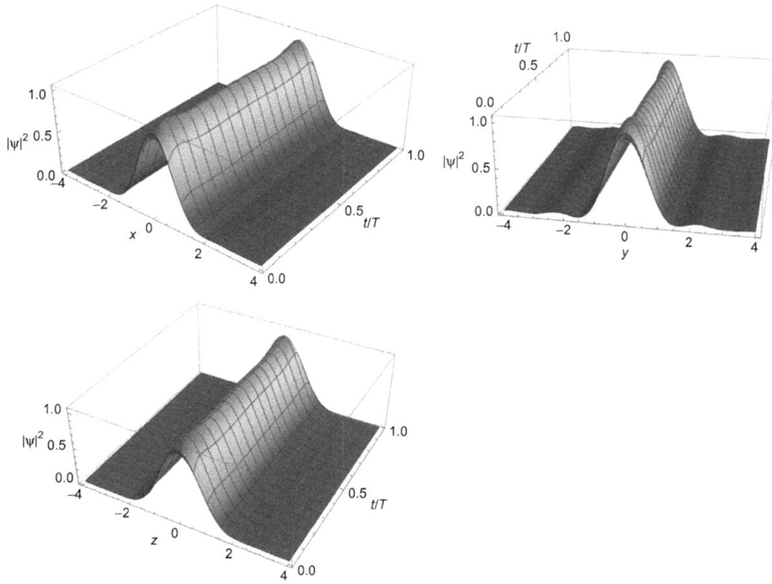

Figure 8 Probability distribution $|\Psi(\bar{r},t)|^2$ (not normalized) of a particle in a Paul trap. The initial distribution is a localized particle as shown in Fig. 7. The particle is trapped in a Paul trap with parameters $A = 1/10$ and $Q = 1$. The number of Sinc points is $N = 4$ for temporal and spatial coordinates. The factor c used in the basis function is $c = 1/100$. The step length is $h = \pi/\sqrt{N}$ in spatial and temporal directions.

$$\Psi^0(x,y,z) = \frac{1}{5000}\cos^2\left(x^2 - 2y^2\right)e^{-(x^2+y^2+z^2)\,4} \tag{71}$$

and the trap parameter chosen in the unstable region with $A = 10^{-3}$ and $Q = 10^{-2}$. The solutions derived under these conditions are singular solutions of the Schrödinger equation in the sense that the life time of the particle is limited. At this point we mention that Sinc methods are able to deal with such kind of solutions in a straight forward and predictable way without any special measures or changes of the algorithm. This kind of insensitivity of the algorithm is a real advantage of the method. The solutions determined under the described conditions are shown in Fig. 9. Note that in Fig. 9 only a domain $B \subset \Omega$ is shown where the probability density is not vanishing.

By these examples computed via use of convolutions based on Sinc approximation we have demonstrated the effectiveness of this method of solution, and in addition, we have demonstrated the existence of novel types of solutions to Schrödinger's equation. These solutions form localized structures in spatial and temporal coordinates. Such localized structures are

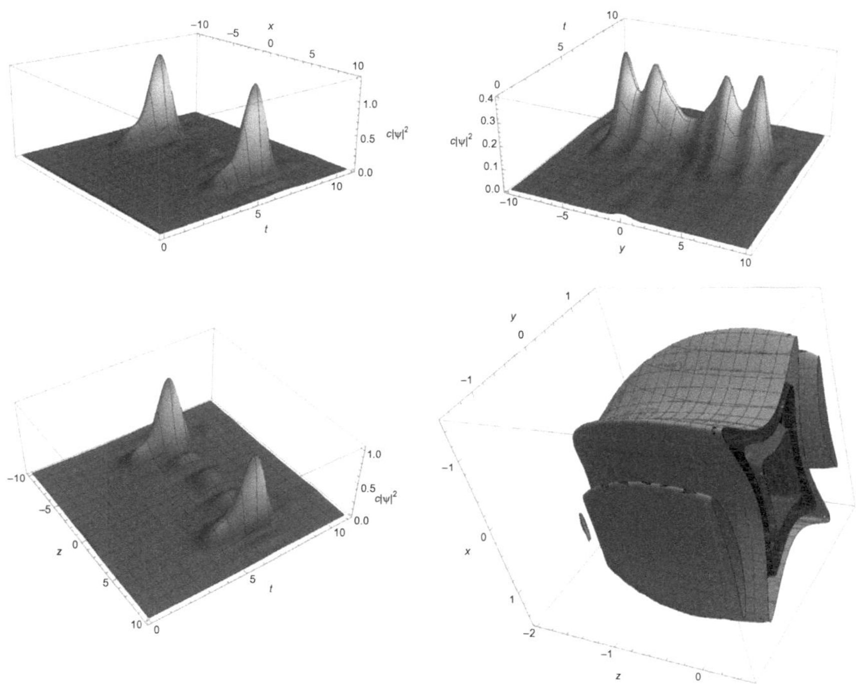

Figure 9 Probability distribution $|\Psi(\bar{r},t)|^2$ (not normalized) of a particle in a Paul trap. The initial distribution is a localized particle similar to that shown in Fig. 7 but with a much smaller amplitude of factor $1/210^{-4}$. The isoprob surfaces of Ψ^0 are shown in the right bottom panel. The solution was calculated on the domain $\Omega = \mathbb{R}^3 \times (0, \infty)$. The subdomain B displayed contains the nonvanishing part of the probability density. The particle is trapped in a Paul trap with parameters $A = 10^{-3}$ and $Q = 10^{-2}$ for a finite time. For this choice of parameters, the mechanical equivalent is unstable.[22] The number of Sinc points is $N = 4$ for temporal and spatial coordinates. The factor c used in the basis function is $c = 1/100$. The step length is $h = \pi/\sqrt{N}$ in spatial and temporal directions.

consistent with the probability interpretation of the density function $\Psi(\bar{r}, t)$. Previously, standard quantum mechanical perturbation methods were used to study spacial and termporal distributions, whereas we can now study such distributions using the convolution methods of this chapter.

REFERENCES

1. Koures, V.; Harris, F. Sinc Collocation in Quantum Chemistry: Solving the Planar Coulomb Schrödinger Equation. *Int. J. Quantum Chem. Symposium* **1996**, *V*(30), 1311–1318.
2. Amore, P. A Variational Sinc Collocation Method for Strong-Coupling Problems. *J. Phys. A Math. Gen.* **2006**, *38*, L349–L355.

3. Ley-Koo, E. Rotations of the Most Asymmetric Molecules via 4-Step and 1-Step Ladder Opertors. To appear.
4. Stenger, F. An Analytic Function Which is an Approximate Characteristic Function. *SIAM J. Numer. Anal.* **1975**, *12*, 239–254.
5. Schulte, M. *Numerical Solution of the Schrödinger Equation on Unbouded Domains;* Inaugural-Dissertation zur Erlangung des akademischen Grades Doktor der Naturwissenshaften im Fachbereich Mathematik und Informatik der Mathematischen Naturwissenschaftlichen Fakultät der Westfälischen Wilhelms-Universität Münster, 2007;1–127, Ph.D. Thesis.
6. Stenger, F. *Numerical Methods Based on Sinc and Analytic Functions*. Springer-Verlag, 1993.
7. Stenger, F. *Handbook of Sinc Numerical Methods;* CRC Press, 2011.
8. Jerke, J. L.; Lee, Y.; Tymczak, C. I. Solving One-Electron System in a Novel Gaussian-Sinc Mixed Basis Set. *Comput. Phys.* **2015**, 1–26, 1405.5073.
9. Berrut, J. P. Barycentrische Formeln zur trigonometrischen Interpolation, (I & II). *Z. Angew. Math. Phys.* **1984**, *35*, 91–105 and 193–205.
10. Han, L.; Xu, J. Proof of Stenger's Conjecture on Matrix $I^{(-1)}$ of Sinc Methods. *J. Comp. Appl. Math.* **2014**, *255*, 805–811.
11. Stenger, F. Numerical Methods Based on the WhittakerCardinal, or Sinc Functions. *SIAM Rev.* **1981**, *23*, 165–224.
12. Koures, V. Solving the Coulomb Shrödinger Equation in $d = 2 + 1$ via Sinc Collocation. *J. Comp. Phys.* **1996**, *128*, 1–5.
13. Arnold, A. Numerically Absorbing Boundary Concitions for Quantum Evolution Equations. *VLSI Des.* **1998**, *6*, 313–319.
14. Schmeisser, G.; Stenger, F. Sinc Approximation with a Gaussian Multiplier. *J. Samp. Th. Sig. Image Poc.* **2007**, *6*, 199–221.
15. Merzbacher, E. *Quantum Mechanics;* Wiley: New York, 1998.
16. Flügge, S. *Practical Quantum Mechanics;* Springer: Berlin, 1999.
17. Vinitsky, S.; Chuluunbaatar, O.; Gerdt, V.; Gusev, A. A.; Rostovtsev, V. Symbolic-Numerical Algorithms for Solving Parabolic Quantum Well Problem with Hydrogen-Like Impurity. In: *Computer Algebra in Scientific Computing*; Gerdt, V., Mayr, E., Vorozhtsov, E. Eds.; Springer: Berlin, 2009, 11th International Workshop, Kobe, Japan, 2009.
18. Schrödinger, E. An Undulatory Theory of the Mechanics of Atoms and Molecules. *Phys. Rev.* **1926**, *28*, 1049–1070.
19. Pöschel, G.; Teller, E. Bemerkungen zur Quantenmechanik des anharmonischen Oszillators. *Z. Phys.* **1933**, *83*, 143–151.
20. Paul, W.; Steinwedel, H. Ein neues Massenspektrometer ohne Magnetfeld. *Z. Naturforsch., A: Phys. Sci.* **1953**, *8*, 448–450.
21. Paul, W.; Raether, M. Das elektrische Massenfilter. *Z. Phys.* **1955**, *140*, 262–273.
22. Paul, W.; Reinhard, H.; von Zahn, U. Das elektrische Massenfilter als Massenspektrometer und Isotopentrenner. *Z. Phys.* **1958**, *152*, 143–182.
23. Leibfried, D.; Blatt, R.; Monroe, C.; Wineland, D. Quantum Dynamics of Single Trapped Ions. *Rev. Mod. Phys.* **2003**, *75*, 281–324.
24. Palmero, M.; Bowler, R.; Gaebler, J.; Leibfried, D.; Muga, J. Fast Transport of Mixed-Species Ion Chains Within a Paul Trap. *Phys. Rev. A.* **2014**, *90*, 053408.
25. Roy, I. G. On Robust Estimation of Discrete Hilbert Transform of Noisy Data. *Geophysics* **2003**, *78*, 239–249.

CHAPTER TWELVE

Approximate Coherent States for Nonlinear Systems

Ricardo Román-Ancheyta*, José Récamier*,[1]

*Instituto de Ciencias Físicas, Universidad Nacional Autónoma de México C.P. 62210 Cuernavaca, Morelos, México

[1]Corresponding author: e-mail address: pepe@fis.unam.mx

Contents

Abstract

On the basis of the f-deformed oscillator formalism, we propose to construct nonlinear coherent states for Hamiltonian systems having linear and quadratic terms in the number operator by means of the following two definitions: (i) as deformed annihilation operator coherent states and (ii) as deformed displacement operator coherent states (DOCS). For the particular cases of the Morse and modified Pöschl-Teller potentials, modeled as f-deformed oscillators (both supporting a finite number of bound states), the properties of their corresponding nonlinear coherent states, viewed as DOCS, are analyzed in terms of their occupation number distribution, their evolution on phase space, and their uncertainty relations.

Advances in Quantum Chemistry, Volume 71
ISSN 0065-3276
http://dx.doi.org/10.1016/bs.aiq.2015.02.006

1. INTRODUCTION

The idea of creating a coherent state for a quantum system was first proposed by Schrödinger, in 1926, in connection with the classical states of the quantum harmonic oscillator; he referred to them as states of the minimum uncertainty product.[1] Almost 40 years later, in 1963 Glauber[2] introduced the field coherent states, these have been realized experimentally with the development of lasers and can be obtained from any one of the following three mathematical definitions: (i) as the right-hand eigenstates of the boson annihilation operator $\hat{a}|\alpha\rangle = \alpha|\alpha\rangle$, with α being a complex number; (ii) as the states obtained by application of the displacement operator $D(\alpha) = \exp(\alpha\hat{a}^\dagger - \alpha^*\hat{a})$ upon the vacuum state; and (iii) as those states with a minimum uncertainty relationship $(\Delta q)^2(\Delta p)^2 = 1/4$, with $q = (\hat{a} + \hat{a}^\dagger)/\sqrt{2}$ and $p = i(\hat{a}^\dagger - \hat{a})/\sqrt{2}$ the position and momentum operators, respectively, and $\Delta q = \Delta p$. The same states are obtained from any one of the three mathematical definitions when one makes use of the harmonic oscillator algebra. At the same time, Klauder[3] developed a set of continuous states in which the basic ideas of coherent states for arbitrary Lie groups were contained. The complete construction of coherent states of Lie groups was achieved by Perelomov[4] and Gilmore[5] almost 10 years later; the basic theme of this development was to connect the coherent states with the dynamical group for each physical problem thus allowing the generalization of the concept of coherent state. For systems far from the ground state or with a finite number of bound states, the harmonic oscillator algebra is no longer adequate, therefore there has arisen an interest in generalizing the concept of coherent state for systems with different dynamical properties.

Nieto and Simmons[6] generalized the notion of coherent states for potentials having unequally spaced energy levels. Their construction is such that the resultant states are localized, follow the classical motion and disperse as little as possible in time. Their coherent states are defined as those which minimize the uncertainty relation equation. They constructed coherent states for the Pöschl-Teller potential, the harmonic oscillator with a centripetal barrier and the Morse potential.

Gazeau and Klauder[7] proposed a generalization for systems with one degree of freedom possessing discrete and continuous spectra. These states present continuity of labeling, resolution of unity, and temporal stability. The key point is to parametrize such states by means of two real values: an amplitude J and a phase γ, instead of using a complex value α.

Man'ko and collaborators[8] introduced coherent states of an *f-deformed* algebra as eigenstates of a deformed annihilation operator $\hat{A} = \hat{a}f(\hat{n})$, where $\hat{a}$ is the usual annihilation operator of the harmonic oscillator algebra and $f(\hat{n})$ is a number operator function that specifies the potential. These states present nonclassical properties such as squeezing and antibunching. More recently, they generalized the displacement operator method for the case of *f-deformed* oscillators introducing a deformed version of the displacement operator[9] with the disadvantage that it is nonunitary and does not displace the deformed operators in the usual form.

On the basis of the f-deformed oscillator formalism, in previous works we put forward the construction of generalized coherent states for nonlinear systems by selecting a specific deformation function in a way such that the energy spectrum of the Hamiltonian it seeks to describe is similar to that of a f-deformed Hamiltonian. In this regard, we examined the trigonometric and the modified Pöschl-Teller potentials, as well as the Morse potential, all of them containing linear and quadratic terms in the number operator.[10] In addition to these, a harmonic oscillator with an inverse square potential in two dimensions was also considered.[11]

Coherent states have been applied not only in optics but in many fields of physics, for instance, in the study of the forced harmonic oscillator,[12] quantum oscillators,[13] quantum nondemolition measurements,[14] nonequilibrium statistical mechanics,[15] and atomic and molecular physics.[16–18] Excellent reviews can be found in Refs. 19–21.

The chapter is organized as follows: In Section 2, we briefly describe the properties of the usual coherent states. In Section 3, we present a methodology to construct deformed Hamiltonians for specific systems. In Section 4, we give two alternative forms to obtain the coherent states pertinent to a nonlinear Hamiltonian. And finally, in Section 5, we present some numerical results.

2. HARMONIC OSCILLATOR OR FIELD COHERENT STATES

The harmonic oscillator coherent states, also called *field coherent states*,[2] are quantum states of minimum uncertainty product which most closely resemble the classical ones in the sense that they remain well localized around their corresponding classical trajectory. When one makes use of the harmonic oscillator algebra, the same coherent states are obtained from the three Glauber's mathematical definitions mentioned above. Such states take the following form in terms of the number state basis:

$$|\alpha\rangle = e^{-\frac{1}{2}|\alpha|^2} \sum_{n=0}^{\infty} \frac{\alpha^n}{\sqrt{n!}} |n\rangle, \tag{1}$$

where α is an arbitrary complex number. For every complex number $\alpha \neq 0$, the coherent state $|\alpha\rangle$ has a nonzero projection on every Fock state $|n\rangle$.

From a statistical point of view, it follows from the above that the occupation number distribution of a coherent state, $P_\alpha(n) = |\langle n|\alpha\rangle|^2$, is characterized by a Poisson distribution:

$$P_\alpha(n) = e^{-|\alpha|^2} \frac{|\alpha|^{2n}}{n!}, \tag{2}$$

with an average occupation number $\bar{n} = |\alpha|^2$ and mean square root deviation

$$\Delta N = \sqrt{\langle\alpha|N^2|\alpha\rangle - \langle\alpha|N|\alpha\rangle^2} = |\alpha| = \sqrt{\bar{n}}.$$

Such states are not orthogonal, and the overlap between two of them is given by

$$\langle\alpha|\beta\rangle = e^{-\frac{1}{2}|\beta - \alpha|^2} e^{\frac{1}{2}(\beta^*\alpha - \beta\alpha^*)} \tag{3}$$

Moreover, they form an overcomplete basis with the closure relationship,

$$\mathcal{I} = \frac{1}{\pi} \int |\alpha\rangle\langle\alpha| d^2\alpha, \tag{4}$$

so that any state of the group generated by the operators $\hat{a}$, $\hat{a}^\dagger$ and $\hat{n} = \hat{a}^\dagger \hat{a}$ can be expanded in terms of coherent states.

3. DEFORMED OSCILLATORS

A deformed oscillator is a nonharmonic system characterized by a Hamiltonian of the harmonic oscillator form:

$$\hat{H}_D = \frac{\hbar\Omega}{2}\left(\hat{A}\hat{A}^\dagger + \hat{A}^\dagger\hat{A}\right), \tag{5}$$

with a specific frequency Ω and deformed boson creation $\hat{A}^\dagger$ and annihilation $\hat{A}$ operators defined as[8]:

$$\hat{A}^\dagger = f(\hat{n})\hat{a}^\dagger = \hat{a}^\dagger f(\hat{n}+1), \qquad \hat{A} = \hat{a}f(\hat{n}) = f(\hat{n}+1)\hat{a}, \tag{6}$$

with $f(\hat{n})$ a number operator function that specifies the deformation. The commutation relations between the deformed operators are:

$$[\hat{A},\hat{A}^{\dagger}]=(\hat{n}+1)f^2(\hat{n}+1)-\hat{n}f^2(\hat{n}),\quad [\hat{A},\hat{n}]=\hat{A},\quad [\hat{A}^{\dagger},\hat{n}]=-\hat{A}^{\dagger}. \tag{7}$$

Notice that the commutation relations between the deformed operators $\hat{A}$, $\hat{A}^{\dagger}$ and the number operator are the same as those among the usual operators $\hat{a}$, $\hat{a}^{\dagger}$ with the number operator. Note that the eigenfunctions of the harmonic oscillator $|n\rangle$ are also eigenfunctions of the deformed oscillator. In terms of the number operator, the deformed Hamiltonian takes the form:

$$\hat{H}_{\mathrm{D}}=\frac{\hbar\Omega}{2}\left(\hat{n}f^2(\hat{n})+(\hat{n}+1)f^2(\hat{n}+1)\right). \tag{8}$$

3.1 Algebraic Hamiltonian for the Morse Potential

The Morse oscillator is a particularly useful anharmonic potential for the description of systems that deviate from the ideal harmonic oscillator conduct and has been used widely to model the vibrations of a diatomic molecule. If we choose the deformation function[22]:

$$f^2(\hat{n})=1-\chi_{\mathrm{a}}\hat{n}, \tag{9}$$

with χ_{a} an anharmonicity parameter and substitute it into Eq. (8), we obtain an algebraic Hamiltonian of the form:

$$\hat{H}_{\mathrm{D}}=\hbar\Omega\left[\hat{n}+\frac{1}{2}-\chi_{\mathrm{a}}\left(\hat{n}+\frac{1}{2}\right)^2-\frac{\chi_{\mathrm{a}}}{4}\right], \tag{10}$$

whose spectrum is in essence identical to that of the Morse potential,[23] i.e.,

$$E_{\mathrm{n}}=\hbar\omega_{\mathrm{e}}\left(n+\frac{1}{2}\right)-\frac{\hbar\omega_{\mathrm{e}}}{2N+1}\left(n+\frac{1}{2}\right)^2, \tag{11}$$

provided that $\omega_{\mathrm{e}}=\Omega$ and $\chi_{\mathrm{a}}=1/(2N+1)$, with N being the number of bound states corresponding to the integers $0\leq n\leq N-1$. For this particular choice of deformation function, the commutator between deformed operators is:

$$[\hat{A},\hat{A}^{\dagger}]=1-\chi_{\mathrm{a}}(2\hat{n}+1)=\frac{2(N-\hat{n})}{2N+1}, \tag{12}$$

from which one can see that the commutator equals a scalar plus a linear function of the number operator.

The action of the deformed operators upon the number states is:

$$\hat{A}|n\rangle = \sqrt{n}f(n)|n-1\rangle = \sqrt{n(1-\chi_a n)}|n-1\rangle \tag{13}$$

and

$$\hat{A}^\dagger|n\rangle = \sqrt{n+1}f(n+1)|n+1\rangle = \sqrt{(n+1)(1-\chi_a(n+1))}|n+1\rangle. \tag{14}$$

The deformed operators change the number of quanta in ± 1 and their corresponding matrix elements depend on the deformation function $f(n)$. Furthermore, from the commutation relations we see that the set $\{\hat{A}, \hat{A}^\dagger, \hat{n}, 1\}$ is closed under the operation of commutation.

3.2 Algebraic Hamiltonian for the Modified Pöschl-Teller Potential

The modified Pöschl-Teller potential can be written as:

$$V(x) = U_0 \tanh^2(ax) \tag{15}$$

where U_0 is the depth of the well, a is the range of the potential, and x is the relative distance from the equilibrium position. The eigenfunctions and eigenvalues are[23]

$$\psi_n^\epsilon(\zeta) = N_n^\epsilon (1-\zeta^2)^{\epsilon/2} F(-n, \epsilon+s+1; \epsilon+1, (1-\zeta)/2) \tag{16}$$

and

$$E_n = U_0 - \frac{\hbar^2 a^2}{8\mu}\left(-(2n+1) + \sqrt{1+\frac{8\mu U_0}{\hbar^2 a^2}}\right)^2, \tag{17}$$

respectively, where N_n^ϵ is a normalization constant, $\zeta = \tanh(ax)$, μ is the reduced mass of the molecule, s is related with the depth of the well so that $s(s+1) = 2\mu U_0/(\hbar^2 a^2)$, $\epsilon = \sqrt{-2\mu(E-U_0)}/\hbar a$, and $F(a,b;c,z)$ stands for the hypergeometric function.[24] If we write the eigenvalues in terms of the parameter s, we obtain

$$E_n = \frac{\hbar^2 a^2}{2\mu}\left(s + 2sn - n^2\right). \tag{18}$$

The number of bound states is determined by the dissociation limit $\epsilon = s - n_{max} = 0$. For integer values of s, the state associated with null energy is not normalizable, whereby the last bound state corresponds to $n_{max} = s - 1$.[25] The modified Pöschl-Teller potential is a nonlinear potential symmetric in the displacement coordinate.

Let us now consider a deformation function of the form[10]:

$$f^2(\hat{n}) = \frac{\hbar a^2}{2\mu\Omega}(2s + 1 - \hat{n}). \tag{19}$$

By substituting it into Eq. (8), we obtain the deformed Hamiltonian

$$\hat{H}_{\mathrm{D}} = \frac{\hbar^2 a^2}{2\mu}\left(-\hat{n}^2 + 2s\hat{n} + s\right), \tag{20}$$

whose spectrum is identical to that of Eq. (18). The harmonic limit is obtained by taking $s \to \infty$, $a \to 0$ with $sa^2 \to \mu\Omega/\hbar$.

For this choice of deformation function, deformed operators, together with the number operator, obey the following commutation relations:

$$[\hat{A}, \hat{A}^\dagger] = \frac{\hbar a^2}{\mu\Omega}(s - \hat{n}), \quad [\hat{A}, \hat{n}] = \hat{A}, \quad [\hat{A}^\dagger, \hat{n}] = -\hat{A}^\dagger, \tag{21}$$

which have the correct harmonic limit and are similar to those of the generators of the *SU*(2) group.

3.3 Algebraic Hamiltonian for the Trigonometric Pöschl-Teller Potential

The trigonometric Pöschl-Teller potential is given by:

$$V(x) = U_0 \tan^2(ax), \tag{22}$$

where U_0 is the strength of the potential and a is its range. This potential supports an infinite number of bound states, its eigenfunctions, and eigenvalues are[6]:

$$\psi_{\mathrm{n}}^{\lambda}(x) = \sqrt{\frac{a(\lambda + n)\Gamma(2\lambda + n)}{\Gamma(n+1)}}(\cos(ax))^{1/2} P^{1/2-\lambda}_{n+\lambda-1/2}(\sin(ax)), \tag{23}$$

$$E_{\mathrm{n}} = \frac{\hbar^2 a^2}{2\mu}(n^2 + 2\lambda n + \lambda) = \hbar\omega\left(n + \frac{1}{2} + \frac{n^2}{2\lambda}\right), \tag{24}$$

where μ is the mass of the particle, $\omega = \hbar\lambda a^2/\mu$ and the parameter λ is related to the potential strength and range by $\lambda(\lambda - 1) = 2\mu U_0/\hbar^2 a^2$. In the harmonic limit $\lambda \to \infty$ and $a \to 0$ with $\lambda a^2 = \mu\omega/\hbar$.

If we choose a deformation function[11]

$$f^2(\hat{n}) = \frac{\hbar a^2}{2\mu\Omega}(\hat{n} + 2\lambda - 1), \tag{25}$$

the deformed Hamiltonian becomes

$$\hat{H}_{\mathrm{D}} = \frac{\hbar^2 a^2}{2\mu}\left(\hat{n}^2 + 2\lambda\hat{n} + \lambda\right), \tag{26}$$

whose eigenvalues are identical to those given by Eq. (24). Once we have given the deformation function, the deformed operators are specified. In this case, the commutation relations are:

$$[\hat{A}, \hat{A}^\dagger] = \frac{\hbar a^2}{\mu\Omega}(\hat{n} + \lambda), \quad [\hat{A}, \hat{n}] = \hat{A}, \quad [\hat{A}^\dagger, \hat{n}] = -\hat{A}^\dagger, \tag{27}$$

which are similar to those of the generators of the $SU(1,1)$ group. We emphasize here that the set of operators $\{\hat{A}, \hat{A}^\dagger, \hat{n}, 1\}$ is closed under the operation of commutation.

4. NONLINEAR COHERENT STATES

In this work, we consider the generalization of two of the known definitions given to construct the field coherent states, namely, (i) as eigenstates of an annihilation operator and (ii) as the states obtained from the application of the displacement operator upon a maximal state.

4.1 Coherent States as Eigenstates of the Deformed Annihilation Operator

Following Man'ko and collaborators, we construct deformed coherent states as eigenstates of the deformed annihilation operator[8]:

$$\hat{A}|\alpha, f\rangle = \alpha|\alpha, f\rangle. \tag{28}$$

In order to solve Eq. (28), let the state $|\alpha, f\rangle$ be written as a weighted superposition of the number eigenstates $\{|0\rangle, |1\rangle, \ldots, |n\rangle, \ldots\}$:

$$|\alpha, f\rangle = N_f \sum_{n=0}^{\infty} c_n^f |n\rangle. \tag{29}$$

On inserting this into Eq. (28), we get the following relation between the coefficients c_n^f and c_{n-1}^f:

$$c_n^f f(n)\sqrt{n} = \alpha c_{n-1}^f. \tag{30}$$

By applying n times the annihilation operator, we get a relationship between c_n^f and c_0^f:

$$c_n^f f(n)!\sqrt{n!} = \alpha^n c_0^f, \tag{31}$$

where $f(n)! = f(n)f(n-1)\cdots f(0)$. Substitution of Eq. (31) into Eq. (28) gives us the following expression for the Annihilation Operator Coherent States (AOCS):

$$|\alpha, f\rangle = N_f \sum_{n=0}^{\infty} \frac{\alpha^n}{\sqrt{n!}f(n)!}|n\rangle. \tag{32}$$

As an example, we replaced the explicit form of the deformation function for the trigonometric Pöschl-Teller potential and obtained the AOCS associated with this system

$$|\alpha, f\rangle = N_f \sum_{n=0}^{\infty} \alpha^n \sqrt{\frac{(2\lambda)^n \Gamma(2\lambda)}{n!\Gamma(2\lambda + n)}}|n\rangle \tag{33}$$

with $N_f = [{}_0F_1(2\lambda; 2\lambda|\alpha|^2)]^{-1/2}$ a normalization constant.

4.2 Coherent States Obtained via the Deformed Displacement Operator

In this section, we construct the Displacement Operator Coherent States (DOCS) for the nonlinear potentials discussed in Section 3. We propose to construct them by application of a deformed displacement operator on the fundamental state of the system, i.e.,

$$|\zeta(\alpha)\rangle = \exp[\alpha\hat{A}^\dagger - \alpha^*\hat{A}]|0\rangle \tag{34}$$

where α is a complex parameter. The deformed displacement operator is obtained from the usual one through the replacement of the harmonic oscillator creation and annihilation operators by their deformed counterparts. Since the commutator between the deformed operators can be a rather complicated function of the number operator, it is not possible, in general, to disentangle the exponential in Eq. (34). However, for the cases we are considering in this work, we have found that the commutator between the deformed operators $\hat{A}$, $\hat{A}^\dagger$ is equal to a scalar plus a linear function of the number operator, and therefore the dynamical group pertinent to these systems is composed by the operators $\{\hat{A}^\dagger, \hat{A}, \hat{n}, 1\}$ which form a Lie algebra. So, the deformed displacement operator can be disentangled[26,27] to get (for the particular case of a Morse potential)[28]

$$\begin{aligned}\hat{D}_f(\alpha) &= \exp[\alpha\hat{A}^\dagger - \alpha^*\hat{A}], \\ &= \exp\left(\zeta\frac{\hat{A}^\dagger}{\sqrt{\chi_a}}\right)\left(\frac{1}{1+|\zeta|^2}\right)^{g(\chi_a\hat{n})/2\chi_a}\exp\left(-\zeta^*\frac{\hat{A}}{\sqrt{\chi_a}}\right),\end{aligned} \tag{35}$$

where, for a given value of $\alpha = |\alpha|e^{i\phi}$, the complex parameter $\zeta = e^{i\varphi}\tan\left(|\alpha|\sqrt{\chi_a}\right)$ is introduced, and $g(\chi_a\hat{n}) = [\hat{A}, \hat{A}^\dagger]$. The DOCS for this particular case can be obtained as:

$$|\zeta\rangle = \hat{D}_f(\alpha)|0\rangle \simeq \sum_{n=0}^{N-1}\binom{2N}{n}^{1/2}\frac{\zeta^n}{(1+|\zeta|^2)^N}|n\rangle \tag{36}$$

where the explicit form of the deformation function has been used and the state is approximate due to the fact that the number of bound states supported by the potential is finite.

5. NUMERICAL RESULTS

5.1 Morse Nonlinear Coherent States, Phase Space Trajectories, and Occupation Number Distribution

In this section, we consider a heteronuclear diatomic molecule HF with 22 bound states. Due to the asymmetry of the molecule, we model it by means of a Morse potential. According to Carvajal et al.,[29] the position and momentum variables for this system can be expressed as a series expansion involving all powers of the deformed creation and annihilation operators. Keeping up to second-order terms we obtain:

$$\hat{x}_D \simeq \sqrt{\frac{\hbar}{2\mu\Omega}}\left(f_{00} + f_{10}\hat{A}^\dagger + \hat{A}f_{01} + f_{20}\hat{A}^{\dagger 2} + \hat{A}^2 f_{02}\right) \tag{37}$$

$$\hat{p}_D \simeq i\sqrt{\frac{\hbar\mu\Omega}{2}}\left(g_{10}\hat{A}^\dagger + \hat{A}g_{01} + g_{20}\hat{A}^{\dagger 2} + \hat{A}^2 g_{02}\right) \tag{38}$$

where the expansion coefficients are functions of the number operator and are given by[30,31]:

$$f_{00}(\hat{n}) = \sqrt{k}\left[f_0 + \ln\left(\frac{(k-2)(k-\hat{n}-1)}{(k-1-2\hat{n})(k-2\hat{n})}\right)(1-\delta_{\hat{n},0})\right] \tag{39}$$

$$f_{10}(\hat{n}) = f_{01}(\hat{n}) = \sqrt{\frac{k-1}{k}}\left(1+\frac{\hat{n}}{k-\hat{n}}\right) \tag{40}$$

$$f_{20}(\hat{n}) = f_{02}(\hat{n}) = \frac{k-1}{2k\sqrt{k}}\left(\frac{-1}{(1-(\hat{n}-1)/k)(1-\hat{n}/k)}\right), \tag{41}$$

$$g_{10}(\hat{n}) = -g_{01}(\hat{n}) = \sqrt{\frac{k-1}{k}}\left(\frac{k-2\hat{n}}{k-\hat{n}}\right), \tag{42}$$

$$g_{20}(\hat{n}) = -g_{02}(\hat{n}) = -\frac{k-1}{k\sqrt{k}}\left(\frac{k-(2\hat{n}-1)}{k(1-(\hat{n}-1)/k)(1-\hat{n}/k)}\right), \tag{43}$$

where in turn,

$$f_0 = \ln k - \left(\sum_{p=1}^{k-2}\frac{1}{p} - \Gamma\right), \tag{44}$$

with

$$\Gamma = \lim_{m\to\infty}\left(\sum_{p=1}^{m}\frac{1}{p} - \ln m\right) = 0.577216 \tag{45}$$

being the Euler constant and k is Child's parameter[32] defined by $k=2N+1=1/\chi_a$.

In what follows, we define the deformed position and momentum operators $\hat{x}_D$ and $\hat{p}_D$ taking $\hbar=\mu=\Omega=1$ in Eqs. (37) and 38. When the deformation function is equal to 1 ($\chi_a=0$), these expressions take the harmonic values. In Ref. 30, we compared the phase space trajectories obtained by averaging the deformed coordinate $\hat{x}_D$ and momentum $\hat{p}_D$ (keeping up to second order terms) with those obtained averaging the Morse coordinate and momentum and we found a very good agreement between them when the averages were taken between nonlinear coherent states obtained as eigenstates of the deformed annihilation operator (AOCS). In order to evaluate the phase space trajectories, here, we will consider the temporal evolution of the nonlinear coherent states obtained by application of the deformed displacement operator on the vacuum state (see Eq. 36). That is, we apply the time evolution operator $\hat{U}(t)=e^{-i\hat{H}_D t/\hbar}$ on the state $|\zeta\rangle$.

$$|\zeta;t\rangle = \hat{U}(t)|\zeta\rangle = e^{-i\Omega t\left[\hat{n}+\frac{1}{2}-\chi_a\left(\hat{n}+\frac{1}{2}\right)^2-\frac{\chi_a}{4}\right]}|\zeta\rangle. \tag{46}$$

The averages are:

$$\langle \zeta, t | \hat{x}_{\rm D} | \zeta, t \rangle = \langle \zeta | \hat{U}(t)^{\dagger} \hat{x}_{\rm D} \hat{U}(t) | \zeta \rangle, \quad \langle \zeta, t | \hat{p}_{\rm D} | \zeta, t \rangle = \langle \zeta | \hat{U}(t)^{\dagger} \hat{x}_{\rm D} \hat{U}(t) | \zeta \rangle. \tag{47}$$

Transformation of the deformed operators yield:

$$\begin{aligned} \hat{U}(t)^{\dagger} \hat{A} \hat{U}(t) &= e^{i\Omega t[\hat{n} - \chi_{\rm a}(\hat{n}^2 + \hat{n})]} \hat{A} e^{-i\Omega t[\hat{n} - \chi_{\rm a}(\hat{n}^2 + \hat{n})]} \\ &= e^{-i\Omega \chi_{\rm a} t \hat{n}^2} e^{i\Omega t(1-\chi_{\rm a})\hat{n}} \hat{A} e^{-i\Omega t(1-\chi_{\rm a})\hat{n}} e^{i\Omega \chi_{\rm a} t \hat{n}^2} \\ &= e^{-i\Omega(1-\chi_{\rm a})t} e^{-i\Omega \chi_{\rm a} t \hat{n}^2} \hat{A} e^{i\Omega \chi_{\rm a} t \hat{n}^2} = e^{-i\Omega t(1-2\chi_{\rm a})} e^{-2i\Omega \chi_{\rm a} t \hat{n}} \hat{A} \end{aligned} \tag{48}$$

and

$$\hat{U}(t)^{\dagger} \hat{A}^{\dagger} \hat{U}(t) = e^{i\Omega t(1-2\chi_{\rm a})} \hat{A}^{\dagger} e^{2i\Omega \chi_{\rm a} t \hat{n}}, \tag{49}$$

notice the presence of the number operator in the exponentials. From these expressions, we can get $\hat{A}^2(t) = \hat{U}(t)^{\dagger} \hat{A}^2 \hat{U}(t)$ and $\hat{A}^{\dagger 2}(t) = \hat{U}(t)^{\dagger} \hat{A}^{\dagger 2} \hat{U}(t)$ and obtain the deformed coordinate and momentum as a function of time.

In Fig. 1, we show the occupation number distributions $P_{\rm n}(\alpha) = |\langle n | \zeta(\alpha) \rangle|^2$ (frames A, C, and E) and the corresponding phase space trajectories (frames B, D, and F) for $\langle \hat{n} \rangle = 0.2$, 2, and 4. In the calculation, a diatomic HF molecule is considered to be modeled by a deformed Morse-like oscillator with $N = 22$ bound states. It can be seen from the figure that for the values used for the parameter α, which fixes the average value of the number operator, the occupation number distributions are such that the states near dissociation are mostly unoccupied so it is to be expected that the contribution from the states in the continuum can be neglected. Concerning the phase space trajectories we see that for a given value of the average number operator, that is, a given value of the average energy, there are several intersecting curves in contrast with the case of a field coherent state where one finds a single trajectory for a given energy. The presence of several trajectories for a given energy is a signature of the nonlinear term in the deformed Hamiltonian. Notice that for a small energy (frame B) the phase space trajectories fill an anular region of phase space with an unaccessible internal region. The width of the anular region narrows when the parameter $|\alpha|$ is decreased, that is, for smaller values of the average number operator. This unaccessible region is lost as the energy is increased (frames D and F). Due to the asymmetry of the Morse potential the deformed coordinate can attain small negative values and much larger positive values along the temporal evolution.

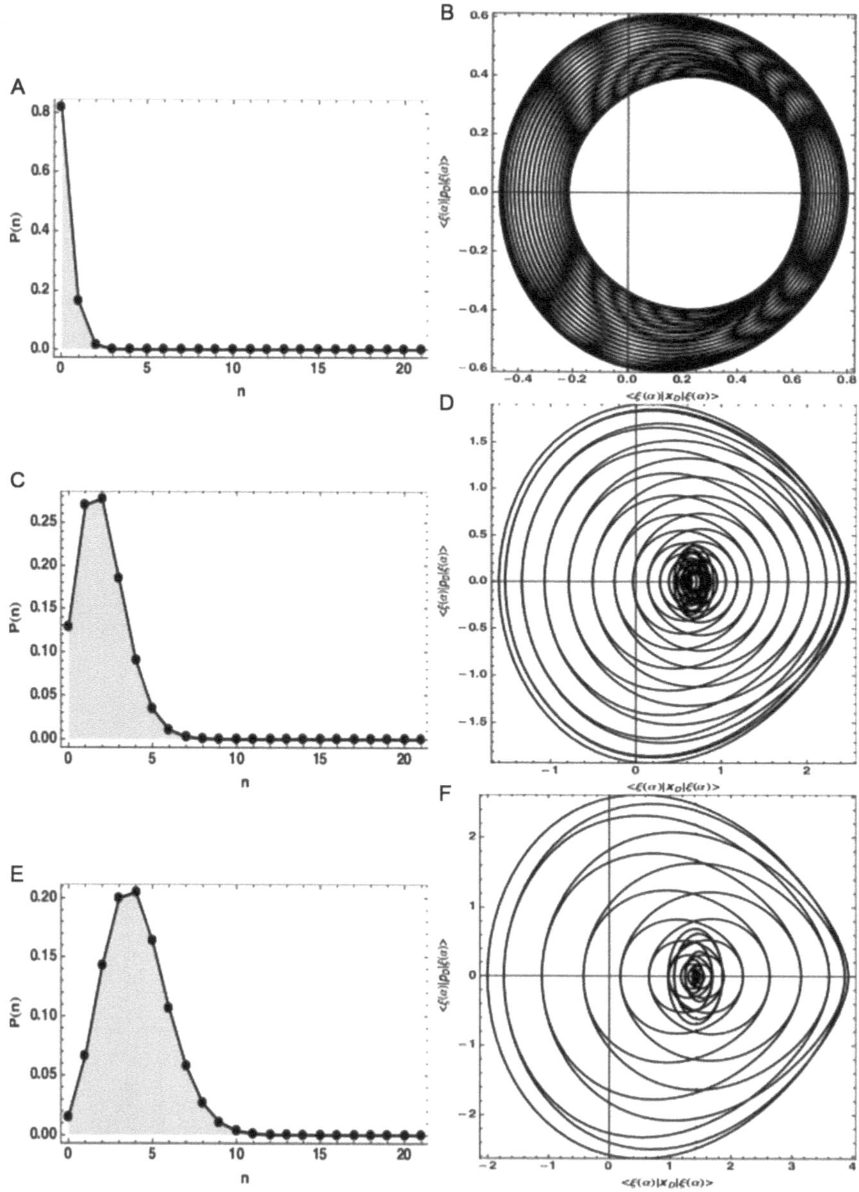

Figure 1 Occupation number distributions (frames A, C, and E) and the corresponding phase space trajectories (frames B, D, and F) of deformed displacement operator coherent states (DOCS) for $\langle \hat{n} \rangle = 0.2$, 2, and 4. In the calculation, a diatomic HF molecule is considered to be modeled by a deformed Morse-like oscillator with $N = 22$ bound states.

In Fig. 2, we show the temporal evolution of the uncertainties in coordinate (frames A, C and E) and momentum (frames B, D and F) of deformed displacement operator coherent states (DOCS) for $\langle\hat{n}\rangle = 0.2$, 2, and 4. The first row corresponds to the case with lowest energy considered $\langle\hat{n}\rangle = 0.2$, the dispersions in the deformed position and momentum are oscillating functions with amplitudes in the range $0.3 \leq \langle(\Delta\hat{O})^2\rangle \leq 0.9$ with $\hat{O} = \hat{x}_D, \hat{p}_D$. Notice that the DOCS are not minimum uncertainty states and notice also that there is squeezing present whenever the dispersion in any coordinate is smaller than 0.5. The second row corresponds to $\langle\hat{n}\rangle = 2$, here the dispersions oscillate with larger amplitude $0.2 \leq (\Delta\hat{O})^2 \leq 4.8$. And finally, in the third row, $\langle\hat{n}\rangle = 4$, the dispersions oscillate with amplitude in the range

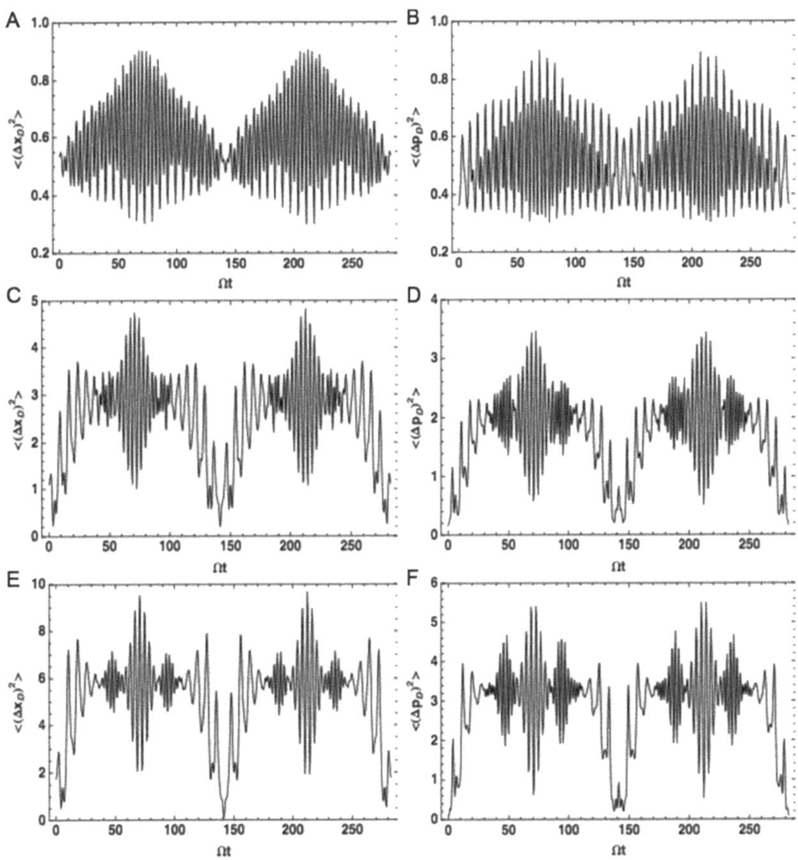

Figure 2 Temporal evolution of the uncertainties in coordinate (frames A, C, and E) and momentum (frames B, D, and F) of deformed displacement operator coherent states (DOCS) for $\langle\hat{n}\rangle = 0.2$, 2, and 4. In the calculation, a diatomic HF molecule is considered to be modeled by a deformed Morse-like oscillator with $N = 22$ bound states.

$0.1 \leq (\Delta\hat{O})^2 \leq 10$. In the cases with $\langle\hat{n}\rangle = 2, 4$, the amplitude of the oscillations in the dispersion in the coordinate is larger than that in the momentum.

In Fig. 3, we show the temporal evolution of the normalized uncertainty product $\Delta_{xp} = 4\langle(\Delta\hat{x}_{\mathrm{D}})^2\rangle\langle(\Delta\hat{p}_{\mathrm{D}})^2\rangle / |\langle[\hat{x}_{\mathrm{D}}, \hat{p}_{\mathrm{D}}]\rangle|^2$ of displacement operator coherent states for $\langle\hat{n}\rangle = 0.2$, 2, and 4 (frames A, B, and C, respectively). We see that there are some specific times at which the dispersion is minimal and these times do not depend on the value of $\langle\hat{n}\rangle$, these correspond to the outermost trajectories shown in Fig. 1. At those times when the trajectories evolve near the origin, so that $\langle\hat{x}_{\mathrm{D}}\rangle$ and $\langle\hat{p}_{\mathrm{D}}\rangle$ are small, the dispersions take their largest values.[30] Most of the time the DOCS are not minimum uncertainty states, the product of the dispersions seems to be a periodic function of time and the amount of dispersion is an increasing function of $\langle\hat{n}\rangle$.

In Fig. 4, we show the dispersions in coordinate and momentum at time $t=0$ as a function of the parameter $|\alpha|$. Here, we see that the dispersion in the deformed coordinate $\hat{x}_{\mathrm{D}}$ starts as that of a minimum uncertainty state and is an increasing function of the parameter $|\alpha|$. The dispersion in the momentum also starts as that of a minimum uncertainty state and is a decreasing function of the parameter $|\alpha|$ so it is squeezed. The normalized uncertainty product remains near a minimum uncertainty state for small values of $|\alpha|$. However, for values of $|\alpha|$ such that the average value of the number operator approaches N the uncertainty product increases rapidly.

5.2 Modified Pöschl-Teller Nonlinear Coherent States, Phase Space Trajectories, and Occupation Number Distribution

In this section, we consider a homonuclear diatomic molecule H_2 supporting 10 bound states. Due to the symmetry of the potential, the deformed coordinate and momentum are written as an expansion involving odd powers of deformed operators[25] as:

$$\hat{X}_{\mathrm{D}} = \sqrt{\frac{\hbar}{2\mu\omega}}\left(\hat{A}F(\hat{n}) + F(\hat{n})\hat{A}^{\dagger} + \hat{A}^{3}G(\hat{n}) + G(\hat{n})\hat{A}^{\dagger 3}\right) \tag{50}$$

$$\hat{P}_{\mathrm{D}} = -i\sqrt{\frac{\hbar\mu\omega}{2}}\left(\hat{A}R(\hat{n}) - R(\hat{n})\hat{A}^{\dagger} + \hat{A}^{3}S(\hat{n}) - S(\hat{n})\hat{A}^{\dagger 3}\right) \tag{51}$$

where we have kept up to third order terms in the deformed operators and the coefficient functions are given by:

$$F(n) = \sqrt{\frac{2\mu\omega}{\hbar}}\frac{\langle n-1|\hat{x}|n\rangle}{f(n)\sqrt{n}}, \tag{52}$$

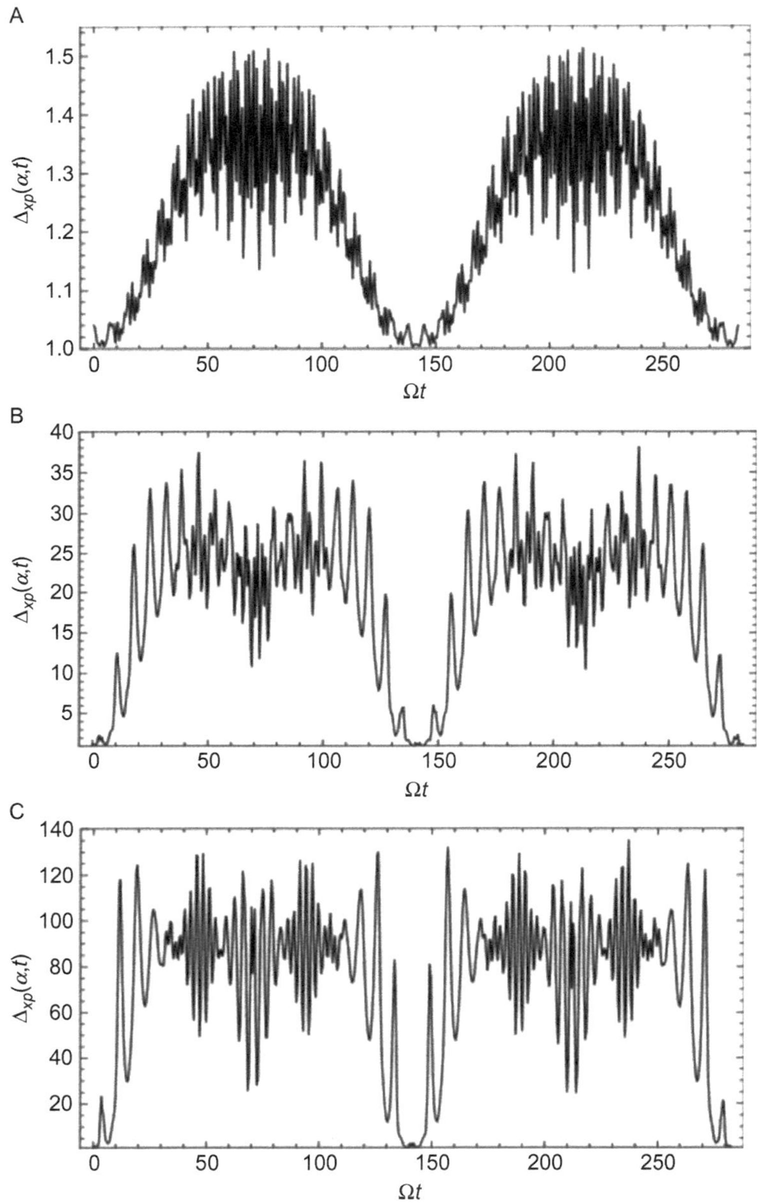

Figure 3 Temporal evolution of the normalized uncertainty product $\Delta_{xp} = 4\langle(\Delta\hat{x}_{\mathrm{D}})^2\rangle\langle(\Delta\hat{p}_{\mathrm{D}})^2\rangle/|\langle[\hat{x}_{\mathrm{D}},\hat{p}_{\mathrm{D}}]\rangle|^2$ of displacement operator coherent states for $\langle\hat{n}\rangle = 0.2$, 2, and 4 (frames A, B, and C, respectively). In the calculation, a diatomic HF molecule is considered to be modeled by a deformed Morse-like oscillator with $N = 22$ bound states.

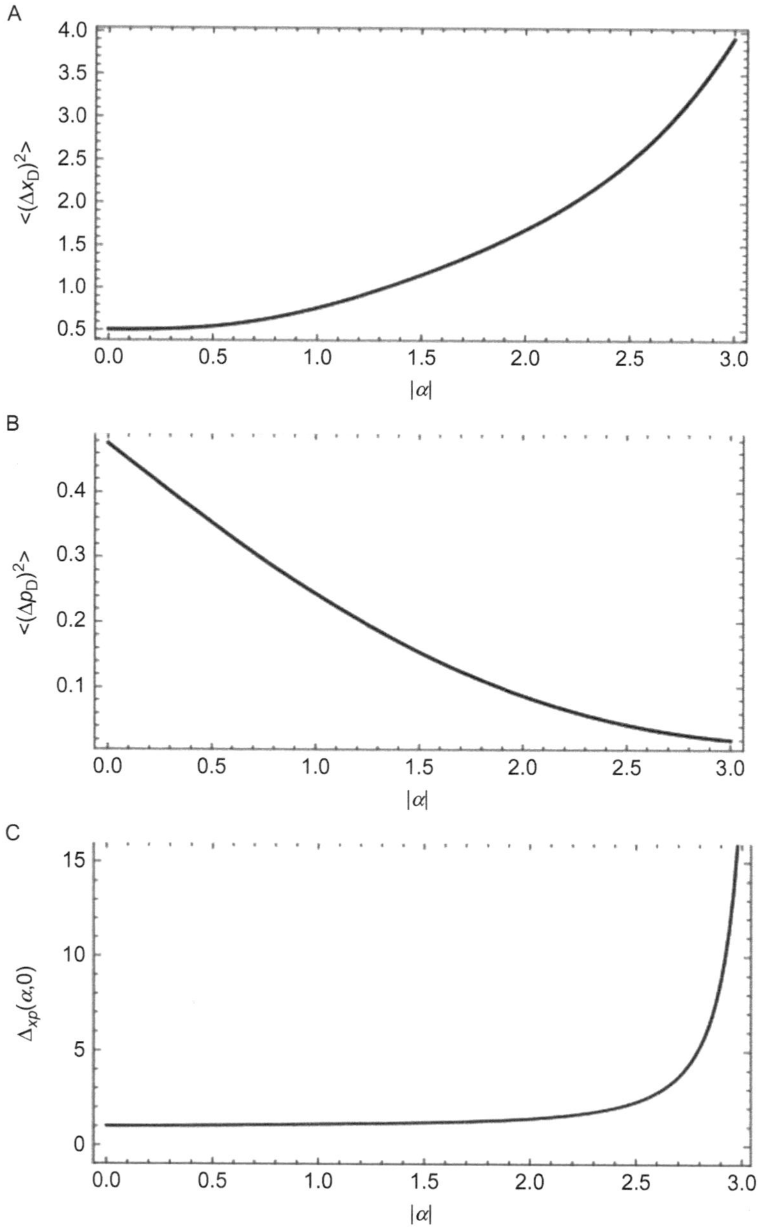

Figure 4 Uncertainty in coordinate $\langle(\Delta\hat{x}_{\mathrm{D}})^2\rangle$ (A), in momentum $\langle(\Delta\hat{p}_{\mathrm{D}})^2\rangle$ (B), and the corresponding normalized uncertainty product $\Delta_{xp}=4\langle(\Delta\hat{x}_{\mathrm{D}})^2\rangle\langle(\Delta\hat{p}_{\mathrm{D}})^2\rangle/|\langle[\hat{x}_{\mathrm{D}},\hat{p}_{\mathrm{D}}]\rangle|^2$ (C) at time $t=0$ as functions of $|\alpha|$ for the DOCS. In the calculation, a diatomic HF molecule is considered to be modeled by a deformed Morse-like oscillator with $N=22$ bound states.

$$G(n)=\sqrt{\frac{2\mu\omega}{\hbar}}\frac{\langle n-3|\hat{x}|n\rangle}{f(n)f(n-1)f(n-2)\sqrt{n(n-1)(n-2)}}, \tag{53}$$

$$R(n)=i\sqrt{\frac{2}{\hbar\mu\omega}}\frac{\langle n-1|\hat{p}|n\rangle}{f(n)\sqrt{n}}, \tag{54}$$

$$S(n)=i\sqrt{\frac{2}{\hbar\mu\omega}}\frac{\langle n-3|\hat{p}|n\rangle}{f(n)f(n-1)f(n-2)\sqrt{n(n-1)(n-2)}}. \tag{55}$$

Here, the matrix elements $\langle n-\beta|\hat{x}|n\rangle$ and $\langle n-\beta|\hat{p}|n\rangle$ are evaluated by numerical integration using the eigenfunctions of the corresponding Scrödinger equation.

The temporal evolution of the deformed coordinate and momentum is calculated taking the expectation values between the states $|\zeta,t\rangle=\hat{U}(t)|\zeta\rangle$ with $|\zeta\rangle$ a nonlinear coherent state obtained by application of the deformed displacement operator on the vacuum state

$$|\zeta\rangle\simeq\frac{1}{(1+|\zeta|^2)^s}\sum_{n=0}^{s-1}\sqrt{\frac{\Gamma(2s+1)}{n!\Gamma(2s+1-n)}}\zeta^n|n\rangle, \tag{56}$$

(the state is approximate since the sum contains a finite number of terms) and

$$\hat{U}(t)=\exp(-\frac{i}{\hbar}\hat{H}_{\mathrm{D}}t)$$

with

$$\hat{H}_{\mathrm{D}}=\frac{\hbar^2a^2}{2\mu}(-\hat{n}^2+2s\hat{n}+s).$$

In Fig. 5, we show the occupation number distributions and the phase space trajectories of DOCS for the particular case of $\langle\hat{n}\rangle\simeq0.1$ and 1.0. Notice that the phase space trajectories are symmetric with respect to the origin, this is a reflection of the symmetry of the potential. For a small value of the parameter $|\alpha|$, we see a behavior similar to that found for the Morse potential (see Fig. 1), that is, the phase space trajectories fill an anular region whose width depends upon the parameter $|\alpha|$ (for smaller values of $|\alpha|$ the width decreases) and there is an unaccessible internal region. For a larger value of the parameter $|\alpha|$, the trajectories in phase space are intersecting curves occupying all the space consistent with the energy, the unaccessible region disappears. The plots of the occupation number distribution manifest the fact that for the cases considered here the population is far from the

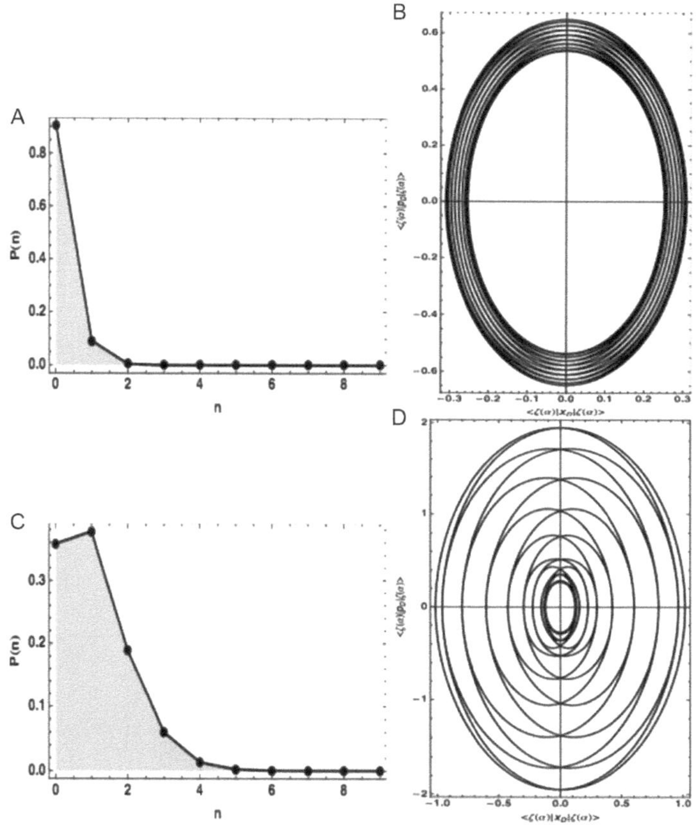

Figure 5 Occupation number distributions (frames A and C) and the corresponding phase space trajectories (frames B and D) of deformed displacement operator coherent states (DOCS) for $\langle \hat{n} \rangle \approx 0.1$, 1.0. In the calculation, a modified Pöschl-Teller potential is modeled by a deformed oscillator with $s = 10$ bound states.

dissociation and the approximation done keeping only the bounded part of the spectrum is justified.

In Fig. 6, we show the temporal evolution of the average value of the deformed coordinate $\langle \hat{X}_{\mathrm{D}} \rangle$, its dispersion $\langle (\Delta \hat{X}_{\mathrm{D}})^2 \rangle$ and the uncertainty product $\langle (\Delta \hat{X}_{\mathrm{D}})^2 \rangle \langle (\Delta \hat{P}_{\mathrm{D}})^2 \rangle$ for $\langle \hat{n} \rangle \simeq 0.1$ (left column) and 1.0 (right column). For these calculations, we considered a system supporting 10 bound states. For a fixed, small value of the parameter $|\alpha|$ (thus a small $\langle \hat{n} \rangle$) the deformed coordinate is an oscillatory function with a slightly varying amplitude, the corresponding dispersion is an oscillatory function whose amplitude is largest when the amplitude of the oscillation is smallest. Notice the presence of squeezing. The corresponding uncertainty product is shown in frame E, it can be seen that the DOCS we have constructed are minimum

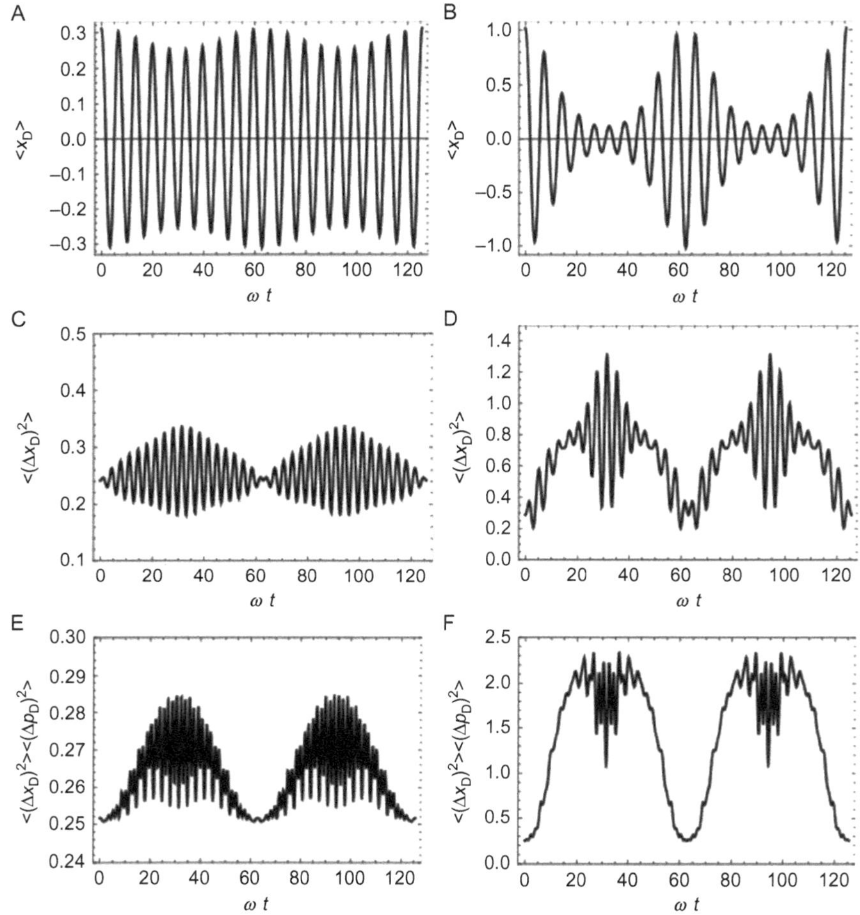

Figure 6 Average coordinate (frames A and B), uncertainty in coordinate (frames C and D), and the corresponding uncertainty product (frames E and F) of deformed displacement operator coherent states (DOCS) for $\langle \hat{n} \rangle \approx 0.1$ (left column) and 1.0 (right column). In the calculation, a modified Pöschl-Teller potential is modeled by a deformed oscillator with $s = 10$ bound states.

uncertainty states at the initial time and there are some specific instants of time when the states return to be of minimum uncertainty, this conduct is periodic. Most of the time the DOCS are not minimum uncertainty states.

For a larger value of the parameter $|\alpha|$ (thus a larger value of the average $\langle \hat{n} \rangle$), we see that the deformed coordinate presents oscillations with varying amplitudes. The largest amplitude corresponds to that of a field coherent state with the same energy. Notice that when the amplitude is largest the dispersion is smallest and corresponds to that of a minimum uncertainty state, this conduct is present at the initial time and repeats itself periodically. Here, we can see also the presence of squeezing. When the amplitude of the oscillations in the deformed coordinate is small we see that the dispersion is large, this is a reflection of the nonlinear terms in the Hamiltonian. The uncertainty product also shows a periodic conduct with instants of time where the state is a minimum uncertainty state.

Finally, in Fig. 7, we see the dispersions in the deformed coordinate and momentum as a function of the parameter $|\alpha|$ evaluated at time $t=0$. Notice

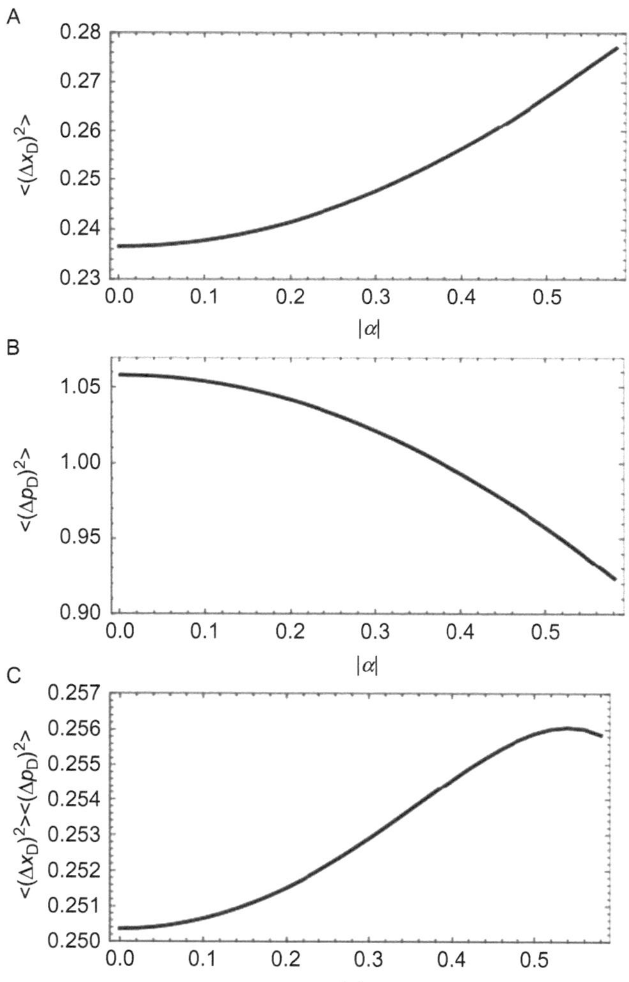

Figure 7 Average coordinate $\langle x_D \rangle$ (A), uncertainty in coordinate $\langle (\Delta x_D)^2 \rangle$ (B), and the corresponding uncertainty product $\langle (\Delta x_D)^2 \rangle \langle (\Delta p_D)^2 \rangle$ (C) at time $t=0$ as functions of $|\alpha|$ for the DOCS. In the calculation, a modified Pöschl-Teller potential is modeled by a deformed oscillator with $s=10$ bound states.

that for small $|\alpha|$ the dispersion in the deformed coordinate is less than 1/2 meaning squeezing, in contrast, that of the deformed momentum is larger than 1/2 and the uncertainty product is near the minimum possible value. As the parameter $|\alpha|$ increases, the dispersion in the deformed coordinate increases so that the state is less squeezed and that of the momentum decreases but is still far from the minimum allowed value (in the range of values for the parameter $|\alpha|$ shown in the figures) so that the uncertainty product separates from the initial value. Because we are dealing with a finite number of bound states s and we are not taking into account the states from the continuum we must use a small enough value of $|\alpha|$ so that the average occupation number is much smaller than s. To complete the examples, we just write the Displaced Operator Coherent States associated we the trigonometric Pöschl-Teller potential as

$$|\zeta(\alpha)\rangle = (1-|\zeta(\alpha)|^2)^{\lambda}\sum_{n=0}^{\infty}\zeta(\alpha)^n\sqrt{\frac{\Gamma(2\lambda+n)}{n!\Gamma(2\lambda)}}|n\rangle. \tag{57}$$

with $\zeta(\alpha) = e^{i\theta}\tanh(|\alpha|/\sqrt{2\lambda})$. This expression should be compared with that of the AOCS (see Eq. 33).

6. DISCUSSION

Based on the f-deformed oscillator formalism we have introduced nonlinear coherent states by generalization of two definitions, as eigenstates of a deformed annihilation operator AOCS and as the states that result by application of a deformed displacement operator on the vacuum state DOCS. We have applied our method to Hamiltonians that contain linear and quadratic terms in the number operator corresponding to the trigonometric and modified Pöschl-Teller Hamiltonians and the Morse Hamiltonian.

The AOCS and the DOCS obtained for the trigonometric Pöschl-Teller potential are exact because the number of bound states supported by the potential is infinite. On the other hand, the AOCS and the DOCS obtained for the Morse and the modified Pöschl-Teller potentials are approximate because the number of bound states is finite. For the numerical results, we have considered only the states obtained by application of a deformed displacement operator on the vacuum state; in Ref. 33, we discussed the nonlinear coherent states obtained as eigenstates of a deformed annihilation operator for the Morse potential. Although from an algebraic-structure point of view the coherent states obtained from each generalization

(DOCS) or (AOCS) are different, as well as their respective statistical behavior,[34] the average values and the phase space trajectories obtained with them are almost identical.[10]

We must mention that the same form of nonlinear coherent states AOCS and DOCS for the Morse and Pöschl-Teller potentials would be obtained, if we had worked with the true ladder operators associated with each potential.[11] This happen because the ladder operators and deformed operators share the same commutation relation for these kind of systems.

ACKNOWLEDGMENTS

We acknowledge useful discussions with O. de los Santos Sánchez and thank Reyes García for the maintenance of our computers. We also thank CONACyT and DGAPA-UNAM for partial support through projects 166961 and IN 108413, respectively.

REFERENCES

1. Mandel, L.; Wolf, E. *Optical Coherence and Quantum Optics*; Cambridge University Press, 1995, pp. 522-456.
2. Glauber, R. J. The Quantum Theory of Optical Coherence. *Phys. Rev.* **1963**, *130*, 2529–2539; Glauber, R. J. Coherent and Incoherent States of the Radiation Field. *Phys. Rev.* **1963**, *131*, 2766–2788; Glauber, R. J. Photon Correlations. *Phys. Rev.* **1963**, *10*, 84.
3. Klauder, J. R. Continuous-Representation Theory. 1. Postulates of Continuous-Representation Theory. *J. Math. Phys.* **1963**, *4*, 1055; Klauder, J. R. Continuous-Representation Theory. 2. Generalized Relation Between Quantum and Classical Dynamics. *J. Math. Phys.* **1963**, *4*, 1058.
4. Perelomov, A. M. Coherent States for Arbitrary Lie Group. *Commun. Math. Phys.* **1972**, *26*, 222.
5. Gilmore, R. Geometry of Symmetrized States. *Ann. Phys.* **1972**, *74*, 391; Gilmore, R. Baker-Campbell-Hausdorff Formulas. *Ann. Phys.* **1974**, *15*, 2090–2092.
6. Nieto, M. M.; Simmons, L. M. Coherent States for General Potentials. *Phys. Rev. Lett.* **1978**, *41*, 207; Nieto, M. M.; Simmons, L. M. Coherent States for General Potentials 1. Formalism. *Phys. Rev. D* **1979**, *20*, 1321; Nieto, M. M.; Simmons, L. M. Simmons, L.M. Coherent States for General Potentials 3. Non-confining One-Dimensional Examples. *Phys. Rev. D* **1979**, *20*, 1342.
7. Gazeau, J. P.; Klauder, J. Coherent States for Systems with Discrete and Continuous Spectrum. *J. Phys. A: Math. Gen.* **1999**, *32* (1), 123–132.
8. Man'ko, V. I.; Marmo, G.; Sudarshan, E. C. G.; Zaccaria, F. F-Oscillators and Nonlinear Coherent States. *Phys. Scr.* **1997**, *55*, 528–541.
9. Aniello, P.; Man'ko, V. I.; Marmo, G.; Solimeno, S.; Zaccaria, F. On the Coherent States, Displacement Operators and Quasidistributions Associated with Deformed Quantum Oscillators. *J. Opt. B: Quant. Semiclass. Opt.* **2000**, *2*, 718.
10. de los Santos-Sánchez, O.; Récamier, J. Nonlinear Coherent States for Nonlinear Systems. *J. Phys. A: Math. Theor.* **2011**, *44* (145307).
11. Román-Ancheyta, R.; de los Sántos-Sánchez, O.; Récamier, J. Ladder Operators and Coherent States for Nonlinear Potentials. *J. Phys. A: Math. Theor.* **2011**, *44*, 435304.
12. Carruthers, P.; Nieto, M. M. Coherent States and Forced Quantum Oscillator. *Am. J. Phys.* **1965**, *33* (7), 537.

13. Glauber, R. J. Classical Behavior of Systems of Quantum Oscillators. *Phys. Lett.* **1966**, *21*, 650.
14. Unruh, W. G. Quantum Nondemolition Measurement and Coherent States. *Phys. Rev. D* **1978**, *17*, 1180–1181.
15. Takahashi, Y.; Shibata, F. Spin Coherent State Representation in Nonequilibrium Statistical Mechamics. *J. Phys. Soc. Jpn.* **1975**, *38* (3), 656–668.
16. Arrechi, F. T.; Thomas, H.; Gilmore, R. Atomic Coherent States in Quantum Optics. *Phys. Rev. A* **1972**, *6* (6), 2211.
17. Wang, Y. K.; Hioe, F. T. Phase-Transition in Dicke Model of Superradiance. *Phys. Rev. A* **1973**, 7 (3), 831–836.
18. Stopera, C.; Grimes, T. V.; McLaurin, P. M.; Privett, A.; Morales, J. A. Some Recent Developments in the Simplest-Level Electron Nuclear Dynamics Method: Theory, Code Implementation and Applications to Chemical Dynamics. *Adv. Quant. Chem.* **2013**, *66*, 113–194.
19. Klauder, J. R.; Skagerstam, B. S. *Coherent States Applications in Physics and Mathematical Physics*; World Scientific: Singapore, 1985.
20. Ali, S.; Antoine, J.; Gazeau, J. *Coherent States, Wavelets and Their Generalizations (Graduate Texts in Contemporary Physics)*; World Scientific: Singapore, 2000.
21. Zhang, W. M.; Feng, D.; Gilmore, R. Coherent States: Theory and Some Applications. *Rev. Mod. Phys.* **1990**, *62*, 867–924.
22. Récamier, J.; Gorayeb, M.; Mochán, W. L.; Paz, J. L. Lie Algebraic Method Applied to a Pulsed Anharmonic Oscillator. *Int. J. Quantum Chem.* **2006**, *106* (15), 3160–3166.
23. Landau, L. D.; Lifshitz, E. M. *Quantum Mechanics (Non Relativistic Theory)*; Oxford: Pergamon, 1977.
24. Abramowitz, M.; Stegun, I. *Handbook of Mathematical Functions*; Dover: New York, NY, 1970, pp. 555.
25. Lemus, R.; Bernal, R. Connection of the Vibron Model with the Modified Pöschl-Teller Potential in Configuration Space. *Chem. Phys.* **2002**, *3* (283), 401–417.
26. Puri, R. *Mathematical Methods of Quantum Optics*; Springer: New York, NY, 2001, pp. 48.
27. Gilmore, R. *Lie Groups, Lie Algebras and Some of Their Applications*; Wiley: New York, 1974, pp. 460.
28. de los Santos-Sánchez, O.; Récamier, J. Phase Space Picture of Morse-Like Coherent States Based Upon the Wigner Function. *J. Phys. A: Math. Theor.* **2012**, *45*, 415310.
29. Carvajal, M.; Lemus, R.; Frank, A.; Jung, C.; Ziemniak, E. An Extended SU(2) Model for Coupled Morse Oscillators. *Chem. Phys.* **2000**, *1-2* (260), 105–123.
30. Récamier, J.; Mochán, W. L.; Gorayeb, M.; Paz, J. L.; Jáuregui, R. Uncertainty Relations for a Deformed Oscillator. *Int. J. Mod. Phys. B* **2006**, *11-13* (20), 1851–1859.
31. de los Santos-Sánchez, O.; Récamier, J. Morse-Like Squeezed Coherent States and Some of Their Properties. *J. Phys. A: Math. Theor.* **2013**, *375303*, 46.
32. Child, M. S.; Halonen, L. Overtone Frequencies and Intensities in the Local Mode Picture. *Adv. Chem. Phys.* **1984**, *57*, 1–58.
33. Récamier, J.; Gorayeb, M.; Mochán, W. L.; Paz, J. L. Nonlinear Coherent States and Some of Their Properties. *Int. J. Theor. Phys.* **2008**, *3* (47), 673–683.
34. Román-Ancheyta, R.; Gutiérrez, C. G.; Récamier, J. Photon-Added Nonlinear Coherent States for a One-Mode Field in a Kerr Medium. *J. Opt. Soc. Am. B* **2014**, *1* (31), 38–44.

CHAPTER THIRTEEN

Electronic Properties in Supercritical Fluids: The Absorption Spectrum of *p*-Nitroaniline in Supercritical Water

Marcelo Hidalgo Cardenuto[*,†], **Kaline Coutinho**[*], **Benedito J.C. Cabral**[*,‡,§], **Sylvio Canuto**[*,1]

[*]Instituto de Física, Universidade de São Paulo, São Paulo, Brazil
[†]Laboratoire de Chimie Théorique, Unité de Chimie Physique Théorique et Structurale, University of Namur, Namur, Belgium
[‡]Grupo de Física Matemática da Universidade de Lisboa, Lisboa, Portugal
[§]Departamento de Química e Bioquímica, Faculdade de Ciências, Universidade de Lisboa, Lisboa, Portugal
[1]Corresponding author: e-mail address: canuto@if.usp.br

Contents

Abstract

Supercritical fluids (SCFs) are of great interest for their remarkable physical–chemical properties. Understanding the properties of SCFs is of industrial and academic interests. Foreign molecules can act as interesting probes giving information from the changes induced by the supercritical environment. The electronic absorption spectrum of *p*-nitroaniline (pNA) in the environment of supercritical water has been addressed. The thermodynamic condition is assured by Monte Carlo simulation, and quantum mechanical calculations are performed on sampled configurations to obtain the spectrum of pNA in supercritical water. Using the same thermodynamic condition of the experiment ($T = 655$ K and $\rho = 0.12$ g/cm^3), we analyze the electronic polarization of the *p*-aniline, the solute–solvent hydrogen bonds, and the red shift of the intense charge transfer π–π^* transition. Compared to normal water, the number of solute–solvent hydrogen bonds in supercritical water decreases to essentially one-fifth, from 2.9 to 0.6. The calculated spectrum in supercritical water is obtained using

Advances in Quantum Chemistry, Volume 71
ISSN 0065-3276
http://dx.doi.org/10.1016/bs.aiq.2015.03.006

time-dependent CAM-B3LYP/aug-cc-pVDZ and shows a maximum located at 305 nm in good comparison with the experimental maximum observed at 309 nm. In comparison to the results in the vapor phase, the experimental red shift of 17 nm is very well described theoretically with the result of 17 ± 3 nm. A discussion is also presented of the inhomogeneous broadening contribution to the half-width at the absorption maximum.

1. INTRODUCTION

The region located in the phase diagram at temperatures and pressures above the critical values corresponds to the supercritical region, where the fluid exhibits novel characteristics, termed as supercritical fluids (SCFs). They are of great importance because of their unusual properties.[1] SCF may exhibit properties that are markedly different from regular liquids.[2–4] Water in supercritical conditions (SCW) has static dielectric constant that varies in the approximate interval between $2 \leq \varepsilon \leq 30$, thus exhibiting dielectric properties equivalent to a range between hexane and methanol. For reference, at the critical point of water ($T_C = 647$ K and $P_C = 221$ atm.) the dielectric constant is close to $\varepsilon = 5.0$ and the critical density is only $\rho_c = 0.32$ g/cm^3.

SCFs are acquiring a great importance for essentially three reasons. First, it is a benign solvent friendly to the environment. Second, it is now widely used in industrial processes not only because of the first point just mentioned but also because tuning the temperature and pressure, the corresponding changes in the fluid may increase solubility, gauge molecular properties, and facilitate chemical reactivity. Third, related to these advantages, there is a need for a better understanding of the physical properties of SCF for industrial purposes and basic sciences.[4] The understanding of the electronic properties of SCF is of great importance and will open wide avenues and enormous possibilities of applications of SCF. From the theoretical point of view, the study of the electronic properties of SCF needs quantum mechanics and, equally important, the inclusion of the thermodynamic condition.

It was only recently that the explicit consideration of the thermodynamic constraint has been included in quantum mechanical calculations.[5–7] This is efficiently obtained by combining statistical mechanics and quantum mechanics and is generally known as QM/MM methods.[8] There are several ways of combining molecular mechanics and quantum mechanics. In the procedure adopted in our laboratory, the MM (molecular mechanics) simulation is used to generate configurations of the system according to the specific thermodynamic condition. These configurations are then submitted to QM (quantum mechanical) calculations. Thus, in principle, any thermodynamic condition can be used but relying on the accuracy and appropriateness of the force field used. In more difficulty situations, the use of first-principle molecular dynamics should be recommended.[9,10] Given these considerations, there are great opportunities for theoretical studies in the SCF region. We are interested in molecular properties and spectroscopy in the supercritical region of water. In this work, we will consider the electronic absorption spectrum of *p*-nitroaniline (pNA) in SCW.

The low-lying absorption spectrum of vapor pNA is characterized by a strong π–π^* transition located at 292 nm.[11] This is known to be a charge transfer transition that upon excitation leads to a considerable increase of the molecular dipole moment. Thus, it is very influenced by a solvent environment that is known to shift the π–π^* transition to longer wavelength, a red shift. The origin of this red shift is easy to rationalize because the increased dipole moment in the excited state leads to a better stabilization in solvents compared to the ground state, decreasing the separation of the two states. This shift is quite appreciable in normal water (NW), where the transition seen in the gas at 292 nm appears at 380 nm.[12,13] As the charge transfer in the excitation involves the nitro (NO_2) and amine groups (NH_2), it is expected that the hydrogen bonds formed between the solvent water and the nitro group (a proton acceptor) and the amine group (a proton donor) have an important effect.[13,14] A more interesting situation appears in the case of SCW because of the change of the hydrogen bond network and the expected reduction in the number of solute–solvent hydrogen bonds.[15] In fact, the structure of hydrogen bonds in SCW is a topic of great interest.[16–19] The electronic absorption spectrum of pNA in SCW was experimentally obtained by Oka and Kajimoto.[13] They have measured the spectrum of pNA in SCW (thermodynamic condition: $T=655$ K and density $\rho=0.12$ g/cm^3) and found the intense π–π^* transition located at 309 nm. This indicates a spectral shift of 17 nm from the gas-phase condition, corresponding in this case to a red shift of 0.24 eV. This is indeed a small shift compared to the case of NW where it is as large as 1.0 eV.

The solvation effects on the low-lying electronic absorption spectrum of pNA in NW have been subjected to several previous theoretical studies and we will resort to some recent ones.[20–22] Most of the interest has been in the description of the spectral shifts associated to the low-lying π–π^* state of pNA in water[20–22] and in some other solvents.[20] However, we are not aware of any study in the supercritical condition of water, although there is a recent study in supercritical CO_2,[23] a solvent that has attracted considerable attention recently.[23–27] This is related to some fundamental questions associated to Lewis basicity and acidity[24–26] and also to the intriguing polar attributes of supercritical CO_2.[9,28]

In this work, we use the sequential QM/MM methodology[29–31] to study the spectral shift of the charge transfer π–π^* transition of pNA in SCW. We will analyze the electronic polarization of pNA due to the solvent, the hydrogen bonds formed between the pNA and the solvent, their influence on the spectral shift, and, finally, we obtain the total spectral shift. Although our emphasis is on SCW as the solvent, we will also perform similar studies in NW for comparison, especially when analyzing the solute–solvent hydrogen bonds.

2. METHODOLOGY

We adopt the sequential QM/MM methodology[29–31] performing Monte Carlo (MC) simulation for generating the fluid structure to be subsequently submitted to QM calculations. Hence, the structure (molecular arrangements of water around the pNA molecule) is obtained by Metropolis MC simulation,[32] and statistically relevant configurations (see below) are sampled for the QM calculations of the electronic absorption spectrum. For proper comparison, the simulations are made at the same experimental thermodynamic conditions where the results for pNA in SCW and NW were obtained. Thus, for NW we use the NPT ensemble with $T=298$ K and $P=1$ atm. The calculated density obtained from the MC simulation is 1.01 ± 0.01 g/cm^3, in agreement with the density of NW. For SCW, the experiments[13] were made at fixed density; thus, we adopt in this case the NVT ensemble with the experimental condition of $T=655$ K and $\rho=0.12$ g/cm^3. The geometry of pNA was fully optimized at the second-order Møller–Plesset with a large basis set, MP2/aug-cc-pVDZ, including diffuse functions for a better description of the dipole moment.[20] To include part of the solvent effects on the geometry of pNA, we have included the solvent influence using the polarizable continuum method (PCM).[33–35] For the case of NW, we used the regular value of $\varepsilon=78.4$ for the dielectric constant. In the case of SCW, we adopted the dielectric constant of $\varepsilon=2.66$ suggested by Pomelli and Tomasi[36] that are in agreement with the corresponding value[37] of the temperature and density used in our simulation. The optimized geometry of pNA is not planar with a calculated dihedral angle of 25° in the amino group, in agreement with crystallography results.[38] After obtaining the geometries, the structure of pNA was held fixed in the MC simulations. The electronic polarization of pNA due to the solvent is included using also the PCM solvent model. As the MC simulation is made before the QM calculations, the inclusion of the electronic polarization is important. This can be made by resorting to an iterative procedure where each MC simulation is followed by a QM calculation of the atomic charges of the solute.[39,40] After convergence, there is an electrostatic equilibrium between the solute and the solvent. This procedure has shown to be accurate but it is cumbersome computationally requiring a series of successive simulations and QM calculations. Hence, in a previous work[41] we have analyzed the combination of QM/MM and the PCM polarization in comparison with this iterative procedure. The success[41] in giving good result for the electronic polarization of the solute by the solvent suggests it can be adopted here. The interatomic potential used is of the Lennard-Jones plus Coulomb type.[32] The potential adopted for water was the SPC/E[42] that has been successfully used before in the study of electronic absorption spectra[43] and nuclear magnetic resonance chemical shielding[44] in SCW environment. For pNA, the Lennard-Jones parameters were obtained from Refs. 45,46, but the atomic charges that compose the Coulomb part of the potential were obtained from the PCM polarization discussed above and using the CHELPG fitting.[47] These calculated charges and the dipole moments obtained with the MP2/aug-cc-pVDZ calculations are shown in Table 1.

The calculated dipole moment for isolated pNA is $\mu = 6.16$ D, in comparison with the reported experimental values of 6.1 D[13,48] and 6.29 D,[11] the latter obtained in benzene solution. These values are lower than previous theoretical estimates.[20–22] The solvent polarization into pNA increases the value of the dipole moment. In SCW, this electronic polarization due to the solvent increases the ground-state dipole moment to 7.33 D and in NW, this value is 8.58 D (Table 2). There is no experimental value for the dipole moment of pNA in SCW. The value in dilute benzene is reported as 7.6 D.[49]

The MC simulations are made in systems composed of the central pNA surrounded by 500 water molecules. After a thermalization stage of 6×10^6 MC steps, a total of 1.5×10^8 steps were performed. Using the autocorrelation function of the energy,[7,50] a total of 300 configurations were separated for the analysis of the solute–solvent hydrogen bonds. For the QM

Table 1 Calculated Atomic Charges of pNA Isolated and in Solvent Environment Using a PCM Representation of the Solvent

Atom	Isolated	SCW ($\varepsilon = 2.66$)	NW ($\varepsilon = 78.4$)
C1, C3	−0.2129	−0.2436	−0.2510
C2	0.1555	0.1580	0.1538
C4, C6	−0.1346	−0.1231	−0.1458
C5	0.2906	0.2995	0.3564
N7	0.5295	0.5755	0.5972
O8, O9	−0.3656	−0.4035	−0.4339
H10, H13	0.1441	0.1597	0.1698
H11, H12	0.1321	0.1398	0.1536
N14	−0.7232	−0.7734	−0.8442
H15, H16	0.3107	0.3409	0.3757
μ (D)	6.16	7.33	8.58

The charges are obtained using a CHELPG fit after MP2/aug-cc-pVDZ calculations. Atomic labels are defined in Fig. 1.

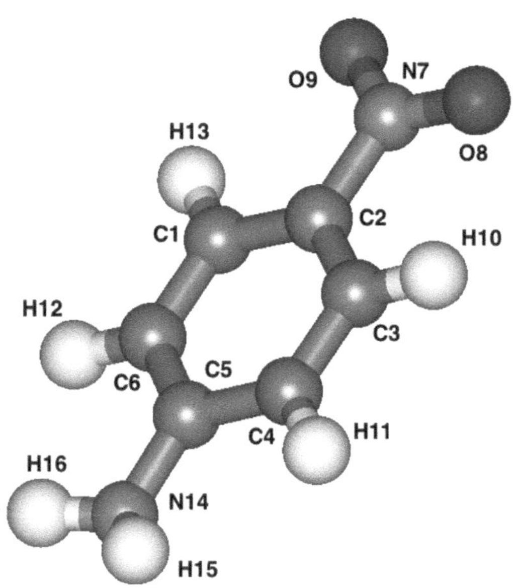

Figure 1 The structure of *p*-nitroaniline.

Table 2 Calculated Electronic Transitions (in nm) and Oscillator Strengths (Within Parentheses) for Isolated pNA Using Different Functionals and the TD-DFT/aug-cc-pVDZ Model

Transition	B3LYP	BHandHLYP	CAM-B3LYP	PBE0
S_0–S_1	332.4 (0.00)	300.8 (0.00)	320.4 (0.00)	325.4 (0.00)
S_0–S_2	320.6 (0.31)	279.4 (0.37)	287.5 (0.36)	308.5 (0.33)
S_0–S_3	291.2 (0.00)	270.2 (0.00)	282.7 (0.00)	286.1 (0.00)

Experimental value of the absorption maximum is 292 nm.

calculation of the spectrum, a total of 100 configurations were selected. It is important to mention that all results reported are statistically converged. The convergence is a consequence of the efficient sampling obtained by using the autocorrelation function of the energy.[7,50]

All MC simulations and statistically analysis were made using the DICE program.[51] The QM calculations were made using the time-dependent density functional theory (TD-DFT).[52,53] We have made some preliminary calculations (Table 2) using different functionals such as B3LYP,[54,55] BHandHLYP,[56] CAM-B3LYP,[57] and PBE0[58,59] with large basis set including functions. Based on the results of Table 2, we selected the CAM-B3LYP/aug-cc-pVDZ for all calculations of the excitation energies. The better accuracy obtained for the CAM-B3LYP[57] functional is not surprising because of the charge transfer nature of the excited state. All QM calculations were performed using the Gaussian-09.[56] The results for the isolated pNA will be discussed in Section 3 along with some earlier theoretical results.[20–22]

3. RESULTS AND DISCUSSION

We will first discuss the hydrogen bonds formed between pNA and water in both thermodynamic conditions, SCW and NW. The solvation of pNA in water is of interest also because of its amphoteric property[60]: it can act both as a proton donor and proton acceptor system. It is a proton donor in the NH_2 and a proton acceptor in the NO_2 group. To identify the formation of hydrogen bonds, the usual procedure is to identify the hydrogen bond shell in the corresponding radial distribution function. Although this geometric procedure gives the correct coordination number, it cannot be assured that all water molecules satisfying this criterion are indeed hydrogen bonded to the solute. This has been extensively discussed in previous works.[50] Hence, we use geometric and energetic criteria. The radial distribution function and the pairwise energy distribution are used for identifying the solute–solvent hydrogen bonds. Specifically, the conditions used here were that (i) the distance between the H atom and the proton acceptor should be less than 3.5 Å, (ii) the angle less than 35°, and (iii) simultaneously, the solute–solvent pairwise interaction should be more attractive than −3.0 kcal/mol. The first geometrical criterion is defined after analysis of the radial distribution function. The hydrogen bond angle is defined as O–$(OH)_w$ in the NO_2 and O_w–NH in the NH_2 groups. The distribution of pairwise energy interaction is shown in Fig. 2. For NW, it is easily discernible the distribution of HB for energies below −3.0 kcal/mol. In the case of SCW, only a bump is seen but located at essentially the same position. This is a consequence of the reduced number of HB in the case of the SCW condition. Figure 2 is used to define the energetic criterion. Table 3 shows the entire statistics obtained from 300 uncorrelated configurations extracted from the MC simulation. In this, we are including the hydrogen bonds formed both in the NH_2 and NO_2 groups. Figure 3 illustrates the configuration space spanned by the water molecules that are hydrogen bonded to pNA in either of these two groups. In the case of SCW, the number of HB is considerably reduced compared to NW, as expected. It can be seen in Table 3 that 55% of the configurations make no HB in the case of SCW, a situation that is statistically negligible in NW. In NW, 28.7% of the configurations make two hydrogen bonds and 34.7% make three hydrogen bonds. On average, considering both the NH_2 and NO_2 groups, pNA makes 2.9 HB with NW but only 0.6 HB in the case of SCW.

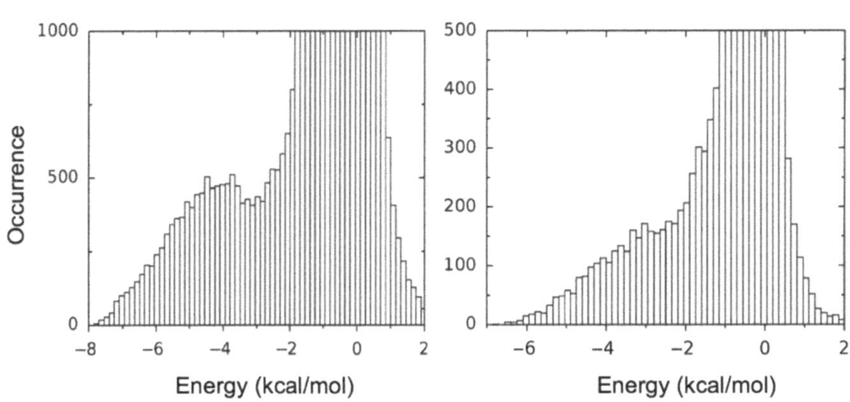

Figure 2 Pairwise energy interaction between *p*-nitroaniline and water in normal (left) and supercritical conditions (right).

Table 3 Statistics of the Hydrogen Bonds Formed Between pNA and Water in the Two Thermodynamic Conditions Obtained from 300 Statistically Uncorrelated Configurations

	SCW		NW	
N_{HB}	Configurations	Occurrence (%)	Configurations	Occurrence (%)
0	166	55	3	1
1	107	36	19	6.3
2	24	8	86	28.7
3	3	3	104	34.7
4	0	0	76	25.3
5	0	0	12	4
Average	0.55		2.89	

The number of hydrogen bonds N_{HB} makes no distinction whether the HB is formed on the NH_2 or NO_2 groups.

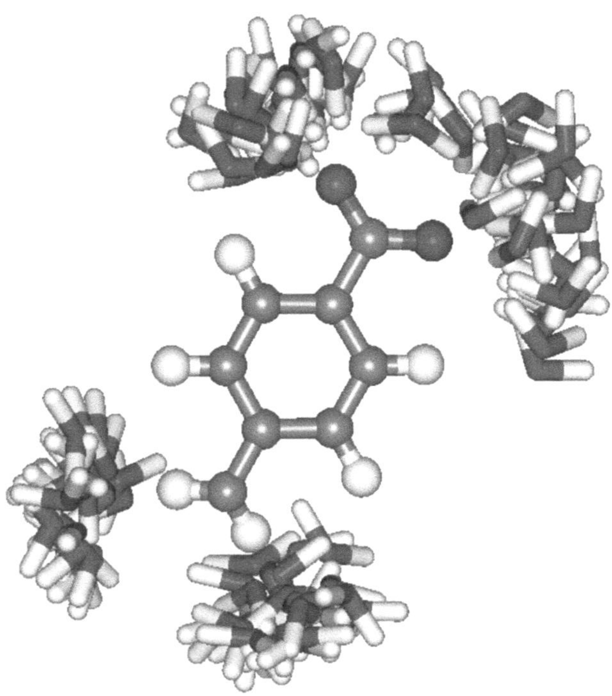

Figure 3 Illustration of the configuration spanned by hydrogen bonds formed in the proton acceptor NH_2 and proton donor NO_2 groups.

Now we discuss the electronic absorption spectrum of pNA. Table 2 shows the calculated results for the transition energies calculated for the isolated molecule in the spectral region of the experimental result. Different functionals are used. It is seen that in all cases, the transition to the first excited state S_0–S_1 has no intensity. Similarly, the S_0–S_3 transition has a zero oscillator strength. These dark states hence do not contribute to the experimental absorption band in gas phase. It is the second calculated state, the S_0–S_2 transition, that carries the intensity for the transition from the ground state. The best result for the S_0–S_2 transition is obtained with the CAM-B3LYP functional, with an intense charge transfer transition calculated at 288 nm, in good agreement with the experimental maximum seen at 292 nm.[11] In this spectral region, the difference of 4 nm in the wavelength corresponds to an energy difference between theory and experiment of only 0.05 eV. This calculated value corresponds to 4.30 eV in comparison with the experimental value of 4.25 eV. We note, in Table 2, that the first excited state, the S_1 (n–π*) state, is well separated from the S_2 (π–π*) state but the separation of this S_2 (π–π*) and the S_3 (n–π*) state is less than 0.1 eV in the CAM-B3LYP results and are amenable to reversal in different QM calculations. Using the effective fragment potential (EFP) in a CIS(D) calculation, Kosenkov and Slipchenko[20] obtained the intense transition to be the S_0–S_3 (π–π*) with the excitation energy of 4.65 eV, that is higher than ours and the experimental values. In turn, Sok et al.[21] using also the EFP in a B3LYP calculation obtained the value of 3.97 eV, lower than experiment. Among the DFT models used, the Coulomb-attenuated functional is more indicated for charge transfer transitions and the better agreement of the CAM-B3LYP functional confirms this and it will be used in all forthcoming calculations. Figure 4 shows the highest occupied molecular orbital and lowest unoccupied molecular orbital involved in the π–π* transition and the charge transfer from the NH_2 to the NO_2 group can be seen.

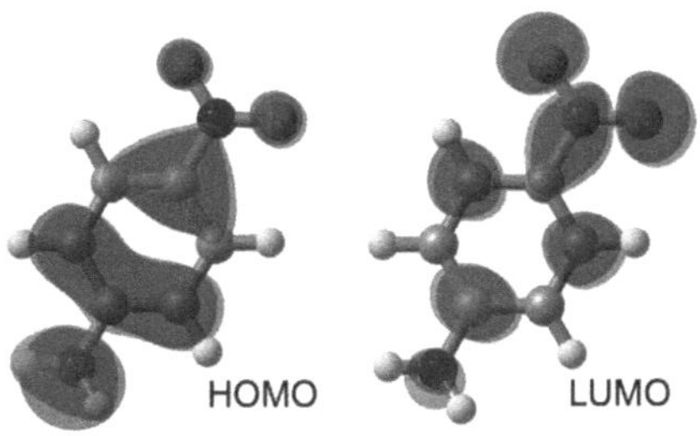

Figure 4 The molecular orbitals of the intense π–π* transition of *p*-nitroaniline.

Now we consider the π–π^* transition in SCW, measured experimentally[13] but where theoretical results are not available for comparison. The calculations are made again with the CAM-B3LYP/aug-cc-pVDZ and two sets of calculations are performed. In the first case, all the solvent molecules are treated as simple point charges using the same atomic charges of the SPC/E potential.[42] This includes only the electrostatic interaction between the pNA solute molecule and the solvent molecules. For this, a total of 350 water molecules are considered in this electrostatic embedding (EE). In the second set of calculations, the water molecules that were identified to be hydrogen bonded to pNA are explicitly considered and the remainder is treated as an EE. This is a reasoning of the fact that solvent molecules that are further away from the pNA molecule interact mostly by the long-range electrostatic interaction.[61] The configurations used for the hydrogen-bonded water molecules obey the statistics shown in Table 3. Figure 5 illustrates the case of one configuration using two HB (one in the NH_2 and the other in the NO_2 groups) plus the EE of all the remaining 348 water molecules. There are 8% of configurations with two solute–water hydrogen bonds. The calculated results for the intense transition are shown in Table 4 with comparison to the experimental result. In all configurations

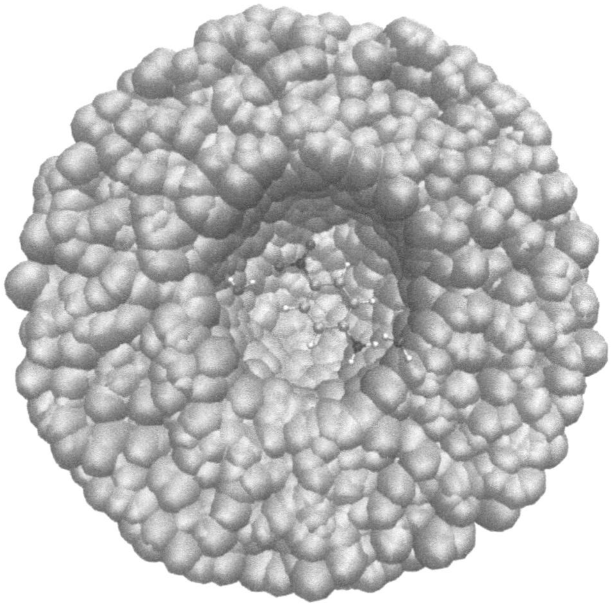

Figure 5 Illustration of *p*-nitroaniline with two explicit hydrogen-bonded water molecules in the electrostatic embedding of the remaining solvent molecules.

Table 4 Calculated Electronic Transitions (in nm) Obtained for pNA Isolated and in SCW Using TD-DFT with CAM-B3LYP/aug-cc-pVDZ

	Calculated	**Experiment**
Isolated pNA	288 (4.30)	292 (4.25) in vapor[11]
pNA + EE	300 (4.13)	
pNA + HB (all) + EE	305 (4.06)	309 (4.01) in SCW[13]
pNA + HB (NO_2) + EE	306 (4.05)	
pNA + HB (NH_2) + EE	304 (4.08)	

Comparison is made with the experimental absorption transition maximum. In parentheses are the corresponding values in eV. HB means the explicit inclusion of the hydrogen-bonded water molecules. EE corresponds to electrostatic embedding. See text for details.

used, the intense transition is still the S_0–S_2 as in the gas phase but in some configurations, due to local symmetry breaking, the S_0–S_1 and S_0–S_3 transitions acquire some intensity, and both being of the n–π* type suffer a blue shift compared to the isolated pNA. Interesting, in some configurations the lowest S_0–S_1 transition gets nonzero intensity and it will contribute to shift the spectral maximum toward the low-energy side. It is seen (Table 4) that the sole inclusion of the electrostatic interaction gives the correct sign and a relatively good value of 12 nm for the experimental shift of 17 nm. But with the inclusion of the explicit water molecules that make hydrogen bonds with pNA a very good agreement with the experimental value is obtained.

The theoretical spectral red shift of the charge transfer π–π* transition obtained from the gas to the SCW condition is now 17 nm (or 0.24 eV) in excellent agreement with the experimental value. The importance of the hydrogen bonds in the spectral shift was anticipated earlier in experimental studies.[13,14] Theoretical results indicate that the presence of hydrogen bonds, either in the proton donor NH_2 or the proton acceptor NO_2, influences the charge transfer and affects the solvatochromic shift. Table 4 shows that using only the HBs that are located in the NO_2 group gives a spectral shift of 18 nm that is very close to the experimental value of 17 nm. In fact, these results are all equivalent, given the statistical error of 3 nm. Similarly, using only the HBs that are located in the NH_2 group gives a spectral shift of 16 nm that is also in very good agreement with experiment. The influence of the hydrogen bonds in the NO_2 group to the spectral shift goes beyond some experimental inferences.[14]

A better comparison between theory and experiment should consider explicitly all the states and their corresponding intensities for obtaining

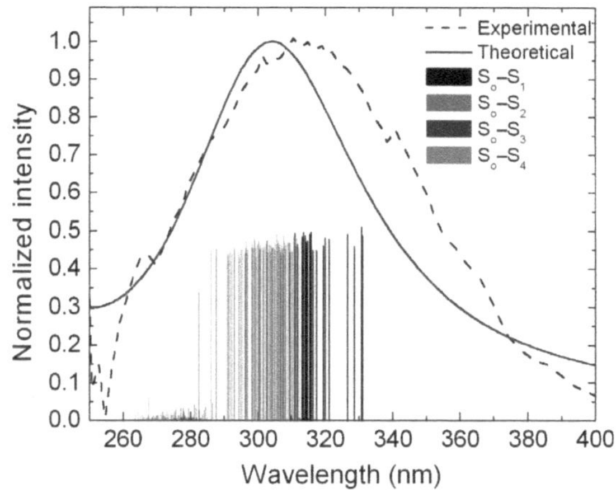

Figure 6 Theoretical absorption spectrum obtained from convolution of the calculated transition energies and intensities and comparison with experiment.

the spectral maximum. Figure 6 makes a direct comparison between the theoretical and experimental spectra. The theoretical spectrum is obtained by convoluting Lorentzians around the individual transitions using the first 10 states calculated in 100 configurations sampled from the MC simulation. Hence, in total, 1000 transitions and their corresponding oscillator strengths are used in the theoretical spectral convolution. This allows an additional comparison, namely the bandwidth obtained from the inhomogeneous broadening in the solvent environment. One can note that similar to the experimental band profile, the theoretical band is also asymmetric. The theoretical band maximum at 305 nm compares well with the experimental value of 309 nm. Note the contribution of some S_0–S_1 transition to the final spectral red shift. The calculated bandwidth at half maximum of 65 nm is in good agreement with the experimental value of 80 nm. The theoretical results include the inhomogeneous part derived from the statistical distribution but lacks possible contribution of the vibrational broadening.

4. SUMMARY AND CONCLUSIONS

Spectroscopy and reactivity of molecules in supercritical environment is of increasing interest and can be studied theoretically by combining methods of statistical mechanics, where the thermodynamic condition is imposed, with conventional quantum chemical methods. Using the challenging pNA as a probe molecule, the spectral shift of the intense charge transfer π–π* transition was studied in supercritical water environment. Using classical MC simulation and TD-DFT, the experimental spectral shift of 0.24 eV from vapor to supercritical water was accurately described using CAM-B3LYP/aug-cc-pVDZ calculations on configurations generated by classical MC simulations. The TD-DFT CAM-B3LYP/aug-cc-pVDZ calculations were made on 100 statistically uncorrelated configurations. These were composed of all water molecules involved in hydrogen bonds with pNA surrounded by an EE composed by all the remaining water molecules up to a distance of nearly 15 Å from the center of mass of the solute. The influence and importance of solute–solvent hydrogen bonds were considered in some detail. Compared to NW, the number of solute–solvent hydrogen bonds in supercritical water ($T=655$ K and $\rho=0.12$ g/cm^3) decreases to essentially one-fifth, from 2.9 to 0.6. The calculated theoretical spectrum shows a maximum located at 305 nm with a width of 65 nm in good comparison with the experimental maximum observed at 309 nm with a width of 80 nm. Compared to measurements made in the vapor phase, this corresponds to a spectral shift of 17 nm that is well described theoretically with the theoretical result of 17 ± 3 nm.

Including explicitly the thermodynamic condition in quantum chemical studies opens large avenues for the study of the electronic structure of SCFs, a region of increasing interest for the industry and the academia.

ACKNOWLEDGMENT

This work has been partially supported by CNPq, CAPES, and FAPESP.

REFERENCES

1. Arai, Y.; Sako, T.; Takebayashi, Y. Eds. *Supercritical Fluids: Molecular Interactions, Physical Properties and New Applications*; Springer: Heidelberg, 2013.
2. See the special issue: Noyori, R., Ed. Supercritical Fluids. *Chem. Rev.* **1999,** *99* (2), 353–634.
3. Beckman, E. J. Supercritical and Near-Critical CO_2 in Green Chemical Synthesis and Processing. *J. Supercrit. Fluids* **2004**, *28*, 121.
4. Besnard, M.; Tassaing, T.; Daten, Y.; Andanson, J.-M.; Soetens, J.-C.; Cansell, F.; Loppinet-Serani, A.; Reveron, H.; Aymonier, C. Bringing Together Fundamental and Applied Science: The Supercritical Fluids Route. *J. Mol. Liq.* **2006**, *125*, 88.
5. Warshel, A.; Levitt, M. Theoretical Studies of Enzymic Reactions: Dielectric, Electrostatic and Steric Stabilization of the Carbonium Ion in the Reaction of Lysozyme. *J. Mol. Biol.* **1976**, *103*, 227.
6. Field, M. J.; Bash, P. A.; Karplus, M. A Combined Quantum Mechanical and Molecular Mechanical Potential for Molecular Dynamics Simulations. *J. Comput. Chem.* **1990**, *11*, 700.
7. Canuto, S. Ed. *Solvation Methods on Molecules and Biomolecules. Computational Methods and Applications*; Springer: Berlin, 2008.

8. See the special issue: Canuto, S., Ed. Combining Quantum Mechanics and Molecular Mechanics. Some Recent Progresses in QM/MM Methods. *Adv. Quantum Chem.* **2010,** *59,* 1–416.
9. Cabral, B. J. C.; Rivelino, R.; Coutinho, K.; Canuto, S. A First Principles Approach to the Electronic Properties of Liquid and Supercritical CO_2. *J. Chem. Phys.* **2015**, *142*, 024504.
10. Martiniano, H. F. M. C.; Cabral, B. J. C. Structure and Electronic Properties of a Strong Dipolar Liquid: Born-Oppenheimer Molecular Dynamics of Liquid Hydrogen Cyanide. *Chem. Phys. Lett.* **2013**, *556*, 119.
11. Millefiori, S.; Favini, G.; Millefiori, A.; Grasso, D. Electronic Spectra and Structure of Nitroanilines. *Spectrochim. Acta A* **1977**, *33*, 21.
12. Thomsen, C. L.; ThØrgensen, J.; Keiding, S. R. Ultrafast Charge-Transfer Dynamics: Studies of *p*-Nitroaniline in Water and Dioxane. *J. Phys. Chem. A* **1998**, *102*, 1062; Kovalenko, S. A.; Schanz, R.; Frarztdinov, V. M.; Henning, H.; Ernsting, N. P. Femtosecond Relaxation of Photoexcited para-Nitroaniline: Solvation, Charge Transfer, Internal Conversion and Cooling. *Chem. Phys. Lett.* **2000**, *323*, 312.
13. Oka, H.; Kajimoto, O. UV Absorption Solvatochromic Shift of 4-Nitroaniline in Supercritical Water. *Phys. Chem. Chem. Phys.* **2003**, *5*, 2535.
14. Kamlet, M. J.; Minesinger, R. R.; Kayser, E. G.; Aldridge, M. H.; Eastes, J. W. Hydrogen Bonding by Hydroxylic Solvents to Aromatic Amines. Effects on Spectra and Relative Basicities of Some N-(4-Nitrophenyl)polymethylenimines. *J. Org. Chem.* **1971**, *36*, 3852.
15. Botti, A.; Bruni, F.; Ricci, M. A.; Soper, A. K. Neutron Diffraction Study of High Density Supercritical Water. *J. Chem. Phys.* **1998**, *109*, 3180.
16. Tassaing, T.; Bellisent-Funel, M. C.; Guillot, B.; Guissani, Y. The Partial Pair Correlation Functions of Dense Supercritical Water. *Europhys. Lett.* **1998**, *42*, 265.
17. Hoffman, M. M.; Conradi, M. S. Are There Hydrogen Bonds in Supercritical Water? *J. Am. Chem. Soc.* **1997**, *119*, 3811.
18. Soper, A. K.; Bruni, F.; Ricci, M. A. Site-Site Pair Correlation Functions of Water from 25 to 400 C: Revised Analysis of New and Old Diffraction Data. *J. Chem. Phys.* **1997**, *106*, 247.
19. Sahle, C. J.; Sternemanna, C.; Schmidt, C.; Lehtola, S.; Jahn, S.; Simonelli, L.; Huotari, S.; Hakala, M.; Pylkkänen, T.; Nyrow, A.; Mende, K.; Tolan, M.; Hämäläinen, K.; Wilke, M. Microscopic Structure of Water at Elevated Pressures and Temperatures. *Proc. Natl. Acad. Sci. U.S.A.* **2013**, *110*, 6301.
20. Kosenkov, D.; Slipchenko, L. V. Solvent Effects on the Electronic Transitions of p-Nitroaniline: A QM/EFP Study. *J. Phys. Chem. A* **2011**, *115*, 392.
21. Sok, S.; Willow, S. Y.; Zahariev, F.; Gordon, M. S. Solvent-Induced Shift of the Lowest Singlet $\pi \rightarrow \pi^*$ Charge-Transfer Excited State of p-Nitroaniline in Water: An Application of the TDDFT/EFP1 Method. *J. Phys. Chem. A* **2011**, *115*, 9801.
22. Eriksen, J. J.; Sauer, S. P. A.; Mikkelsen, K. V.; Christiansen, O.; Jensen, H. J. A.; Kongsted, J. Failures of TDDFT in Describing the Lowest Intramolecular Charge-Transfer Excitation in Para-Nitroaniline. *Mol. Phys.* **2013**, *111*, 1235.
23. Hidalgo, M.; Rivelino, R.; Canuto, S. Origin of the Red Shift for the Lowest Singlet $\pi \rightarrow \pi^*$ Charge-Transfer Absorption of p-Nitroaniline in Supercritical CO_2. *J. Chem. Theory Comput.* **2014**, *10*, 1554.
24. Raveendran, P.; Wallen, S. L. Cooperative C—H···O Hydrogen Bonding in CO_2—Lewis Base Complexes: Implications for Solvation in Supercritical CO_2. *J. Am. Chem. Soc.* **2002**, *124*, 12590.
25. Rivelino, R. Lewis Acid–Base Interactions in Weakly Bound Formaldehyde Complexes with CO_2, HCN, and FCN: Considerations on the Cooperative H-Bonding Effects. *J. Phys. Chem. A* **2008**, *112*, 161.

26. Altarsha, M.; Ingrosso, F.; Ruiz-Lopez, M. F. A New Glimpse into the CO_2-Philicity of Carbonyl Compounds. *ChemPhysChem* **2012**, *13*, 3397.
27. Foltran, S.; Méreaux, R.; Tassaing, T. On the Interaction Between Supercritical CO_2 and Epoxides Combining Infrared Absorption Spectroscopy and Quantum Chemistry Calculations. *Phys. Chem. Chem. Phys.* **2011**, *13*, 9209.
28. Raveendran, P.; Ikushima, Y.; Wallen, S. L. Polar Attributes of Supercritical Carbon Dioxide. *Acc. Chem. Res.* **2005**, *38*, 478.
29. Coutinho, K.; Rivelino, R.; Georg, H. C.; Canuto, S. The Sequential QM/MM Method And Its Applications To Solvent Effects In Electronic And Structural Properties Of Solutes. In: *Solvation Effect on Molecules and Biomolecules. Computational Methods and Application*; Canuto, S., Ed.; Springer: Berlin, 2008; pp 159–189.
30. Coutinho, K.; Canuto, S. Solvent Effects from a Sequential Monte Carlo-Quantum Mechanical Approach. *Adv. Quantum Chem.* **1997**, *28*, 89.
31. Coutinho, K.; Canuto, S.; Zerner, M. C. A Monte Carlo-Quantum Mechanics Study of the Solvatochromic Shifts of the Lowest Transition of Benzene. *J. Chem. Phys.* **2000**, *112*, 9874.
32. Allen, M. P.; Tildesley, D. J. *Computer Simulation of Liquids*; Oxford University Press: New York, 1987.
33. Miertũs, S.; Scrocco, E.; Tomasi, J. Electrostatic Interaction of a Solute with a Continuum. A Direct Utilization of Ab Initio Molecular Potentials for the Prevision of Solvent Effects. *Chem. Phys.* **1981**, *55*, 117.
34. Cancés, E.; Mennucci, B.; Tomasi, J. A New Integral Equation Formalism for the Polarizable Continuum Model: Theoretical Background and Applications to Isotropic and Anisotropic Dielectrics. *J. Chem. Phys.* **1997**, *107*, 3032.
35. Cossi, M.; Barone, V.; Mennucci, B.; Tomasi, J. Ab Initio Study of Ionic Solutions by a Polarizable Continuum Dielectric Model. *Chem. Phys. Lett.* **1998**, *286*, 253.
36. Pomelli, C. S.; Tomasi, J. Ab Initio Study of the SN2 Reaction $CH_3Cl + Cl^- \rightarrow Cl^- + CH_3Cl$ in Supercritical Water with the Polarizable Continuum Model. *J. Chem. Phys. A* **1997**, *101*, 3561.
37. Archer, D. G.; Wang, P. The Dielectric Constant of Water and Debye-Huckel Limiting Law Slopes. *J. Phys. Chem. Ref. Data* **1990**, *19*, 371.
38. Trueblood, K. N.; Goldfish, E.; Donohue, J. A Three-Dimensional Refinement of the Crystal Structure of 4-Nitroaniline. *Acta Crystallogr.* **1961**, *14*, 1009.
39. Georg, H. C.; Coutinho, K.; Canuto, S. Converged Electronic Polarization of Acetone in Liquid Water and the Role in the n–p* Transition. *Chem. Phys. Lett.* **2006**, *429*, 119.
40. Georg, H. C.; Coutinho, K.; Canuto, S. Solvent Effects on the UV-Visible Absorption Spectrum of Benzophenone in Water: A Combined Monte Carlo Quantum Mechanics Study Including Solute Polarization. *J. Chem. Phys.* **2007**, *126*, 034507.
41. Manzoni, V.; Lyra, M. L.; Gester, R. M.; Coutinho, K.; Canuto, S. Study of the Optical and Magnetic Properties of Pyrimidine in Water Combining PCM and QM/MM Methodologies. *Phys. Chem. Chem. Phys.* **2010**, *12*, 14023.
42. Berendsen, H. J. C.; Griegera, J. R.; Straatsma, T. P. The Missing Term in Effective Pair Potentials. *J. Phys. Chem.* **1987**, *91*, 6269.
43. Fonseca, T. L.; Coutinho, K.; Canuto, S. Probing Supercritical Water with the n-π^* Transition of Acetone: A Monte Carlo/Quantum Mechanics Study. *J. Chem. Phys.* **2007**, *126*, 034508.
44. Fonseca, T. L.; Coutinho, K.; Canuto, S. The Isotropic Nuclear Magnetic Shielding Constants of Acetone in Supercritical Water: A Sequential Monte Carlo/Quantum Mechanics Study Including Solute Polarization. *J. Chem. Phys.* **2008**, *129*, 034502.
45. Jorgensen, W. L.; Maxwell, D. S.; Tirado-Rives, J. Development and Testing of the OPLS All-Atom Force Field on Conformational Energetics and Properties of Organic Liquids. *J. Am. Chem. Soc.* **1996**, *118*, 11225.

46. Reis, H.; Papadopoulos, M. G.; Gryzbowski, A. Computer Simulation of the Linear and Nonlinear Optical Susceptibilities of p-Nitroaniline in Cyclohexane, 1,4-Dioxane, and Tetrahydrofuran in Quadrupolar Approximation. II. Local Field Effects and Optical Susceptibilities. *J. Phys. Chem. B* **2006**, *110*, 18537.
47. Breneman, C. M.; Wiberg, K. B. Determining Atom-Centered Monopoles from Molecular Electrostatic Potentials. The Need for High Sampling Density in Formamide Conformational Analysis. *J. Comput. Chem.* **1990**, *11*, 361.
48. Liptay, W.; Eberlein, W.; Weidenberg, H.; Elflein, O. Die Beeinflussung der optischen Absorption von Molekülen durch ein äußeres elektrisches Feld. V. Übergangsmomentrichtungen und Dipolmomente der niedrigen Elektronenzustände in 4-Nitranilin, 1-Nitro-3,5-diaminobenzol, 3,5-Dinitroanilin, Carbazol und 3,6-Dinitrocarbazol. *Ber. Bunsen-Ges. Phys. Chem.* **1967**, *71*, 548.
49. Breitung, E. M.; Vaughn, W. E.; Mahon, R. J. Measurement of Solute Dipole Moments in Dilute Solution: A Simple Three-Terminal Cell. *Rev. Sci. Instrum.* **2000**, *71*, 224.
50. Fileti, E. E.; Coutinho, K.; Malaspina, T.; Canuto, S. Electronic Changes due to Thermal Disorder of Hydrogen Bonds in Liquids: Pyridine in an Aqueous Environment. *Phys. Rev. E* **2003**, *67*, 61504.
51. Coutinho, K., Canuto, S. *DICE: A Monte Carlo Program for Molecular Liquid Simulation*, v: 2.9; University of São Paulo, 2010.
52. Runge, E.; Gross, E. K. U. Density-Functional Theory for Time-Dependent Systems. *Phys. Rev. Lett.* **1984**, *52*, 997.
53. Dreuw, A.; Head-Gordon, M. Single-Reference Ab Initio Methods for the Calculation of Excited States of Large Molecules. *Chem. Rev.* **2005**, *105*, 4009.
54. Becke, A. D. Density-Functional Thermochemistry. III. The Role of Exact Exchange. *J. Chem. Phys.* **1993**, *98*, 5648.
55. Lee, C.; Yang, W.; Parr, R. G. Development of the Colle-Salvetti Correlation-Energy Formula into a Functional of the Electron Density. *Phys. Rev. B* **1988**, *37*, 785.
56. Frisch, M. J.; Trucks, G. W.; Schlegel, H. B.; Scuseria, G. E.; Robb, M. A.; Cheeseman, J. R.; Scalmani, G.; Barone, V.; Mennucci, B.; Petersson, G. A.; Nakatsuji, H.; Caricato, M.; Li, X.; Hratchian, H. P.; Izmaylov, A. F.; Bloino, J.; Zheng, G.; Sonnenberg, J. L.; Hada, M.; Ehara, M.; Toyota, K.; Fukuda, R.; Hasegawa, J.; Ishida, M.; Nakajima, T.; Honda, Y.; Kitao, O.; Nakai, H.; Vreven, T.; Montgomery, J. A., Jr.; Peralta, J. E.; Ogliaro, F.; Bearpark, M.; Heyd, J. J.; Brothers, E.; Kudin, K. N.; Staroverov, V. N.; Kobayashi, R.; Normand, J.; Raghavachari, K.; Rendell, A.; Burant, J. C.; Iyengar, S. S.; Tomasi, J.; Cossi, M.; Rega, N.; Millam, J. M.; Klene, M.; Knox, J. E.; Cross, J. B.; Bakken, V.; Adamo, C.; Jaramillo, J.; Gomperts, R.; Stratmann, R. E.; Yazyev, O.; Austin, A. J.; Cammi, R.; Pomelli, C.; Ochterski, J. W.; Martin, R. L.; Morokuma, K.; Zakrzewski, V. G.; Voth, G. A.; Salvador, P.; Dannenberg, J. J.; Dapprich, S.; Daniels, A. D.; Farkas, O.; Foresman, J. B.; Ortiz, J. V.; Cioslowski, J.; Fox, D. J. *Gaussian 09, Revision A.02*; Gaussian Inc.: Wallingford, CT, 2009.
57. Yanai, T.; Tew, D. P.; Handy, N. C. A New Hybrid Exchange–Correlation Functional Using the Coulomb-Attenuating Method (CAM-B3LYP). *Chem. Phys. Lett.* **2004**, *393*, 51.
58. Perdew, J. P.; Burke, K.; Ernzerhof, M. Generalized Gradient Approximation Made Simple. *Phys. Rev. Lett.* **1996**, *77*, 3865.
59. Adamo, C.; Barone, V. Toward Reliable Density Functional Methods Without Adjustable Parameters: The PBE0 Model. *J. Chem. Phys.* **1999**, *110*, 6158.
60. Rivelino, R.; Canuto, S. An Ab Initio Study of the Hydrogen-Bonded H_2O:HCN and HCN:H_2O Isomers. *Chem. Phys. Lett.* **2000**, *322*, 207.
61. Almeida, T. S.; Coutinho, K.; Cabral, B. J. C.; Canuto, S. Electronic Properties of Liquid Ammonia: A Sequential Molecular Dynamics/Quantum Mechanics Approach. *J. Chem. Phys.* **2008**, *128*, 014506.

CHAPTER FOURTEEN

On a Hyperbolic Solution to the Nonlinear Schrödinger Equation for a Square Well Potential Coupled to a Contact Impurity at the Delocalization Threshold

Ricardo Méndez-Fragoso*, Remigio Cabrera-Trujillo[†,1]

*Facultad de Ciencias, Universidad Nacional Autónoma de México, Circuito Exterior S/N Delegación Coyoacán, Ciudad Universitaria, México

[†]Instituto de Ciencias Físicas, Universidad Nacional Autónoma de México, Cuernavaca, Morelos, México

[1]Corresponding author: e-mail addresses: rich@ciencias.unam.mx; trujillo@fis.unam.mx

Contents

Abstract

The behavior of an ultracold matter wave-packet under confinement conditions by an impurity near the delocalization threshold is of fundamental importance in atomic physics, particularly in atom chip technology. We study the effect of a waveguide impurity on an ultracold matter wave-packet at the threshold of delocalization. The impurity is modeled by a 1D short range square well potential with depth V_0 and width $2R_0$ coupled to a contact impurity at the center of the bend. We report the ground state exact solution for the time-independent nonlinear Schrödinger equation that describes a Bose–Einstein condensate at the delocalization threshold, obtaining the density profile and the maximum nonlinear coupling constant, g_{max}. This allows us to obtain the maximum number of atoms, N_{max}, that the defect potential can localize for the ground state. We find that for small impurity reduced size, $\xi = \sqrt{V_0}R_0$, the maximum number of particles and nonlinear coupling constant, g_{max}, that characterize a contact impurity, become constant. For high values of ξ, we recover the Thomas–Fermi result. With

Advances in Quantum Chemistry, Volume 71
ISSN 0065-3276
http://dx.doi.org/10.1016/bs.aiq.2015.03.007

the analytic solution, we report the full width at half maximum for the wave-function and density profile finding a large tunneling probability for small confining conditions. Implications of these findings to atom chips are discussed.

1. INTRODUCTION

It is with great pleasure that we celebrate the scientific achievements of our dear friend and colleague Frank E. Harris. His numerous contributions to the fields of quantum chemistry and mathematical physics are reflected in his number of publications and citations, and its impact is reflected in the wide range of fields this AQC volume contain in his honor.

In our particular case, we would like to have this small contribution in a field that involves the nonlinear Schrödinger Equation (NLSE) which for the case of a square well potential its solutions are given in terms of Jacobi elliptic functions and hyperbolic functions.[1,2] It is interesting to note that the standard reference book *Mathematical Methods for Physicist*,[3] in which Frank is a coauthor, has not a chapter dedicated to the Jacobi elliptic functions. We hope that this contribution will motivate such an endeavor on his part.

The study of the collective dynamics of macroscopic ensembles of atoms occupying the same single-particle quantum state has been possible due to the experimental realization of Bose–Einstein condensates (BEC).[4-6] The technological development of an atom chip,[7-9] has enabled the cooling and trapping of a BEC in a waveguide which is created by magnetic fields generated above patterned microwire circuits.[10] Thus, atom chips are capable of generating a myriad of trap geometries on an ultracold atom gas.[11] In an atom chip, the BEC is loaded and trapped in the transverse ground state and is allowed to propagate freely along the third dimension. An ultracold matter wave-packet is transported through quasi 1D waveguides due to the strong confinement in the two transverse dimensions.[12]

The collective dynamics of a BEC at temperature $T = 0$ is well described by a mean-field approximation which results in the NLSE for the single-particle orbitals

$$i\frac{\partial \Psi}{\partial t} = -\frac{1}{2}\frac{\partial^2 \Psi}{\partial x^2} + V\Psi + g|\Psi|^2\Psi. \tag{1}$$

Here, g is the nonlinear coupling constant that in the quasi 1D is given by $g = 2(N-1)a_s/a_\perp$ where a_s is the s-wave scattering length, N the number of atoms, and $a_\perp$ the transversal length of the waveguide. We are using the fact that in 1D, $g = g_{3D}/2\pi$ with g_{3D} being the 3D coupling constant. In the case of $g \propto N$, the above NLSE is known as the Gross–Pitaevskii equation (GPE).

With the substitution $\Psi = e^{-i\mu t}\phi$, the previous equation takes the time-independent form

$$\mu\phi = -\frac{1}{2}\frac{\partial^2\phi}{\partial x^2} + V\phi + g\phi^3, \tag{2}$$

where μ is the chemical potential. Equation (2) is written in oscillator units (o.u.) with characteristic length $a_\perp = \sqrt{\hbar/m\omega_\perp}$ for a chosen transverse frequency $\omega_\perp$ and m is the mass of the individual atoms that compose the ultracold gas. Therefore the energy is given in units of $\hbar\omega_\perp$, time in units of $1/\omega_\perp$, such that in S.I. units $g^{SI} = 2\hbar^2 a_s(N-1)/m = \hbar^2 g/(a_\perp m)$.

As mentioned previously, there exist analytic solutions to Eq. (2) for the case of a square well potential[13,14] and for a contact impurity (Dirac's delta function).[14,15] These solutions are found in terms of Jacobi elliptic functions.[2]

A local modification in a transverse waveguide, such as a constriction[16–19] or a local curvature[20–22] can be treated by a quasi 1D defect potential. The defect potential will have a number of atoms trapped as a wave-packet goes through the defect.[23,24] As the number of atoms increases in the impurity, the energy of the system will increase until becoming null respect to the transverse trap energy[20] such that the system is no longer bound and the system becomes delocalized.[13,25] Thus, in turn, gives us a critical or maximum value for g, i.e., a g_{max}, with a maximum number of trapped atoms N_{max}. This corresponds to determining the limit at which the system transitions from a bound to a scattering state.

Analytic approximations for g_{max} were derived by Carr et al.,[13] Leboeuf and Pavloff,[20] and recently by Mendez-Fragoso and Cabrera-Trujillo.[14] Carr et al.[13] solved the 1D NLSE for a finite square-well finding localized solutions when the 1D eigenenergy, $\mu < 0$. The transition at $\mu = 0$ was found in terms of an approximate expression for g_{max} and is given by

$$\frac{g_{max}}{2\sqrt{V_0}} \approx \frac{1}{2}\left[\sqrt{2}\left(\frac{e^{-\sqrt{2}\xi}}{1 + 2e^{-\sqrt{2}\xi} - e^{-2\sqrt{2}\xi}}\right) + 2\xi\right], \tag{3}$$

where the potential well has depth V_0 and width $2R_0$, and $\xi = \sqrt{V_0}R_0$ is the impurity reduced size.

More recent analytic work by Seaman et al.[15] showed that, for a potential $V(x) = -\beta\delta(x)$, the value of $g_{\max} = 4\beta$. Finally, in the work of Mendez-Fragoso and Cabrera-Trujillo an exact expression that interpolates both Carr's and Seaman's results was provided[14] with the proper behavior for small and large confinement conditions. All of these solutions are given in terms of Jacobi elliptic functions. One special case of the Jacobi elliptic functions is when the elliptic modulus, m, goes to zero, such that the solutions become hyperbolic functions.

Our main goal in this work is to present a new analytic solution near the delocalization threshold for the case of a matter wave-packet that can be trapped by an attractive defect described by 1D potential with an attractive contact impurity of variable strength based only on hyperbolic functions.

Our results, in this work, provide an analytic expression for the ground state which is simple and shows the general properties of the ultracold matter wave-packet at the delocalization threshold corroborating the proper behavior for low confinement conditions based only on a hyperbolic function.

This work is organized with the following layout. In Section 2, we provide the analytic solution for the ground state of a square well potential with a contact impurity at the center of the constriction when at the delocalization threshold. In Section 3, we discuss the width behavior of the confinement solution, and in Section 4, we provide the conclusions of our results.

2. GROUND STATE ANALYTIC SOLUTION

The attractive impurity is modeled by a finite square well potential with a contact impurity at the center by

$$V(x) = \begin{cases} -V_0 + U_0\delta(x), & |x| < R_0, \text{region II} \\ 0, & |x| \geq R_0, \text{regions I and III} \end{cases} \tag{4}$$

with V_0 the square well depth and $2R_0$ its width. Here, U_0 is the strength of the contact impurity. The solutions to Eq. (2) for this potential for $U_0 = 0$ can be found in the classical reference on differential equations by Kamke[1] and when $V_0 = 0$ by Seaman et al.[15] In the context of a BEC they were first reported by Carr et al.[13] The solution for this potential when $U_0 = 0$ are given in terms of Jacobi elliptic functions and hyperbolic functions[2] depending on the region of the potential well and the value of the chemical potential.

The solution we are seeking must fulfill the matching conditions

$$\phi_{II}(x = R_0) = \phi_{III}(x = R_0) \tag{5}$$

and

$$\frac{d\phi_{II}}{dx}(x = R_0) = \frac{d\phi_{III}}{dx}(x = R_0) \tag{6}$$

with a similar condition at $x = -R_0$ between regions I and II. The solution is also restricted to the condition that the wave function must be normalized. Furthermore, in the case $U_0 \neq 0$, we have to take into account the discontinuity of the wave function at the origin due to the delta impurity. In this case, the wave-function must satisfy

$$\phi'(0) = U_0\phi(0). \tag{7}$$

In Eq. (2), g is a free parameter which is dependent on the number of atoms in the waveguide or confining cavity.[13] However, we would like to obtain the nonlinear coupling constant, g_{max}, for which a state is near the delocalization threshold. Thus, we would like to find the g_{max} for the ground state when $\mu = 0$, thus determining the maximum number of atoms that the impurity can hold.

The reader can note that in the case $\mu = 0$ the wave function in Eq. (2) for $|x| \geq R$ (regions I and III) is given by

$$\begin{aligned} \phi_I &= 1/(a - \sqrt{g}x) \quad , x < -R_0 \quad \text{and} \\ \phi_{III} &= 1/(a + \sqrt{g}x) \quad , x > R_0 \end{aligned} \tag{8}$$

where a is the matching parameter consistent with the boundary conditions [Eqs. (5) and (6)].

For region *II*, we propose a solution in terms of a hyperbolic function. The solution that would satisfy Eq. (2) with the proper behavior at the origin is the hyperbolic tangent function for the potential given in Eq. (4). So, we propose the solution as

$$\phi_{II} = A\tanh(\alpha x + \beta) \tag{9}$$

By applying the boundary conditions [Eqs. (5) and (6)], we find that the solution is

$$\phi_{\mathrm{I}} = \frac{1}{\sqrt{\frac{2g}{V_0}} - \sqrt{g}(x + R_0)} \qquad , x < -R_0$$
$$\phi_{\mathrm{II}} = -\sqrt{\frac{V_0}{g}} \tanh\left[\sqrt{V_0}(|x| - R) + \tanh^{-1}\left(-\frac{1}{\sqrt{2}}\right)\right] \qquad , |x| < R_0 \tag{10}$$
$$\phi_{\mathrm{III}} = \frac{1}{\sqrt{\frac{2g}{V_0}} + \sqrt{g}(x - R_0)} \qquad , x > R_0.$$

Furthermore, we have to take into account the discontinuity of the wave function at the origin due to the contact impurity. In order to satisfy Eq. (7), the contact potential strength, U_0, should be given by

$$U_0 = -2\sqrt{V_0}\ \mathrm{cosech}\left[2\left(\sqrt{V_0}R_0 + \tanh^{-1}\left(\frac{1}{\sqrt{2}}\right)\right)\right]. \tag{11}$$

This means that the contact impurity strength is coupled to the square well potential via the impurity reduced size $\xi = \sqrt{V_0}R_0$. Furthermore, the contact impurity is also attractive. That means that there will be always a ground state and therefore a finite number of trapped atoms in the limit of $\xi \to 0$.

To find $g_{\max}$, we proceed to normalize the wave-function. For $U_0 < 0$, we have the result

$$\frac{g_{\max}}{2\sqrt{V_0}} = \sqrt{V_0}R_0 + \tanh\left[-\sqrt{V_0}R_0 + \tanh^{-1}\left(-\frac{1}{\sqrt{2}}\right)\right] + \sqrt{2}. \tag{12}$$

Note that $g_{\max}$ does not depend explicitly on U_0.

The approximation for $V_0 \to \infty$ and $R_0 \to 0$ such that $\xi \to 0$ produce the following result

$$\frac{g_{\max}}{2\sqrt{V_0}} = \frac{1}{\sqrt{2}} \quad \text{or} \quad g_{\max} = \sqrt{2V_0} \tag{13}$$

and the contact strength becomes

$$\frac{U_0}{\sqrt{V_0}} = -\frac{1}{\sqrt{2}} \quad \text{or} \quad U_0 = -\sqrt{\frac{V_0}{2}} \tag{14}$$

such that $g_{\max} = -2U_0$, in agreement with Leboeuf and Pavloff,[20] but for the case of only an attractive contact impurity.

In the case of $V_0 = 0$ (no square well potential), our solution is not valid since the square well potential vanishes together with the contact impurity strength such that our solution, Eq. (10), is not valid anymore.

In Fig. 1, we show the nonlinear coupling constant, g_{max}, given by Eq. (12) (solid line). In the same figure, we compare with the analytic result of Carr *et al.*[13] (long-dashed line). For high values of ξ, we notice that both ours and Carr's solutions agree very well in the TF limit. For low values of ξ, our results separate from that of Carr's *et al.* for $\xi \sim 5$. For lower values of ξ, the nonlinear coupling constant, g_{max}, is a factor of two higher than Carr's approximated expression. It is interestingly to note, however, the similarity between both results.

In Fig. 2, we show the wave-function for different values of the impurity reduced size ranging from $\xi = 0.1$ (solid line) to $\xi = 10$ (dot short-dashed line) as a function of x/R_0. For large values of ξ, the wave-function takes the shape of the square well potential and the effect of the contact impurity has disappeared at the origin. As ξ decreases, the characteristic peak at $x = 0$ appears, as consequence of the contact impurity. For $\xi \ll 1$, the value of the contact strength intensity diminishes, i.e., $U_0 \to 0$ and there is not anymore contact impurity again. Our solution, thus is not valid for $\xi = 0$ since there is no square well and contact impurities.

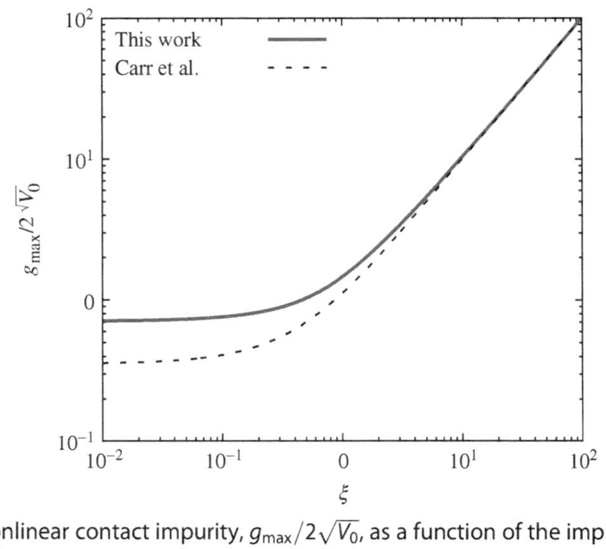

Figure 1 Nonlinear contact impurity, $g_{max}/2\sqrt{V_0}$, as a function of the impurity reduced size, $\xi = \sqrt{V_0}R_0$, as given by Eq. (12) (solid line). We also show Carr's et al. approximation, Eq. (3) (long-spaced-dashed line).

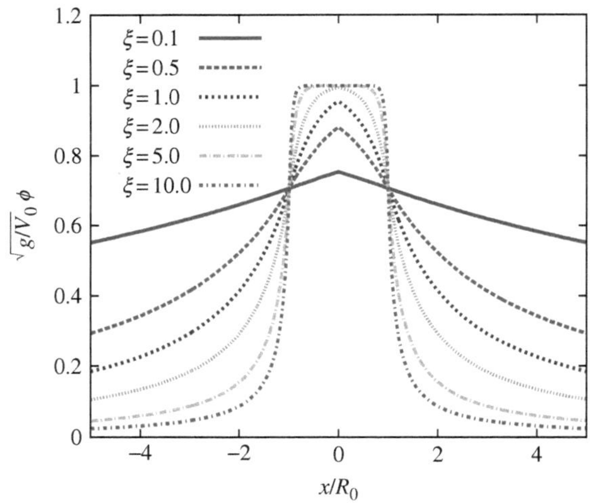

Figure 2 Scaled wave-function $\phi(x)$, as given by Eq. (10), as a function of $\chi = x/R_0$ for several values of the confinement impurity reduced size ξ. Note the effects of the square well impurity and the contact impurity at the center (see text for details).

3. FULL WIDTH HALF MAXIMUM

One of the properties we can immediately check with our solutions is how big the wave-packet would be. Our solution at the threshold of delocalization, are square-integrable, however, $\langle x^n \rangle$ diverges for $n > 1$. Nevertheless, we can still say how spread out is the wave-packet by reporting the Full Width at Half Maximum (FWHM) for the wave-function. From the analytic solution, we determine x_{FWHM} for which the wave-function is half the value than at the maximum peak. Thus, this defines $\Delta x_{\text{FWHM}} = 2x_{\text{FWHM}}$.

For the ground state at the threshold of delocalization, we find that the FWHM for the wave-function is

$$\frac{\Delta x_{\text{FWHM}}}{R_0} = 2 + \frac{4}{\xi}\left[\frac{1}{\tanh\left(\xi + \tanh^{-1}\left(\frac{1}{\sqrt{2}}\right)\right)} - \frac{1}{\sqrt{2}}\right] \tag{15}$$

and for the density profile

$$\frac{\Delta x_{\text{FWHM}}}{R_0} = 2 + \frac{2\sqrt{2}}{\xi}\left[\frac{1}{\tanh\left(\xi + \tanh^{-1}\left(\frac{1}{\sqrt{2}}\right)\right)} - 1\right] \quad (16)$$

Our results, as expected, tend to the width of the square well ($2R_0$) as $\xi \to \infty$ as in the TF approximation for both, wave-function and density profile.

In Fig. 3, we show the FWHM for the wave-function as a function of the reduced impurity size $\xi = \sqrt{V_0}R_0$ for the ground state at the threshold of delocalization (short-dashed line). For $\xi > 2$, we find that the wave-packet is confined within the walls of the square well impurity. For $\xi < 2$, the BEC starts to spill over and occupies a very large size respect to the confining impurity size, thus getting completely delocalized. For the case of the density profile, we show in the same figure the result of Eq. (16) (long-dashed line). For $\xi > 1$, we notice that the density profile width is the same as the square well potential. For small values of ξ, the density profile starts to spread, although with a lower width that the wave-function.

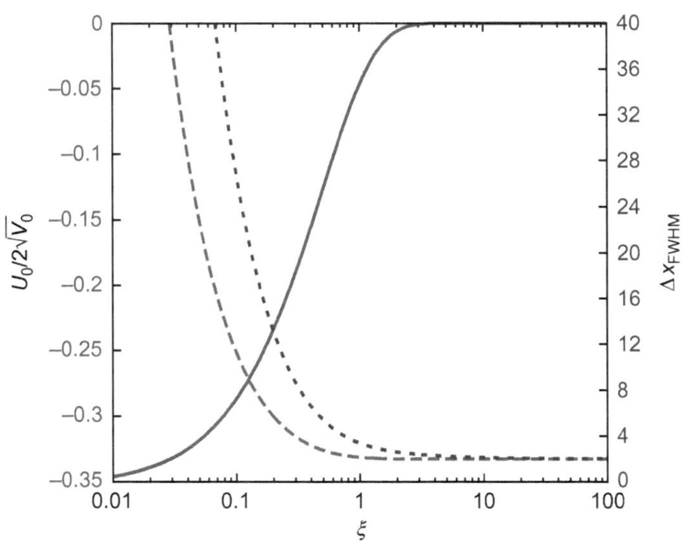

Figure 3 Attractive contact potential strength $U_0/2\sqrt{V_0}$ as a function of the impurity reduced size $\xi = \sqrt{V_0}R_0$ (red (gray in the print version) solid line). Full width at half maximum, Δx_{FWHM}, for the wave-function as a function of the impurity reduced size ξ (blue (black in the print version) short-dashed line) and for the density profile (brown (gray in the print version) long-dashed line).

In the same Fig. 3, we show the value of the contact impurity strength, $U_0/2\sqrt{V_0}$, as a function of the impurity reduced size, ξ, Eq. (11) (solid line). Also, note that for values of $\xi > 2$ the contact impurity disappears and only remains the pure square well potential. For $\xi < 2$, the contact impurity starts to increase its attractive strength until it reaches the limit value $1/2\sqrt{2}$ for $\xi \ll 1$. Note that the value $\xi = 0$ we do not have a physical solution.

4. CONCLUSIONS

The maximum number of atoms that a square well potential trap can hold is studied analytically in this work for the ground state of an ensemble of cold atoms. The maximum number of atoms is proportional to the nonlinear coupling term, g_{max}, and we find its analytic expression when the ground state is at the threshold of delocalization for a square well potential coupled with a delta (contact) impurity at the origin. The presence of the contact impurity has an effect on the shape of the wave function at the origin and on the maximum number of atoms that the well can hold. When the delta impurity is attractive, the number of trapped atoms is higher than the pure square potential impurity, since there is always a bound state. The behavior for $\xi \gg 1$ is as predicted by the Thomas–Fermi approximation and the contribution by the delta function impurity is negligible. However, for small impurity reduced sizes, $\xi = \sqrt{V_0}R_0$, we find that our solution for g_{max} resembles closely that found approximately by Carr et al.[13] for just a square well potential.

Finally, the FWHM of the wave-function and the density profile are reported as a guide to determine how much spread out is the wave-packet at the delocalization threshold. The pure square well potential has the maximum tunneling of the wave-function for small reduced impurity size and keeps the potential shape as $\xi \rightarrow \infty$ as predicted by the Thomas–Fermi approximation. As has been shown, one of the main advantages of our analytic solution that we found in this contribution is that it allows us to calculate system properties that can be measured experimentally.[14] Work is in progress to incorporate other effective potentials with analytic solutions that can be realized experimentally in atom chips.

ACKNOWLEDGMENTS

The authors acknowledge financial support from the grant PAPIIT IN-110-714 and SC15-1-IR-25 to RCT and the grants PAPIIT IA-102-414, SC15-1-S-81 and PAPIME PE-106-615 to R.M.F., We thank Reyes Garcia at the ICF-UNAM computer center for technical assistance.

REFERENCES

1. Kamke, E. *Differentialgleichungen, Lösungsmethoden Und Lösungen Leipzig*; 1959.
2. Abramowitz, M.; Stegun, I. A. *Handbook of Mathematical Functions*; Dover: New York, 1972.
3. Arfken, G.; Weber, H.; Harris, F. *Mathematical Methods for Physicists: A Comprehensive Guide*, 7th ed.; Academic Press, 2013.
4. Griffin, A.; Snoke, D.; Stringaro, S. *Bose-Einstein Condensation*; Cambridge University Press: New York, 1995.
5. Parkins, A. S.; Walls, D. F. The Physics of Trapped Dilute-Gas Bose-Einstein Condensates. *Phys. Rep.* **1998**, *303*(1), 1–80. URL: http://www.sciencedirect.com/science/article/pii/S0370157398000143.
6. Dalfovo, F.; Giorgini, S.; Pitaevskii, L. P.; Stringari, S. Theory of Bose-Einstein Condensation in Trapped Gases. *Rev. Mod. Phys.* **1999**, *71*(3), 463–512. URL: http://dx.doi.org/10.1103/RevModPhys.71.463.
7. Ott, H.; Fortagh, J.; Schlotterbeck, G.; Grossmann, A.; Zimmermann, C. Bose-Einstein Condensation in a Surface Microtrap. *Phys. Rev. Lett.* **2001**, *87*, 230–401. URL: http://link.aps.org/doi/10.1103/PhysRevLett.87.230401.
8. Reichel, J.; Hänsel, W.; Hänsch, T. W. Atomic Micromanipulation with Magnetic Surface Traps. *Phys. Rev. Lett.* **1999**, *83*, 3398–3401. URL: http://link.aps.org/doi/10.1103/PhysRevLett.83.3398.
9. Brugger, K.; Calarco, T.; Cassettari, D.; Folman, R.; Haase, A.; Hessmo, B.; Krger, P.; Maier, T.; Schmiedmayer, J. Nanofabricated Atom Optics: Atom Chips. *J. Mod. Opt.* **2000**, *47*(14-15), 2789–2809. URL: http://www.tandfonline.com/doi/abs/10.1080/09500340008232197.
10. Fortágh, J.; Zimmermann, C. Magnetic Microtraps for Ultracold Atoms. *Rev. Mod. Phys.* **2007**, *79*(1), 235–289.
11. Henderson, K.; Ryu, C.; MacCormick, C.; Boshier, M. G. Experimental Demonstration of Painting Arbitrary and Dynamic Potentials for Bose-Einstein Condensates. *New J. Phys.* **2009**, *11*(4), 043030. URL: http://stacks.iop.org/1367-2630/11/i=4/a=043030.
12. Leanhardt, A. E.; Chikkatur, A. P.; Kielpinski, D.; Shin, Y.; Gustavson, T. L.; Ketterle, W.; Pritchard, D. E. Propagation of Bose-Einstein Condensates in a Magnetic Waveguide. *Phys. Rev. Lett.* **2002**, *89*(4), 040401. http://dx.doi.org/10.1103/PhysRevLett.89.040401.
13. Carr, L. D.; Mahmud, K. W.; Reinhardt, W. P. Tunable Tunneling: an Application of Stationary States of Bose-Einstein Condensates in Traps of Finite Depth. *Phys. Rev. A* **2001**, *64*(3), 033603. http://dx.doi.org/10.1103/PhysRevA.64.033603.
14. Méndez-Fragoso, R.; Cabrera-Trujillo, R. Confinement Effects on an Ultra-Cold Matter Wave-Packet by a Square Well Impurity Near the De-Localization Threshold Analytic Solutions, Scaling, and Width Properties. *Eur. J. Phys. D* **2015**, *69*, 139. http://dx.doi.org/10.1140/epjd/e2015-50736-1.
15. Seaman, B. T.; Carr, L. D.; Holland, M. J. Effect of a Potential Step Or Impurity on the Bose-Einstein Condensate Mean Field. *Phys. Rev. A* **2005**, *71*(3), 033609.
16. Jääskeläinen, M.; Stenholm, S. Adiabatic Propagation in Potential Structures. *Phys. Rev. A* **2002**, *66*(2), 023608.
17. Lahaye, T.; Cren, P.; Roos, C.; Guéry-Odelin, D. Propagation of Guided Cold Atoms. *Commun. Nonlinear Sci. Numer. Simul.* **2003**, *8*(3-4), 315–328. http://dx.doi.org/10.1016/S1007-5704(03)00032-7. Chaotic transport and compexity in classical and quantum dynamics, URL: http://www.sciencedirect.com/science/article/pii/S1007570403000327.
18. Leboeuf, P.; Pavloff, N.; Sinha, S. Solitonic Transmission of Bose-Einstein Matter Waves. *Phys. Rev. A* **2003**, *68*(6), 063608. http://dx.doi.org/10.1103/PhysRevA.68.063608.

19. Koehler, M.; Bromley, M. W. J.; Esry, B. D. Transverse Excitations of Ultracold Matter Waves Upon Propagation Past Abrupt Waveguide Changes. *Phys. Rev. A* **2005**, *72*(2), 023603. http://dx.doi.org/10.1103/PhysRevA.72.023603.
20. Leboeuf, P.; Pavloff, N. Bose-Einstein Beams: Coherent Propagation through a Guide. *Phys. Rev. A* **2001**, *64*(3), 033602. http://dx.doi.org/10.1103/PhysRevA.64.033602.
21. Bromley, M. W. J.; Esry, B. D. Propagation of Ultracold Atoms through Bends in Waveguides. *Phys. Rev. A* **2003**, *68*, 043609.
22. Bromley, M. W. J.; Esry, B. D. Classical Aspects of Ultracold Atom Wave Packet Motion Through Microstructured Waveguide Bends. *Phys. Rev. A* **2004**, *69*(5), 053620.
23. Ernst, T.; Brand, J. Resonant Trapping in the Transport of a Matter-Wave Soliton Through a Quantum Well. *Phys. Rev. A* **2010**, *81*(3), 033614.
24. Gattobigio, G. L.; Couvert, A.; Georgeot, B.; Guéry-Odelin, D. Interaction of a Propagating Guided Matter Wave with a Localized Potential. *New J. Phys.* **2010**, *12*(8), 085013. URL: http://stacks.iop.org/1367-2630/12/i=8/a=085013.
25. Cabrera-Trujillo, R.; Bromley, M. W. J.; Esry, B. D. Scaling Laws for the Non-Linear Coupling Constant of a Bose-Einstein Condensate at the Threshold of Delocalization. ArXiv (2012) 1202.4801v1.

CHAPTER FIFTEEN

Multiresolution Approach for Laser-Modified Collisions of Atoms and Ions

Fco. Javier Domínguez-Gutiérrez*,†,1, Predrag S. Krstić*, Remigio Cabrera-Trujillo†

*Institute for Advanced Computational Science, Stony Brook University, Stony Brook, New York, USA
†Instituto de Ciencias Físicas, Universidad Nacional Autónoma de México, Cuernavaca, Morelos, México
[1]Corresponding author: e-mail address: dominijavier@gmail.com

Contents

Abstract

We expand the time-evolution functionality of the Multiresolution ADaptive Numerical Environment for Scientific Simulation (MADNESS), band-limited, gradient-corrected, symplectic propagator approach to solve numerically the time-dependent Schrödinger equation for a single-electron ion-atom collisions in a strong femtosecond laser field. We apply this method to $H^+ + H(2s)$ collision system in the range of 100 ev–25 keV collision energies modified by the presence of a 800 nm laser field of terawatt intensity.

We calculate the $n = 1 - 6$ state charge exchange and $n = 3 - 6$ state excitation cross-sections, as well as the total ionization laser-free cross-section, comparing our results with existing theoretical, and experimental data reported in the literature. The presence of the laser field increase the charge exchange and excitation cross-sections by a factor of ~ 15 for collision energies $1 - 15$ keV.

Advances in Quantum Chemistry, Volume 71
ISSN 0065-3276
http://dx.doi.org/10.1016/bs.aiq.2015.03.008

1. INTRODUCTION

Ion-atom collision systems at low-to-intermediate energies with the target initially in the ground state have been extensively studied in the last century, both theoretically and experimentally, bringing a good understanding of the transition dynamics.[1-6] However, experiments of ion-atom collisions in a laser field with the atomic target initially in a metastable state are scarce.[7-10] These were studied theoretically by a few authors. For example, the H^+ + $H(n = 2)$ system at intermediate energies.[11-14] Thus, Chen et al.[11,12] used the two-centered atomic orbital close-coupling (AOCC) expansion method to study the excitation to $n = 3$ and $n = 4$ states of atomic hydrogen as well as to study the ionization of H(2*s*) in collision with H^+. Pindzola et al.[13] used the time-dependent lattice (TDL) method and AOCC with pseudostates, as well as the classical trajectory Monte Carlo (CTMC) methods to study the $n = 3 - 5$ state excitation and $n = 1 - 5$ state charge exchange for proton colliding at intermediate collision energies with H(2*s*). These authors mutually agreed well for the state selective charge exchange and excitation cross-sections. Janev et al. calculated charge exchange, excitation, and ionization cross-sections of excited states of H at low-to-intermediate energies using hidden crossing theory.[14]

On the other hand, studies of the laser-assisted ion-atom collision systems have been reported by a number of authors in the literature[15-22] obtaining enhancement of the charge exchange and excitation cross-sections at low collision energies in presence of an ultrafast and intense laser pulse. For example, the total and state-to-state charge exchange and excitation cross-sections could be increased by an order of magnitude in low-collision energy range by bringing quasimolecular adiabatic states into resonance or accidental (local) resonance.[18,21]

In the present work, we study the H^+ + H(2*s*) system at 0.1 − 25 keV collision energies by calculating the $n = 1 - 6$ state charge exchange, $n = 3 - 6$ state excitation, and total ionization cross-sections. We also analyze the effects of an intense, few cycles, and near-infrared (NIR) laser pulse on the charge exchange, excitation and ionization processes in H^+ + H(2*s*) collisions, demonstrating a new capability to treat this complex transition dynamics with unsurpased accuracy. This is achieved by numerically solving the time-dependent Schrödinger equation (TDSE) using, for the first time, the time-dependent implementation for ion-atom collisions in the (MADNESS)[23-29] with classical, predefined time-dependent projectile trajectories.[30,31] MADNESS uses highly accurate multiresolution analysis, separated representations and adaptive numerical mesh to solve the TDSE. Its functionality was expanded by Vense et al.[30] to solve the TDSE to treat atomic transition dynamics in a time-dependent laser field. It is worth mentioning that the MADNESS–TDSE has the capability of the straightforward generalization to a multielectron collision system, although practical utilization of this feature might be currently a formidable computational task.

This chapter is organized as follows: In Section 2, we define the Hamiltonian of the collision system, with classical, predefined time-dependent trajectories for the internuclear motion coupled to the electronic degrees of freedom and to the laser field. We describe our implementation of the MADNESS–TDSE approach to study the ion-atom collision dynamics and present the numerical solutions for a few collision systems. In Section 3, we report our results for charge exchange, excitation and ionization, and compare them to existing data in the literature. Finally, in Section 4, we give our conclusions. Atomic system of units is used throughout the text, unless otherwise stated.

2. THEORY

We use the time dependent implementation of MADNESS to calculate the effects of the laser pulse on the charge exchange and excitation collision processes of the H^+ + $H(2s)$ collisions, by considering the system adiabatic quasi-molecular ground state $2s\sigma$ electronic potential energy curves (PEC) (Fig. 1) to describe the classical time-dependent trajectories of the projectile at low collision energies. The MADNESS and its numerical tools provide a high-level environment to solve numerical problems in N-dimensions (partial differential and integral equations) developed by a group of computational chemists, applied mathematicians, and computer scientists, led by Robert Harrison.[29] This software package uses the multiwavelet bases to permit adaptive decomposition of functions and operators with efficient accommodation of boundary conditions keeping track of the numerical precision, providing hardware optimization and the message parsing interface at modern, distributed memory computers.[23] MADNESS is built to be applied primarily in computational chemistry, atomic and molecular physics, material science, and nuclear structure, including Hartree–Fock and density functional theory, multiconfiguration interaction, and other quantum-chemistry approaches.[23-28] Details of MADNESS, its refinement mesh and applications, can be found in Refs. 23-30.

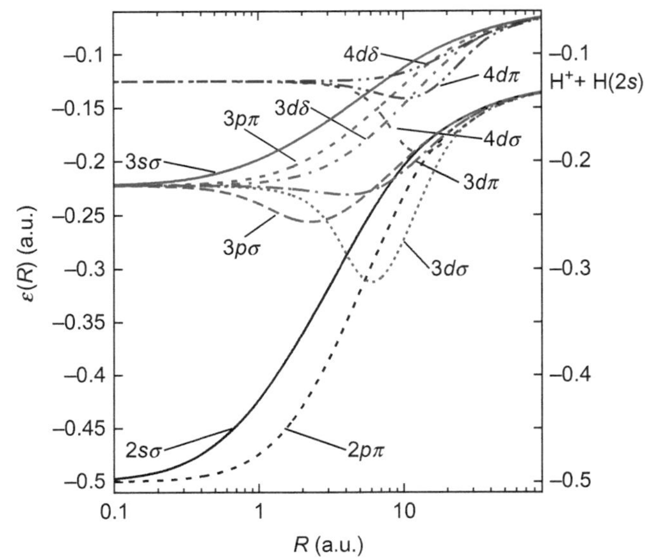

Figure 1 Adiabatic electronic potential energy curves (PEC) of the H_2^+ quasi-molecule. We follow the $2s\sigma$ PEC to obtain the classical, predefined time-dependent trajectories of the projectile for the $H^+ + H(2s)$ systems at low collision energies.

MADNESS has been applied previously to the study of atomic transition dynamics in an attosecond laser field, in the length gauge and in the dipole approximation.[30] In this work, we expand the MADNESS functionality to solve the TDSE, $\mathbf{H}_{\text{Elec}}\Psi = i\frac{\partial}{\partial t}\Psi$, for a single-electron heavy-particle collision system with the following electronic Hamiltonian $\mathbf{H}_{\text{Elec}} = \mathbf{T} + \mathbf{V}$, with

$$\mathbf{T} = -\frac{1}{2}\nabla^2, \text{ and } \mathbf{V} = -\frac{1}{|\vec{r}|} - \frac{1}{|\vec{r} - \vec{R}(t)|} - \vec{r} \cdot \vec{\mathcal{E}}(t). \tag{1}$$

$\mathbf{T}$ is the kinetic energy operator of the active electron, the first and second terms of $\mathbf{V}$ are the interaction potentials of the electron with the projectile and target nuclei, respectively, and $\vec{R}(t)$ is the classical, predefined time-dependent trajectory of the projectile. Finally, the third term of $\mathbf{V}$ is the semi-classical potential of the electron coupling with the electric field of the laser, $\vec{\mathcal{E}}(t)$, in the dipole approximation and in length gauge. The electron coordinate origin is set at the target nuclei.

2.1 The Laser Pulse Model

We model the laser pulse by a Gaussian envelope as

$$\vec{\mathcal{E}}(t) = \mathcal{E}_0 \exp\left[-\left(\frac{\omega t}{2n_{cy}}\right)^2\right]\cos(\omega t + \phi)\hat{k}; \quad -\frac{2\pi n_{cy}}{\omega} < t < \frac{2\pi n_{cy}}{\omega}. \tag{2}$$

We start counting the collision time at $t_0 = -z_0/v_p$, when the laser field amplitude is assumed negligible; $n_{cy} = 2$ is the number of cycles in a pulse, $\mathcal{E}_0 = 0.1$ a.u. is the amplitude of the electric field corresponding to the laser field power flux of $I = 3.15 \times 10^{14}$ W/cm^2; the laser frequency is $\omega = 0.0569$ a.u., i.e., a photon energy ~ 1.55 eV; $\phi = -\pi/2$ is the relative laser field phase. The polarization vector of the laser field is placed in the plane of the collision along the z-axis direction, parallel to the initial collision velocity. We choose these parameters as close as possible to the experimental set of Sansone et al.[32,33]

2.2 Classical Time-Dependent Trajectories

For higher collision energies ($E_p > 1$ keV, with E_p the collision energy), the projectile trajectory can be considered as a straight line defined by $\vec{R}(t) = (b, 0, z_0 + v_p t)$, where b is the impact parameter, z_0 is the initial position of the projectile and v_p is the constant projectile velocity.

In order to account for the proper dynamics at low collision energies (tipically less than 1 keV), we calculate the classical, predefined time-dependent trajectories, $\vec{R}(t) = (R_x, R_y, R_z)$, for the projectile by considering the total Hamiltonian of nuclei and electron, $\mathbf{H}_{Tot} = \mathbf{H}_{Elec} + \mathbf{H}_{Nuc}$. The electronic Hamiltonian, $\mathbf{H}_{Elec}$, is given by Eq. (1) and the nuclei interaction Hamiltonian, $\mathbf{H}_{Nuc}$, is given by

$$\mathbf{H}_{Nuc} = \frac{\vec{P}^2(t)}{2\mu} + \frac{1}{|\vec{R}(t)|}, \tag{3}$$

where $\vec{P}(t) = (P_x, P_y, P_z)$ is the momentum of the ion projectile and $\vec{R}(t) = (R_x, R_y, R_z)$ is the internuclear distance between the nuclei in the relative coordinate system sitting in the target. Notice that we are not coupling the projectile trajectories to the electric field as $m_p \gg m_e$. The total energy of the system is given by

$$E_{\mathrm{Tot}} = \frac{\langle \Psi | \mathbf{H}_{\mathrm{Tot}} | \Psi \rangle}{\langle \Psi | \Psi \rangle}, \quad \text{with} \quad |\Psi\rangle = \sum_i c_i(R) |\phi_i\rangle, \tag{4}$$

where E_{Tot} is conserved for the whole system. For the case of classical trajectories, we have expanded the electronic wave-function $|\Psi\rangle$ as a linear combination of adiabatic molecular states $|\phi_i\rangle$ of the system for each internuclear distance R. Therefore, the average over the electronic states defines the following Hamiltonian

$$\mathcal{H}(R,P) = E_{\mathrm{Tot}} = \frac{P^2}{2\mu} + \frac{1}{|\vec{R}(t)|} + \sum_{i=1}^{\infty} |c_i|^2 \epsilon_i(R), \text{ with } \epsilon_i(R) = \frac{\langle \phi_i | \mathbf{H}_{\mathrm{Elec}} | \phi_i \rangle}{\langle \phi_i | \phi_i \rangle}. \tag{5}$$

Here, we have assumed orthogonality between the molecular states, and $\epsilon_i(R)$ are the electronic potential energy curves (PEC) for the molecule $H^+ + H$ in the state i (Fig. 1). Equation (5) gives the classical, averaged over the electronic states, projectile trajectory. However, $c_i = c_i(t)$ are unknown transition amplitudes into various adiabatic states, and assuming that the system remains mostly all the time in the initial electronic state j (in our case $2s\sigma$), we assume $|c_i|^2 \simeq \delta_{ij}$ which yields[31]

$$\mathcal{H}(R,P) = \frac{P^2}{2\mu} + \frac{1}{|\vec{R}(t)|} + \epsilon_j(R). \tag{6}$$

Now, we are able to calculate the classical, predefined time-dependent projectile trajectories by numerical integration of the classical Hamilton equations

$$\frac{\mathrm{d}}{\mathrm{d}t} R_\zeta(t) = \frac{\partial}{\partial P_\zeta} \mathcal{H}(R,P); \quad \frac{\mathrm{d}}{\mathrm{d}t} P_\zeta(t) = -\frac{\partial}{\partial R_\zeta} \mathcal{H}(R,P), \tag{7}$$

with $\zeta = \{x, y, z\}$.

2.3 MADNESS Spatial Representation and Time Evolution

In this work, we use the smooth potential defined by MADNESS to model the Coulomb potential, as well as the parameters used to solve the TDSE as in Refs. 23 and 30. Thus, for the potential $\mathbf{V}$ in Eq. (1), we use the smoothed, electron–nucleus potential model, $V_\xi(x)$, which has been successfully used in other applications of MADNESS.[23,25,30] This potential model is defined as

$$V_\xi(x) = \frac{Z}{\xi}\frac{\mathrm{erf}(x)}{x} + \frac{\mathrm{e}^{-x^2}}{\sqrt{\pi}}, \quad \text{with} \quad \lim_{\xi\to 0} V_\xi(x) = -\frac{1}{r}, \tag{8}$$

where Z is the nuclear charge, $x = r/\xi$, controlled by the cut parameter ξ.

In order to perform the dynamics of the initial stationary state of the target, we apply the fourth-order, gradient-corrected, symplectic propagator operator developed by Chin and Chen,[34] and applied in MADNESS–TDSE by Vence et al.,[30] which is expressed as

$$U(\mathrm{d}t, t) = \mathrm{e}^{-\frac{i}{6}\mathbf{V}(t+\mathrm{d}t)\mathrm{d}t}\mathrm{e}^{-\frac{i}{2}\mathbf{T}\mathrm{d}t}\mathrm{e}^{-\frac{2i}{3}\mathbf{V}(t+\frac{\mathrm{d}t}{2})\mathrm{d}t}\mathrm{e}^{-\frac{i}{2}\mathbf{T}\mathrm{d}t}\mathrm{e}^{-\frac{i}{6}\mathbf{V}(t)\mathrm{d}t} + O(\mathrm{d}t^5). \tag{9}$$

T and **V** are defined in Eq. (1). The operator $U(\mathrm{d}t,t)$ reduces the error of time discretization and the computational time by allowing larger time steps than the second-order accurate Trotter splitting.[35] The complete description of the ion-atom collision dynamics is obtained by applying successively the time evolution operator to the instantenuos stationary state, i.e., $\Psi(\vec{r}, t+\mathrm{d}t) = U(\mathrm{d}t, t)\Psi(\vec{r}, t)$.

High frequencies associated to the large spatial gradients require small time-step, coupled to the spatial step. Thus the TDSE is limited by the condition $\mathrm{d}t \propto \Delta x^2$, where Δx is the smallest spatial mesh spacing.[30]

2.4 Charge Exchange and Excitation Probabilities

The total wave function $\Psi(E_\mathrm{p},b,t_\mathrm{f})$ of the collision dynamics provides the electronic density at the final projectile position, at time t_f. Therefore, we project the nlm-stationary hydrogenic orbitals on the total wave function $\Psi(E_\mathrm{p},b,t_\mathrm{f})$ to calculate the nl-state excitation and charge exchange probabilities as follows

$$P_{nl}(E_\mathrm{p}, b) = \sum_m \left| \int_V \phi^*_{nlm}(\vec{r})\Psi(E_\mathrm{p}, b, t_\mathrm{f})\mathrm{d}V \right|^2, \tag{10}$$

where $\phi_{nlm}(\vec{r})$ are the stationary hydrogenic orbitals calculated by MADNESS,[36] $P_{nl}(E_\mathrm{p},b)$ is the probability of charge exchange or excitation process for each nl state, and the volume V of integration is chosen over the complete numerical grid. For the nl-state excitation probabilities, we place the stationary hydrogen orbitals $\phi_{nlm}(\vec{r})$ at the center of the numerical grid. But, for the nl-state charge exchange probabilities the ϕ_{nlm} orbitals are placed at the final position of the projectile with center $\vec{R}_p = (b, 0, -z_0)$. To take into account the momentum of the electron caused by motion of the projectile, we combine the electron translation factor with the nlm-state stationary hydrogenic orbitals as

$$\phi_{nlm}(\vec{r}) = \phi_{nlm}(\vec{r} - \vec{R})\exp(i\vec{v}_p(t_f) \cdot \vec{r}), \tag{11}$$

for classical, time-dependent trajectories, $\vec{v}_p(t_f)$ is the velocity at the final time of the dynamics. The phase-dependent factor in Eq. (11) implies that the electron is moving with the projectile, gaining the momentum and energy from that motion. For the straight line trajectories case, the velocity is constant and it is given by $\vec{v}_p(t_f) = v_p\hat{k}$.

2.4.1 Cross-sections

We calculate the nl-state charge exchange and excitation cross-sections integrating transition probabilities over the impact parameters, i.e.,

$$\sigma_{nl} = 2\pi \int_0^{b_{max}} bP_{nl}(E_p, b)\mathrm{d}b, \tag{12}$$

where b_{max} is the maximum impact parameter within the numerical box such that Eq. (12) converges. Here, $P_{nl}(E_p,b)$ is either the charge exchange probability or excitation probability, as given by Eq. (10).

2.4.2 Total Ionization Probability

To describe the ionization process, we take into account the sum of the probabilities for the nl-state charge exchange $P_C(E_p,b)$, the nl-state excitation processes $P_E(E_p,b)$, and the unitary of the total wave function. Thus the total ionization probability is

$$P_I(E_p, b) = 1 - P_E(E_p, b) - P_C(E_p, b), \tag{13}$$

and is also a function of the impact parameter b and projectile energy E_p. For the calculation of the total ionization cross-section, $P_{nl}(E_p,b)$ is replaced in Eq. (12) with $P_I(E_p,b)$.

3. RESULTS

We first present our current results for the n-state charge exchange and excitation cross-sections for the $H^+ + H(1s)$ and $H^+ + H(2s)$ laser-free collision systems. Since there is much more data in the literature for ion collisions with ground state atoms, we study the $H^+ + H(1s)$ system for testing our procedures.[5,11-14,22,37-39] After that, we study the effects of an intense (3.15×10^{14} W/cm^2), near-infrared (800 nm), and ultra-fast (1 fs) laser pulse on the charge exchange, excitation, and ionization processes of $H^+ + H(2s)$ in the collision energy range $0.1 - 25$ keV.

3.1 Laser-Free Collisions

We test our procedures at the example of $H^+ + H(1s)$ colliding at 25 keV along a straight line trajectory. The initial position of the projectile H^+ is set at $z_0 = -35$ a.u. The numerical grid is defined in a box of 200 a.u. lateral dimension. For the MADNESS parameters, we use $k = 8$ (the number of scaling functions), $\xi = 0.3$ (cut parameter of the Coulomb potential), and the numerical tolerance $\epsilon = 1 \times 10^{-4}$. With these parameters, we get the energy of the hydrogen ground state $E_0 = -0.4999$ a.u. We direct the impact parameter in the x direction, using 50 uniformly spaced values for b, from 0.2 to $b_{\max} = 10$ a.u. with $\Delta b = 0.2$ a.u.

We present our numerical results for charge transfer to $n = 2$ and for excitation to $n = 2$ in Fig. 2A and B, respectively, and compare them to previously published finite difference (FD) data.[22] The two sets of data show quite good agreement.

For studying the transition dynamics of $H^+ + H(2s)$ collision system at 25 keV, we set the initial position of the projectile at $z_0 = -200$ a.u., with the numerical grid defined in a cube of lateral dimension 500 a.u., centered at the target atom. The MADNESS parameters used are $k = 8$, $\epsilon = 1 \times 10^{-4}$, and $\xi = 0.25$. The energy of the initial state of the target is obtained as $E_0 = -0.1248$ a.u., in quite good agreement with the exact value of $E_{2s} = -0.125$ a.u. The impact parameter b is in the x direction, ranging from 0.5 to $b_{\max} = 100$ a.u. in steps of $\Delta b = 1.0$ a.u. for $b \leq 20$, and $\Delta b = 5.0$ a.u. for $b > 20$ a.u. By starting the dynamics from a metastable state, we show the final state $n = 1 - 6$ charge exchange in Fig. 2C and excitation probabilities in Fig. 2D, weighted by the impact parameter, as function of b. We obtain

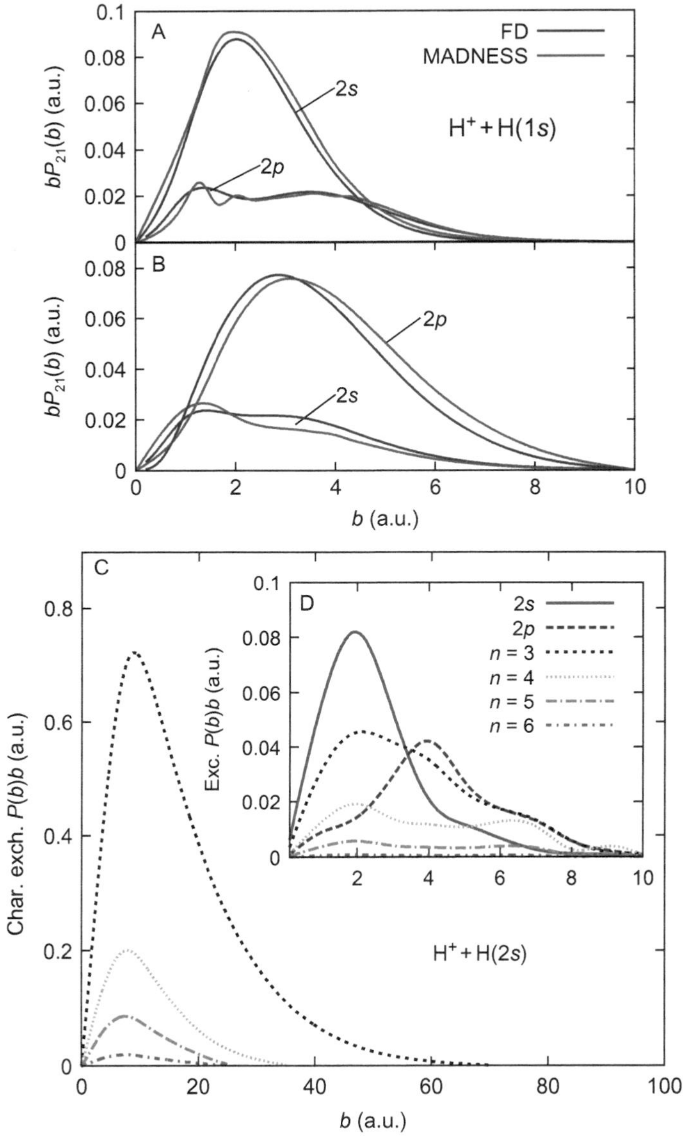

Figure 2 (A) $n = 2$ State charge exchange and (B) excitation probabilities $P_{2l}(b)$ weighted by the impact parameter as a function of b for the $H^+ + H(1s)$ collision system at 25 keV. We compare the results obtained by using MADNESS–TDSE to previously published FD data.[22] (C) $n = 3$–6 State excitation and (D) $n = 2$–6 state charge exchange probabilities weighted by the impact parameter as a function of b for 25 keV collision energy for $H^+ + H(2s)$.

that the charge exchange process populates the $n = 2$ and 3 states of the projectile with highest probability.

3.1.1 State-to-State Charge Exchange and Excitation Cross-sections

In order to assess our numerical results, we compare the charge exchange and excitation cross-sections to $n = 1 - 3$, for the $H^+ + H(1s)$ collision system with the results obtained by a FD approach[22] in Table 1. We also compare with the results of Winter.[5] Our total charge exchange cross-section, summed over all n-states in Table 1 shows a good agreement with the experimental measurement ($3.557 \pm 0.3 \times 10^{-16}$ cm^2) of McClure et al.[39]

In Table 2, we present the results for $n = 1 - 6$ states charge exchange and excitation cross-sections for $H^+ + H(2s)$ collision system and compare them with the results of Pindzola et al.[13]

Table 1 State Charge Exchange and Excitation Cross-sections for $H^+ + H(1s)$ Obtained by the MADNESS for a Collision Energy of 25 keV

	Charge Exchange			Excitation		
State	**MADNESS**	**FD**	**Ref. 5**	**MADNESS**	**FD**	**Ref. 5**
$n = 1$	3.05251	2.985	3.048	–	–	–
$n = 2$	0.53165	0.532	0.570	0.59093	0.643	0.641
$n = 3$	0.14525	–	0.159	0.14278	–	0.072
Total	3.72941	3.820	3.942	–	–	–

We compare our results with those obtained by a FD approach[22] and the results reported by Winter[5] (units are in 10^{-16} cm^2).

Table 2 Same as Table 1 but for the $H^+ + H(1s)$ Collision System

	Charge Exchange		Excitation	
State	**MADNESS**	**Ref. 13**	**MADNESS**	**Ref. 13**
$n = 1$	0.32749	0.326	0.17482	–
$n = 3$	0.35489	0.312	14.76969	21.662
$n = 4$	0.16291	0.138	3.74984	5.091
$n = 5$	0.13588	0.068	1.55075	2.212
$n = 6$	0.09722	–	0.80642	–

We compare our results to those of Pindzola et al.[13] (units are in 10^{-16} cm^2).

3.2 Laser-Assisted H^+ + H(2*s*) Collision System

We apply the MADNESS to study the effect of an intense, ultrafast and near-infrared laser pulse on the charge exchange, excitation and ionization processes of the $H^+ + H(2s)$ collision system. At energies $E_p \geq 2$ keV, the projectile trajectories are defined by straight lines. For low collision energies ($E_p < 2$ keV), we use classical, predefined time-dependent trajectories obtained by numerical integration of Eq. (7) for the ion H^+ by following the $2s\sigma$ electronic PEC (Fig. 1) of the molecule $H^+ + H(2s)$. The starting position of each trajectory is $\vec{R}(0) = (b, 0, z0)$, varying through the set of used impact parameters. Along these trajectories we use the numerical grid, the initial position of the projectile, and the MADNESS parameters as in Section 3.1.

3.2.1 n = 3 − 6 State Excitation Cross-sections

In Fig. 3A, we show the $n = 3 - 6$ state excitation cross-sections as function of the collision energy E_p for the laser-free case (filled square symbols connected by solid line), comparing our numerical results obtained by MADNESS–TDSE with the theoretical results reported by Janev et al.[14] (short-dashed line), Chen et al.[12] (dotted-dashed line), and Pindzola et al.[13] (dotted line). In this figure, the $n = 4$ and $n = 6$ states results have been multiplied by 0.1 to have a better visualization of the complete data. We also notice that our results for $E_p \leq 1$ keV with classical, time-dependent projectile trajectories are in a good agreement with the results in the literature. In Fig. 3B, we compare the $n = 3 - 6$ state excitation cross-section in absence of the laser (filled square symbols connected by solid line) with the same results for the laser case (circle symbols connected by dashed line). The laser results are multiplied by 0.1 to compare to the laser absence case, noticing that the laser-assisted collisions results have been modified by an increase factor of $\sim 10 - 30$ at $E_p < 15$ keV. This means that the excitation process has been increased with respect to the laser-free collision for the $n \geq 4$ state cross-sections. This increase is due to the fact that the photon energy 1.55 eV is close to the energy difference of the $n = 4$ and $n = 6$ states of the hydrogen atom.

3.2.2 n = 1 − 6 State Charge Exchange Cross-sections

For laser-free collisions, as we expect, the main contribution of the charge exchange process is to $n = 2$ and 3 states as suggested by the results of Fig. 2D. In Fig. 4A and B, we show the $n = 1 - 6$ state charge

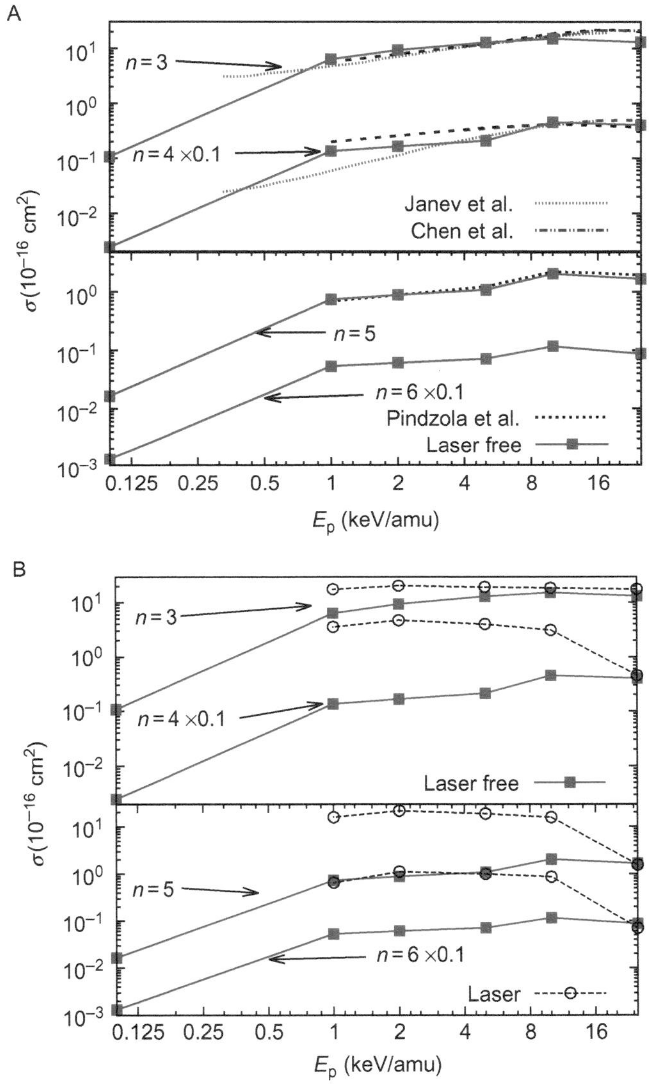

Figure 3 (A) $n = 3 - 6$ State excitation cross-sections (filled square symbols connected by solid line) obtained by using the MADNESS for laser-free collisions. These are compared to theoretical results reported by Janev et al.[14] (short-dashed line), Chen et al.[12] (dotted-dashed line), and Pindzola et al.[13] (dotted line). (B) The laser-free (filled square points connected by solid line) and the laser-assisted (circle symbols connected by dashed line) collisions excitation cross-section. Note that we have multiplied by 0.1 the results for the laser-assisted collisions results for a better visibility.

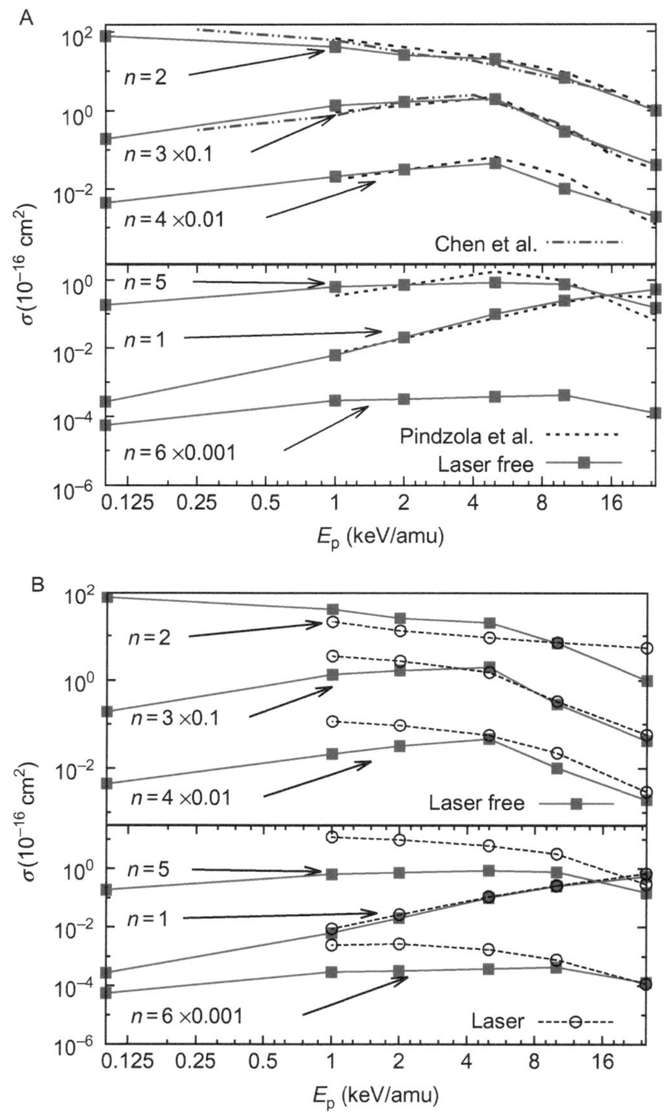

Figure 4 (A) $n = 1 - 6$ State charge exchange cross-sections for the laser-free collisions (filled square symbols connected by solid line) by using MADNESS. These are compared to those reported by Chen et al.[12] (dotted-dashed line) and Pindzola et al.[13] (dotted line). (B) Same as Fig. 3B but for charge exchange cross-sections.

exchange cross-section (filled square symbols connected by solid line), obtained by numerical solution of the TDSE with MADNESS, as function of the collision energy E_p. To improve the visibility of the state by state in the figure, the $n = 3$ states and $n = 4$ results have been multiplied by 0.1 and 0.01, respectively; the $n = 6$ states are multiplied by 0.001 to avoid crossings between the $n = 1$ and 5 curves at high collision energies ($E_p > 8$ keV). We also compare our results for the laser-free case to the theoretical results of Chen et al.[11] (dotted-dashed line) and Pindzola et al.[13] (dotted line) obtaining a good agreement with them. In Fig. 4B, we compare the results for laser-assisted collisions (circle symbols connected by dashed line) to those of the laser-free case (filled square symbols connected by solid line). The laser-assisted results are multiplied by 0.1 for better visibility in the figure. We also show that the laser field has increased the $n = 1 - 6$ state cross-sections by a factor of $\sim 10 - 20$ for collision energies $1 - 10$ keV.

3.2.3 Total Ionization Cross-section

We calculate the total ionization probability using Eq. (13). To illustrate the convergence with the increase of impact parameter, we show in Fig. 5A the total ionization probability, P_I, for 25 keV multiplied by b, as

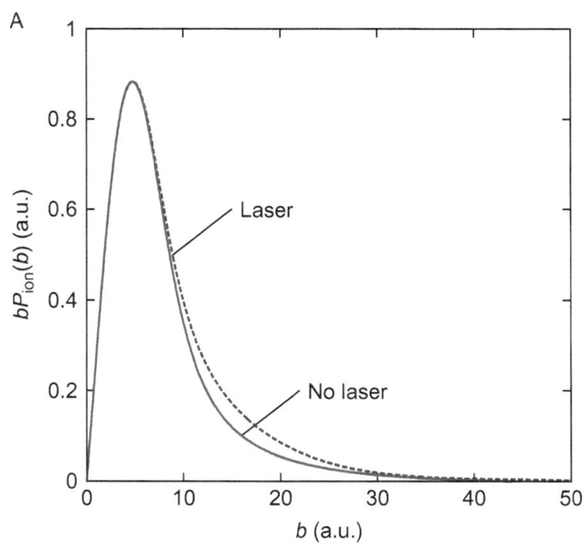

Figure 5 (A) Total ionization probability P_I calculated by Eq. (13), multiplied by b and as function of the impact parameter at 25 keV. We notice that the laser effect on the ionization process is almost negligible at high energies.

(Continued)

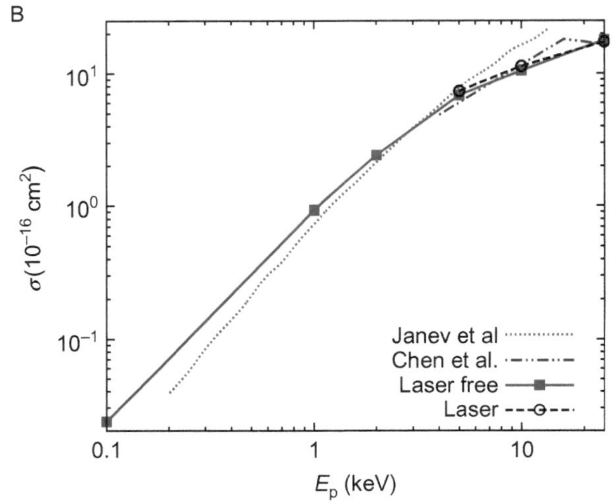

Figure 5—Cont'd (B) Total ionization cross-section (filled square symbols connected by solid line) as function of the collision energy compared to the theoretical results of Janev et al.[14] (short-dashed line) and Chen et al.[12] (dotted-dashed line) for the laser-free case. We also compare to the laser-assisted results (circle symbols connected by dashed line) noticing a small decrease of the total ionization cross-section (see text).

a function of the impact parameter. The laser field modification of the ionization probability is small toward higher impact energies. In Fig. 5B, we present the total ionization cross-section as function of the collision energy, E_p, obtained by means of Eq. (12). We also compare the laser-free results (filled square symbols connected by solid line) to those of Janev et al.[14] (short-dashed line) and Chen et al.[12] (dotted-dashed line) reporting a good agreement with them.

We note that the minimum number of photons to ionize H(2*s*) with the used field is $n_{ph} = 3$. Since the laser field amplitude is $\mathcal{E}_0 = 0.1$, the ionization cross-section is proportional to $I^{n_{ph}} = (0.01)^3 = 1 \times 10^{-6}$. This small factor in the three photon ionization probability is responsible for the small effects of the laser observed in the ionization probabilities for the considered cases.

4. CONCLUDING REMARKS

We have successfully applied, for the first time, the multiresolution approach (MADNESS–TDSE) to study laser-free and laser-assisted ion-atom collision systems in a metastable state at low-to-intermediate energies, $0.1 - 25$ keV. We accurately calculate state resolved cross-sections for charge exchange to the projectile and for excitation of the target, for up to $n = 6$ relevant final states, starting from the excited, $2s\sigma$ state of the target. The obtained results for the laser-free collisions agree well with the data in the literature.

For laser-assisted collisions with a femtosecond laser field of terawatt intensities, we find that the laser effect on the excitation and charge exchange processes of $H^+ + H(2s)$ system increases the state-to-state cross-section by a factor between 10 and 30 times for collision energies $1 - 15$ keV.

ACKNOWLEDGMENTS

We acknowledge the support from grant PAPIIT IN-110-714 and SC15-1-IR-25 to RCT and CONACyT for Ph.D. scholarship to F.J.D.G. We also would like to thank Dr. Robert Harrison for helpful discussions on MADNESS and to Dr. Scott Thornton and the staff at the Institute for Advance Computational Science at Stony Brook University (IACS-SBU) for technical assistance. Simulations were performed at high-performance computer cluster "Handy" of the IACS-SBU.

REFERENCES

1. Fritsch, W.; Lin, C. D. Excitation and Charge Transfer to 2s and 2p States in 1-20 keV $H^+ - H$ Collisions. *Phys. Rev. A* **1982**, *26*, 762–769.
2. Schultz, D. R.; Strayer, M. R.; Wells, J. C. Lattice, Time-Dependent Schrödinger Equation Solution for Ion-Atom Collisions. *Phys. Rev. Lett.* **1999**, *82*, 3976–3979.
3. Barnett, C. F. *Collisions of H, H2, He and Li Atoms and Ions with Atoms and Molecules*; 1990, Oak Ridge National Laboratory Report 6068.
4. Toshima, N. Convergence and Completeness of the Pseudostate Expansion for Proton-Hydrogen Collisons in Two Center Close-Coupling Calculations. *Phys. Rev. A* **2009**, *80*, 032701.
5. Winter, T. G. Electron Transfer, Excitation, and Ionization in p+H(1s) Collisions Studied with Sturmian Bases. *Phys. Rev. A* **2009**, *80*, 032701.
6. Cornelius, K. R.; Wojtkowski, K.; Olson, R. E. State-Selective Cross Section for Electron Capture Collisions. *J. Phys. B* **2000**, *33*, 2017–2035.
7. Bayfield, J. E. Measurement of the Total Cross Section for Charge Transfer into the Metastable State H(2s) for Proton Collisions with Atomic Hydrogen. *Phys. Rev.* **1969**, *185*, 105–112.
8. Williams, D. G.; Lee, A. R.; Butcher, E. C. A Photodetector for Measurement of Metastable H(2s) in Ion-Atom Collisions. *Aust. J. Phys.* **1987**, *40*, 47.
9. Gonzalez-Urena, A.; Vetter, R. Reactive Collisions with Excited-State Atoms. *J. Chem. Soc., Faraday Trans.* **1995**, *91*, 389–398.
10. Weiner, J.; Bagnato, V. S.; Zilio, S.; Julienne, P. S. Experiments and Theory in Cold and Ultracold Collisions. *Rev. Mod. Phys.* **1999**, *71*, 1–85.
11. Esry, B. D.; Chen, Z.; Lin, C. D.; Piacentini, R. D. Close-Coupling Calculations of Electron Capture Cross Sections from the n=2 States of H by Protons and Alpha Particles. *J. Phys. B Atomic Mol. Phys.* **1993**, *26* (9), 1579.

12. Chen, Z.; Esry, B. D.; Lin, C. D.; Piacentini, R. D. Excitation and Ionization of H(2s) by Proton Impact. *J. Phys. B Atomic Mol. Phys.* **1994**, *27* (12), 2511.
13. Pindzola, M. S.; Lee, T. G.; Minami, T.; Schultz, D. R. Excitation and Charge Transfer in p + H(2s) Collisions. *Phys. Rev. A* **2005**, *72*, 062703.
14. Janev, R. K.; Krstić, P. S. Excitation and Ionization Processes in Slow Collisions of Ground-State and Excited Hydrogen Atoms with Protons. *Phys. Rev. A* **1992**, *46*, 5554–5573.
15. Krstić, P. S.; Janev, R. K. Nonadiabatic Transitions Between Quasi-Stationary States in a Laser Field. *Phys. Rev. A* **1986**, *34*, 157.
16. Janev, R. K.; Krstić, P. S. Two-State Model for Non-Adiabatic Transitions with Gaussian Interactions. *J. Phys. B Atomic Mol. Phys.* **1990**, *24*, L39.
17. Kirchner, T. Manipulating Ion-Atom Collisions with Coherent Electromagnetic Radiation. *Phys. Rev. Lett.* **2002**, *89*, 093203.
18. Kirchner, T. Laser-Field-Induced Modifications of Electron-Transfer Processes in Ion-Atom Collisions. *Phys. Rev. A* **2004**, *69*, 063412.
19. Niederhausen, T.; Thumm, U. Capture and Ionization in Laser-Assisted Proton˘Hydrogen Collisions. *Phys. Rev. A* **2006**, *73*, 041404.
20. Lu, Z.; Chen, D.; Fan, R.; Xia, Y. Femtosecond Laser Field Induced Modifications of Electron-Transfer Processes in Ne^+ –He Collisions. *Appl. Phys. Lett.* **2012**, *100* (1), 014105.
21. Domínguez-Gutiérrez, F. J.; Cabrera-Trujillo, R. Pulse Duration Effects on Laser-Assisted Electron Transfer Cross Section for He^{2+} Ions Colliding with Atomic Hydrogen. *Eur. Phys. J. D* **2014**, *68* (8), 226.
22. Domínguez-Gutiérrez, F. J.; Cabrera-Trujillo, R. Comparison of Laser-Assisted Charge Transfer of Symmetric and Asymmetric Colliding Systems. *J. Phys. Conf. Ser.* **2014**, *512* (1), 012033.
23. Harrison, R.; Fann, G.; Yanai, T.; Gan, Z.; Beylkin, G. Multiresolution Quantum Chemistry: Basic Theory and Initial Applications. *J. Chem. Phys.* **2004**, *121*, 11587.
24. Alpert, B.; Beylkin, G.; Gines, D.; Vozovoi, L. Adaptive Solution of Partial Differential Equations in Multiwavelet Bases. *J. Comput. Phys.* **2002**, *182*, 149.
25. Fann, G.; Beylkin, G.; Harrison, R.; Jordan, K. Singular Operators in Multiwavelet Bases. *IBM J. Res. Dev.* **2004**, *48*, 161–171.
26. Beylkin, G.; Fann, G.; Harrison, R. J.; Kurcz, C.; Monzòn, L. Multiresolution Representation of Operators with Boundary Conditions on Simple Domains. *Appl. Comput. Harmon. Anal.* **2012**, *33*, 109. http://dx.doi.org/10.1016/j.acha.2011.10.001.
27. Yanai, T.; Harrison, R. J.; Handy, N. C. Multiresolution Quantum Chemistry in Multiwavelet Bases: Time-Dependent Density Functional Theory with Asymptotically Corrected Potentials in Local Density and Generalized Gradient Approximations. *Mol. Phys.* **2005**, *103*, 413–424.
28. Natarajan, B.; Genovese, L.; Casida, M. E.; Deutsch, T.; Burchak, O. N.; Philouze, C.; Balakirev, M. Y. Wavelet-Based Linear-Response Time-Dependent Density-Functional Theory. *Chem. Phys.* **2012**, *402*, 29.
29. Harrison, R. *Implementation Notes*. Online, 2011. Available from: http://code.google.com/p/m-a-d-n-e-s-s/downloads/list.
30. Vence, N.; Harrison, R.; Krstić, P. Attosecond Electron Dynamics: A Multiresolution Approach. *Phys. Rev. A* **2012**, *85*, 033403.
31. Krstić, P. S. Coupled Channel Representation of the Advanced Adiabatic Approach. *J. Phys. B Atomic Mol. Phys.* **2004**, *37* (11), L217.
32. Sansone, G.; Kelkensberg, F.; Perez-Torres, J. F.; Morales, F.; Kling, M. F.; Siu, W.; Ghafur, O.; Johnsson, P.; Swoboda, M.; Benedetti, E.; Ferrari, F.; Lepine, F.; Sanz-Vicario, J. L.; Zherebtsov, S.; Znakovskaya, I.; L'Huillier, A.; Ivanov, M. Y.;

Nisoli, M.; Martin, F.; Vrakking, M. J. J. Electron Localization Following Attosecond Molecular Photoionization. *Nature* **2010**, *465*, 763.

33. Nisoli, M.; Sansone, G. New Frontiers in Attosecond Science. *Prog. Quantum Electron.* **2009**, *33*, 17–59.
34. Chin, S.; Chen, C. Fourth Order Gradient Symplectic Integrator Methods for Solving The Time-Dependent Schrödinger Equation. *J. Chem. Phys.* **2001**, *114*, 7338.
35. Tannor, D. J. *Introduction to Quantum Mechanics: A Time-Dependent Perspective*; University Science Book, 2007.
36. Galassi, M.; et al. GNU Scientific Library Reference Manual, 3rd Ed.; ISBN 0954612078. http://www.gnu.org/software/gsl/.
37. Fite, W. L.; Stebbings, R. F.; Hummer, D. G.; Brackmann, R. T. Ionization and Charge Transfer in Proton-Hydrogen Atom Collisions. *Phys. Rev.* **1960**, *119*, 663–668.
38. Ludde, H. J.; Dreizler, R. M. Direct and Capture Processes in Proton-Hydrogen Scattering. II. Total Cross Sections for Bombarding Energies of 1 to 50 keV. *J. Phys. B Atomic Mol. Phys.* **1982**, *15* (16), 2703.
39. McClure, G. W. Electron Transfer in Proton-Hydrogen-Atom Collisions: 2-117 keV. *Phys. Rev.* **1966**, *148*, 47–54.

INDEX

Note: Page numbers followed by "*f*" indicate figures, "*t*" indicate tables and "*b*" indicate boxes.

E

F

G

N

O

P

Q

R

S

T

W

MIX
Papier | Fördert gute Waldnutzung
FSC® C083411

Kolibri 360 GmbH
Ferdinand-Jühlke-Straße 7
99095 Erfurt, Deutschland
produktsicherheit@kolibri360.de